Studies in Computational Intelligence

Volume 691

More information about this series at http://www.springer.com/series/7092

About this Series

The series "Studies in Computational Intelligence" (SCI) publishes new developments and advances in the various areas of computational intelligence—quickly and with a high quality. The intent is to cover the theory, applications, and design methods of computational intelligence, as embedded in the fields of engineering, computer science, physics and life sciences, as well as the methodologies behind them. The series contains monographs, lecture notes and edited volumes in computational intelligence spanning the areas of neural networks, connectionist systems, genetic algorithms, evolutionary computation, artificial intelligence, cellular automata, self-organizing systems, soft computing, fuzzy systems, and hybrid intelligent systems. Of particular value to both the contributors and the readership are the short publication timeframe and the worldwide distribution, which enable both wide and rapid dissemination of research output.

Izzat M. Alsmadi • George Karabatis
Ahmed AlEroud
Editors

Information Fusion for Cyber-Security Analytics

Springer

Editors
Izzat M. Alsmadi
Department of Computing and Cyber Security
University of Texas A&M
San Antonio, TX, USA

Ahmed AlEroud
Department of Computer Information Systems
Yarmouk University
Irbid, Jordan

George Karabatis
Department of Information Systems
University of Maryland
Baltimore County (UMBC)
Baltimore, MD, USA

ISSN 1860-949X ISSN 1860-9503 (electronic)
Studies in Computational Intelligence
ISBN 978-3-319-44256-3 ISBN 978-3-319-44257-0 (eBook)
DOI 10.1007/978-3-319-44257-0

Library of Congress Control Number: 2016954920

Printed on acid-free paper

This Springer imprint is published by Springer Nature
The registered company is Springer International Publishing AG
The registered company address is: Gewerbestrasse 11, 6330 Cham, Switzerland

Preface

In spite of the increasing efforts in designing preventive security measures, new attack types arise on a regular basis. The reasons for these include: programming errors, design flaws, insider threats, and the inadequate security tools being used by organizations. Additionally, attackers keep evolving attack strategies, resulting in new attack variations being undetected at a system's real-time execution. Therefore, academic efforts with supporting material are needed to advance the existing attack prediction models, recognize the threats and vulnerabilities in the existing techniques, and learn how to create new intrusion detection systems in the future.

To this end, Internet communications and distributed networked environments have become rich media for electronic data transfer. Due to a huge amount of data transmission, it becomes vital to build effective security policies and threat detection systems that are capable of analyzing network data. As such, providing appropriate protection mechanisms and techniques is significant to combat cyber-threats and to preserve information systems' integrity, confidentiality, and availability. This book discusses several trending topics in the area of information security. Since there is an increase in the volume of malicious cyber-attacks which demands a collaborative effort between security professionals and researchers to design and utilize cyber-defense systems, the first part of this book discusses the recent attack prediction techniques that infuse one or more aspects of information to create attack prediction models. The second part is dedicated to new trends on cybersecurity such as graph data analytics for cybersecurity, unwanted traffic detection and control based on trust management software-defined networks, security in wireless sensor networks and their applications, and emerging trends in security system design using the concept of social behavioral biometric.

By creating this book, from the perspective of information-based security systems, we hope to close the gap in most of the existing systems which mainly focus on low-level data analytics to predict attacks. In addition, we hope to make readers gain a clear understanding of recent techniques in cybersecurity.

San Antonio, TX, USA — Izzat M. Alsmadi
Baltimore, MD, USA — George Karabatis
Irbid, Jordan — Ahmed AlEroud

Acknowledgments

Many people have contributed to this book. The authors would like to thank Dr. Yaser Jararweh from Jordan University of Science and Technology who made great suggestions about the topics covered in the book. The authors would also like to thank the review team including Dr. Ahmed Manasrah from Yarmouk University in Jordan, Dr. Mohammed Akour who is also a faculty member at Yarmouk University, Dr. Josephine Namayanja who is a faculty member at the University of Massachusetts Boston, Dr. Mohammed Naji Alkabi who is a faculty member at Zarqa Private University in Jordan, and Dr. Yan Zheng who is a member at the University of Espoo, Finland. The authors would like to thank the State of Maryland TEDCO (MII) and Northrop Grumman Corporation, USA.

Contents

Chapter 1
Using Contextual Information to Identify Cyber-Attacks

Ahmed AlEroud and George Karabatis

> *"An important challenge in cyber-attack detection is to develop learning methods that can integrate and fuse a much broader array of contextual information. Traditional statistical methods break down, because the broader the array of information, the more training examples are required to achieve good performance. We need to develop methods for breaking the learning problem up into modules that can be learned separately and then combined."*
>
> Thomas G. Dietterich

Abstract A recent trend is toward utilizing knowledge-based intrusion detection systems (IDSs). Knowledge-based IDSs store knowledge about cyber-attacks and possible vulnerabilities and use this knowledge to guide the process of attack prediction. Since an IDS contains information about these vulnerabilities, it can discover attempts to exploit them. One significant limitation of knowledge-based IDSs is the lack of contextual information used to detect attacks. Contextual information is not only information about the configuration on the targeted systems and their vulnerabilities. It also covers any relevant preconditions the attacks require to proceed successfully and the possible contextual semantic relationships between the activities of attackers in terms of time of these activities and the targeted locations. To overcome these limitations, we introduce a novel contextual framework which consists of several attack prediction models that can be utilized in conjunction with IDSs to detect cyber-attacks. We utilized extractable contextual elements from network data to create several knowledge-based, context-aware prediction models that are applied in conjunction with other intrusion detection techniques to assist in identifying known and unknown attacks. The created prediction models are utilized for several tasks including (1) expanding the predictions

A. AlEroud (✉)
Department of Computer Information Systems, Yarmouk University, Irbid 21163, Jordan
e-mail: Ahmed.aleroud@yu.edu.jo

G. Karabatis
Department of Information Systems, University of Maryland, Baltimore County (UMBC), 1000 Hilltop Circle, Baltimore, MD 21250, USA
e-mail: georgek@umbc.edu

I.M. Alsmadi et al. (eds.), *Information Fusion for Cyber-Security Analytics*, Studies in Computational Intelligence 691, DOI 10.1007/978-3-319-44257-0_1

of other intrusion detection techniques using pre-identified contextual relationships between attacker activities, (2) filtering the nonrelevant predictions based on the situation of the hosts targeted by attacks, and (3) predicting the occurrence of unknown attacks. Our framework focuses on the significant dimensions in data; thus, it can be utilized to detect cyber-attacks while keeping the computational overhead as low as possible.

1.1 Significance of the Problem

Providing appropriate protection techniques is significant to combat cyber threats and preserve the integrity, confidentiality, and availability of information systems. The increasing volume of malicious cyber-attacks demands a collaborative effort between security professionals and researchers to design and develop effective cyber-defense systems. Cyber-attacks continue to rise worldwide in a manner that costs companies millions of dollars each year and leads to loss or misuse of information assets. Therefore, companies are under pressure to avoid the occurrence of these attacks or decrease the damage they cause to cyber systems.

There are various types of cyber-attacks, including infected Web pages, viruses, worms, spam botnets, and other unauthorized use of computer systems to modify or access data. Any computer system that does not have a proper security infrastructure can be compromised. Therefore, the cybersecurity authorities aim to design defensive security techniques to protect these systems. However, in spite of the increasing efforts in creating security countermeasures, new attack types arise on a regular basis. The reasons for these include programming errors, design flaws, insider threats, and inadequate security tools. Additionally, attackers keep evolving attack strategies, resulting in new attack variations being undetected at real time.

While it is theoretically possible to combat all types of cyber-attacks, most of the existing techniques provide reactive rather than proactive solutions to these attacks. The proactive techniques aim to eliminate the vulnerabilities in computer systems; however, avoiding all types of vulnerabilities is not possible in practice. Over the past decades, intrusion detection systems (IDSs) have been employed as one of the major reactive techniques against computer attacks. IDSs are an important component of defensive measures for protecting computer systems and networks from abuse. IDSs utilize logic operations, statistical techniques, and machine learning approaches to discern between different types of network activities.

Although modern IDSs are definitely useful and they keep on improving, they still generate a high amount of false alarms, fail to identify unknown attacks, and exhibit a low degree of reliability. Most of the existing IDSs depend on data analytics techniques that work on raw network data at a very low abstraction level to detect cyber-attacks.

1.1.1 Background

Safeguarding computer systems against attacks is one of the most challenging tasks that cannot be easily measured. Most security mechanisms can be breached due to unknown system vulnerabilities that exist and novel hacks applied by attackers to initiate an intrusion. The latter has been defined as any action the user of an information system takes when he/she is not legally allowed to [1]. Powell et al. have also defined an intrusion as a malicious, externally induced fault resulting from a successful attack [2]. Halme et al. have referred to an intrusion attempt as a sequence of actions by means of which an intruder attempts to gain control of a system [3]. The intrusion detection as a process involves determining that an intrusion has been attempted to gain unauthorized access to a system. Krugel et al. consider responding to malicious actions that targeted computing and network resources as part of intrusion detection process [4].

Systems with the capability of detecting intrusions are called intrusion detection systems (IDSs). The role of IDSs is to differentiate between intrusions and normal system execution. The existing intrusion detection techniques combat cyber-attacks at two levels of protection, the network level and the host level. The network-based IDSs monitor the features of network connections in order to detect cyber-attacks. Conversely, host-based IDSs monitor the status of workstations and the internals of a computing system using intrusion detection techniques to discover possible attacks at the host level. There have been also other classifications of IDSs [5–7] from several perspectives, such as the data the system analyzes (log file data, network data), the time of analysis (online, offline), and the distribution mode utilized in the analysis process (centralized, distributed). Machine learning researchers classify IDSs into three major categories: signature-based, anomaly-based, and hybrid-based IDSs. A signature-based IDS measures the similarity between the events under analysis and the patterns of known attacks. Alarms are generated if previously known patterns are detected. For instance, the Snort IDS [8] is one of the most well-known signature-based intrusion detection systems. Snort performs real-time traffic analysis, content searching, and content matching to discover attacks using pre-identified attack signatures. While these systems are accurate in identifying known attacks, they cannot recognize new types of attacks. In anomaly-based IDSs, normal profiles are created and used by an anomaly detection technique to detect anomaly patterns that deviate from such profiles. The anomaly-based IDSs rely on statistical techniques to create normal profiles. Overall, the main advantage of these systems is their ability to detect unknown attacks that do not have existing signatures; however, their major limitation is the difficulty of accurately defining normal profiles. Intuitively, activities that deviate from normal profiles are not necessarily attacks. Failure to identify the boundaries of normal activity in network data leads to incorrect prediction of normal activities as attacks; thus, a high false positive rate is very possible. The hybrid-based IDSs combine signature-based and anomaly-based detection techniques to discover attacks. The major disadvantage of hybrid-based approaches is the computational overhead of using both signature matching and anomaly detection to analyze incoming network connections. Although IDSs have shown a good level of success in detecting intrusion attempts to networks, they show a visible deficiency in their effectiveness. Yet again, intrusion detection technologies have several research challenges that need to be addressed:

First, **lack of information about relationships between entities at the prediction time**. The existing IDSs do not have the capability to analyze information in a relational manner. Primarily, the information about an entity relationship with other entities is not available at the prediction time. Integrating the relationships the entity has with other entities in the intrusion detection process is very significant to identify relevant events that occur in specific context. In general, the events that target the system are not independent. Although the behavior of attackers is not predictable, the sequence of events is initiated by them to reach to a specific objective; therefore, it is very likely to discover several forms of relationships between such events. Detecting these relationships helps in predicting cyber-attacks at their early stages.

Second, **lack of awareness of the current situation**. Awareness of existing conditions is a prerequisite knowledge for an IDS to filter out nonrelevant predictions. For example, specific contextual environments inherently are not conducive to certain types of attacks (i.e., when a computer system is patched against a certain attack); therefore, it is safe not to search for these attacks, saving time and resources. Additionally, IDSs have no knowledge about how situations evolve. They usually work as sensors that cannot recognize multistep attacks; instead they can only identify individual suspicious behaviors. Moreover, they cannot determine the impact of the current situation on the targeted hosts.

Third, **analyzing data at low abstraction level is not sufficient**. Current intrusion detection techniques do not fully acquire the high level of abstraction required to discover attacks. In particular, the analysis is performed without incorporating the contextual factors that affect the analysis results. For instance, intrusion detection based on per-packet inspection is a time-consuming task. A recent trend is to analyze IP flows instead of packet-based analysis. All packets with the same source/destination IP address, source/destination ports, protocol interface, and class of service are grouped into a flow, and then packets and bytes are tallied. Flow-based intrusion detection is still performed at a relative low level leading to high false alarm rates; therefore, one needs to consider the contextual factors that connect one flow to another such as their source and targets along with the time they occur. Analysis performed solely on raw data is clearly insufficient. In order to identify cyber-attacks, data needs to be analyzed at several abstraction layers. The data analyzed at the lower layer does not only overburden cybersecurity systems but also overwhelm human decision-makers. Furthermore, data analyzed at the lower level might not contain certain evidence about the intentions and the objectives of an attacker. Therefore, one must provide means to derive higher-level contexts from lower-level sensors. In attack prediction process, high-level context is the contextual information that is further processed to create context-aware attack prediction models.

Fourth, **existing intrusion detection approaches lack semantic inference needed to identify cyber-attacks**. The alerts generated by the current IDSs are too elementary. Therefore, it is a tedious and time-consuming task for the security operators to analyze semantic relationships between alerts. Recent trends focus on alert correlation and causality analysis to discover these relationships. However, these techniques work at the syntactic not the semantic level. New approaches are needed to convert raw alerts into knowledge with appropriate evidence from which decisions can be taken. Furthermore, when the amount of alerts to be analyzed is very large, semantic reasoning approaches are expected to discover indirect relationships between alerts produced by IDSs. Consequently, utilizing semantic-based approaches for context reasoning helps domain experts make accurate decisions about security threats.

Fifth, **analyzing unknown events**. Uncertainty in the network data may lead to incorrect decisions. While signature-based IDSs are designed to detect known attacks using rule-based signatures [9], simple modifications on attack signatures can lead to unknown attacks that cannot be identified by the existing systems. Detection of unknown attacks in computer networks is a long-standing issue on the wish list of security specialists. In reality, analyzing unknown situations is considered as a crucial challenge for all types of IDSs [10]. Identifying vulnerability paths that lead to these attacks is yet another challenge.

There have been attempts to address the problem of detecting unknown attacks using machine learning and anomaly detection techniques; however, the existing approaches have no capability to effectively capture all normal activities in order to declare other activities as unknown attacks. Although anomaly detection techniques have the capability to identify unknown attack patterns, they produce large and unmanageable amounts of false alarms [11].

Failure to differentiate among unknown attacks and normal activity is a major reason of such a high false alarm rate. In order to identify unknown attacks, it is significant to allow IDSs to maintain the understanding about the evolution of known attacks, so that they can maintain a holistic view in a bigger context. To achieve this objective, links need to be created between known attacks to generate possible paths of unknown attack. The latter represent combinations of known and unknown steps that might be carried out by attackers to initiate unknown attacks.

Based on these challenges, IDSs need to integrate situation-based and event-based information to cover several aspects of a particular situation, such as (1) information about semantic relationships between events that are relevant to that situation, (2) information about the characteristics of the targeted systems in order to exclude some predictions of IDSs that are not applicable in that situation, (3) information about the activities that target that system, (4) relationships between such activities in regard to their sources and time of occurrence, and (5) information to enable an IDS to have better knowledge about how attackers evolve a set of activities to generate unknown attacks. In summary, an IDS must be "context aware" and empowered with automated techniques for capturing and utilizing contextual information to effectively detect cyber-attacks.

1.1.2 Contextual Information Fusion in IDSs

Context has been utilized in different computing areas where it is vital to be aware of the current situation. Generally, the purpose of creating context-aware IDSs is to decrease the dependency on human experts who perform correlation between runtime activities to determine the current situation and react accordingly.

According to several studies, elements for the description of context information fall into five categories *individuality*, *activity*, *location*, *time*, and *relations* [12, 13] as shown in Fig. 1.1. The *activity* predominantly defines the relevance of context elements in specific situations, and it covers all tasks the entity may be involved in. The *location* and *time* primarily drive the creation of relations between activities that target that entity. The *individuality* category contains properties and attributes describing the entity itself. The *relations* category represents information about any possible relationship between activities that target such entity. Based on these categories, the aspects of contextual information create a situation that consists of several circumstances that can lead to/avoid a completion of a specific task.

In order for an IDS to be aware of context, infusion of these five context aspects in the intrusion detection process is essential. First, **location information** reveals the physical or virtual information about location. The IDS has to be aware of the location of victims and attackers. This is very significant to identify relationships between activities that target the system based on the source location of such activities.

Furthermore, semantic correlation with respect to source and target location is necessary to discover multistep attacks.

Second, the IDS has to be aware of **time information** which refers to the time of events that target a particular entity. For instance, the occurrence of two activities in

Fig. 1.1 Contextual information categories

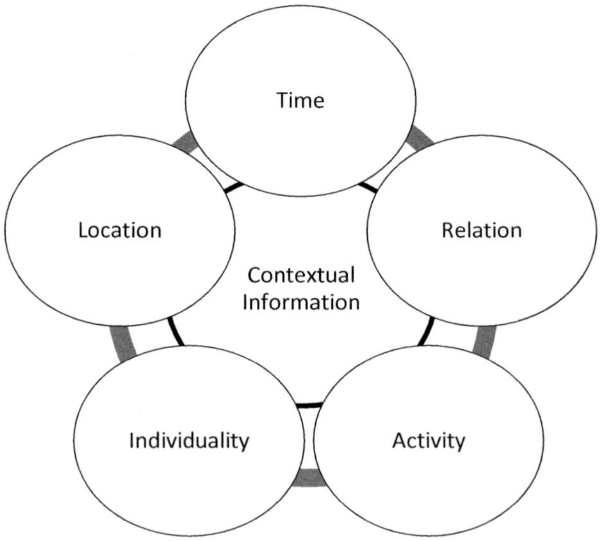

several time intervals indicates a possible relationship between them. Moreover, the current situation of an entity is also part of time context. The configuration of computer systems changes from time to time; it is essential to capture this change in order to identify the dynamic properties of the system from time to time. For instance, when a specific workstation is updated to fix security vulnerabilities, the time of such an update should be added to that workstation's profile. This makes it easier for an IDS to be aware of the current configuration on the target systems and the relevancy of activities based on their time of occurrence.

Third, **activity information** describes events that are applicable to the system. This category of contextual information is the major element in the intrusion detection process. The information in this category covers all events that occur during the system execution time. The set of activities that target the system can lead to one or more cyber-attacks. In general, the activity element of context is important to create event- and/or situation-based prediction models to detect these cyber-attacks.

For event-based prediction models, the activity aspect of context needs to be profiled and used to predict future attacks based on their history of occurrence. For situation-based prediction models, activity contextual features are needed to identify relationships between suspicious events, given a specific situation.

Fourth, the **relations** category of contextual information is significant to identify dependency between multiple events. The relation aspect of context is identified over other categories of contextual information such as time, location, and activity. It is very important to capture contextual relationships and use them in attack prediction. As part of demonstrating the relation aspect of context, if two alerts are related in terms of time of occurrence, targeted locations, and activities that lead to them, such a relation needs to be captured using a specific modeling approach (e.g., a graph with node and edges). In intrusion detection process, contextual relationships are significant to analyze situations rather than just single events.

Fifth, the environmental characteristics of computing entities are captured through the **individuality** aspect of context. For instance, the current characteristics of computer systems, their applications, and the patches applied are considered significant to realize the impact of the activities in progress on the targeted system. Some suspicious events are deemed as nonrelevant when the system is patched against them.

Utilizing context is of utmost importance in improving the effectiveness of the intrusion detection process. As part of this work, we propose a substantial change in approaching these challenges by taking advantage of these categories of contextual information to create a framework which intelligently assists intrusion detection techniques to predict related suspicious activities, identify their actual impact on the targeted system based on the current situation, filter out nonrelevant threats based on the current situation, and be able to detect modifications the attacker can make on a set of known activities to initiate unknown attacks. While we devise an approach that would be classified as research in data mining, databases, statistics, and machine learning, our methodology significantly enhances these techniques through working situations rather than single events using relational databases as evidenced by our recent research, which has revealed encouraging results attributed to the use of context [14–20]. We will use the following attack scenario to explain some motivations of utilizing contextual information in the intrusion detection process.

Attack Scenario 1.1
This attack scenario describes an attempt by an attacker to overflow a buffer and then proceed to execute arbitrary code on a host running Microsoft Internet Information Server (IIS).

The attacker may be trying to overflow a buffer via exploiting vulnerabilities in nsiislog.dll and idq.dll files. The vulnerability arises when requests for "Server Side Includes" are not properly checked by the Web server or when IIS is vulnerable to a buffer overflow when handling ISAPI (Internet Server Application Programming Interface) extensions. An unchecked buffer in the code that handles idq.dll ISAPI extensions in the indexing service for IIS allows a remote attacker to overflow a buffer and execute code by sending a specially crafted indexing service request. An attacker could exploit this vulnerability to gain complete control over the affected server. The alerts that could be raised by an IDS which are identified by their identification number (Alert ID) are shown in Table 1.1. The log shows that some activities that targeted a specific host with IP (154.241.88.201) at a specific time lead to three alerts. The first one "bare byte encoding" with ID [119:4:1] is raised by an IDS in response to an attacker attempt to encode a Web request using a nonstandard format. Microsoft IIS servers are able to use non-ASCII characters as values when decoding UTF-8 values. This is a nonstandard behavior for a Web server, and it violates RFC (Request for Comments) recommendations. All non-ASCII values should be encoded with a %. Usually, bad encoding is an indication of initial steps in a buffer-overflow attack against a Web server. The IDS also raises another two alerts to indicate a suspicious attempt to compromise WEB-IIS isapi. idq ([1:2229:5]) and WEB-IIS nsiislog.dll ([1:2129:13) files. The role of an IDS stops at this point. It is then the responsibility of domain experts to investigate and analyze this set of security alerts in search of possible intrusions. First, how are these alerts related given their time of occurrence, the source, target locations, and the set of activities that lead to them? Second, the domain expert needs to

Table 1.1 IDS alert log: attack scenario 1.1

Alert ID	Alert description
[119:4:1]	[**] (http_inspect) bare byte unicode encoding [**] [Priority: 3] 11/08-13:01:52.972141 **10.2.190.254:50559**-> **154.241.88.201:80**
[1:2229:5]	[**]WEB-IIS isapi.idq attempt [**] 11/08-13:01:58.165085 **10.2.190.254:36887** -> **154.241.88.201:80**
[1:2129:13]	[**]] WEB-IIS nsiislog.dll access [**] 11/08-13:05:08.477276 **10.2.190.254:55772** -> **154.241.88.201:80**

Fig. 1.2 Relationships between alerts based on attack scenario 1.1

investigate if these alerts are relevant to the target host according to the current configuration on that host and if one (or more) of these alerts is (are) false positive (s). Intuitively, in order to identify the relationships between these alerts, the domain expert needs to investigate their description and correlate the activities that lead to them, the time of these alerts, and the locations targeted. In other words, a domain expert needs to identify the context under which these alerts occur. The identification of relationships is significant in order to infer if these alerts co-occur on a regular basis. Since these relationships cannot be identified manually due to large amounts of alerts contrasted with the attack scenario above, one solution is to automatically find the similarity between alerts. A feature-based similarity metric can be used to identify the relationship between these alerts. A quick look at the description feature of these alerts in Table 1.1 along with their time of occurrence indicates low or no similarity between the bad encoding (bare byte encoding) alert and the buffer-overflow attempt against WEB-IIS nsiislog.dll (see Fig. 1.2). The source of low similarity is the difference in their signatures. This difference is clearly observed from the description of both alerts.

Now let us consider the alerts about bad encoding (bare byte encoding) and the buffer overflow against WEB-IIS isapi.idq. Although their signatures are dissimilar, they occur close to each other in terms of time of occurrence; thus, from the time aspect of context, they are similar. Let this similarity be denoted by sim_1. On the contrary, we expect some similarity between buffer-overflow WEB-IIS isapi.idq and WEB-IIS nsiislog.dll alerts based on their description. Therefore, the set of activities that lead to these two alerts are also related. Let this similarity be denoted by sim_2. The following graph depicts the alerts (represented as nodes) along with the similarity values assigned on edges that connect alerts.

While there is no direct relationship between the bad encoding (bare byte encoding) and the buffer-overflow (WEB-IIS nsiislog.dll) alerts, based on this similarity-based graph, an inference mechanism can detect an indirect relationship between these two alerts through the buffer-overflow (WEB-IIS ISAPI.idq) alert. Therefore, based on such an inference mechanism, bare byte encoding and WEB-IIS nsiislog.dll are semantically related, and they can be observed together in similar or close context. Using this relationship, when an IDS raises an alert about bare byte encoding, it needs to consider the contextual relationship with other buffer-overflow attempts so that all three alerts can be predicted together as a possible attack scenario. The produced contextual relationships can be utilized to generate unknown attack paths an attacker might follow to create unknown variations of this buffer-overflow attempt (more details will be given later). One last question is whether the discovered alerts are false positives according to the

Table 1.2 Information on the targeted host: attack scenario 1.1

Target host	OS	Alert	Description	Category	Patch
`154.241.88.201`	Windows	`WEB-IIS nsiislog.dll access`	This event is generated when an attempt is made to access nsiislog.dll on a host running Microsoft Internet Information Server (IIS)	Web application attack	MS03-022

contextual situation on the target host. Based on the current configuration or the individuality-based characteristics shown in Table 1.2, assume that the target host has been recently patched against the vulnerability causes WEB-IIS nsiislog.dll buffer-overflow attempt. This can be used to infer that the buffer-overflow (WEB-IIS ISAPI nsiislog.dll) alert is a false positive. Therefore, it must be filtered out. While an IDS should be aware of the individuality-based contextual information or the current configuration on the targeted hosts, it is only aware of what is known about the targeted host. For instance, based on the attack scenario described above, the IDS has no idea if patches are applied against bad encoding and WEB-IIS ISAPI.idq buffer-overflow attempts; therefore, the target host is considered incorrectly vulnerable to this type of alerts.

1.2 Statement of the Problem and Research Questions

We can clearly see from the discussion and the examples above that contextual information is a critical part of the current state awareness during the attack prediction process. Thus, we need to utilize contextual information to address several problems in the area of intrusion detection. The goal of this work is therefore motivated by finding solutions to each of these problems guided by the following research questions.

- Given a large amount of intrusion detection data, how important are the discovered contextual relationships between suspicious activities to accurately predict contextually related cyber-attacks? Creating techniques that automatically discover contextual relationships between suspicious activities is a critical factor in the attack detection process. The techniques we utilize to discover such relationships aim to address several significant challenges including the early detection of cyber-attacks, discovering attacks that occur in several steps, and lowering the possibility of false negatives (i.e., predicting attacks as benign activities).
- Using the information extracted about one or more aspects of context, is it effective to implement a multi-evidence intrusion detection process? Instead of relying on a single signature-based intrusion detection technique, we used the categories of contextual information described earlier to build a coherent

analysis of evidence to recognize attacks. Therefore, more informative and comprehensive decisions are taken based on context. It is important to notice that context-aware IDSs need to utilize multiple dimensions of contextual information as opposed to current attack graphs and anomaly detection techniques. Information extracted about each dimension is supposed to contribute to the final decision about suspicious activities. Using multiple dimensions of contextual information during attack prediction process addresses one of the key problems in the existing rule-based IDSs, which only rely on signature matching to discover attacks without being aware of the surrounding context.

- Using contextual information, can we predict how the attacker exploits the relationships between two known attacks to initiate an unknown one? Addressing this challenge means ultimately discovering how situations evolve using contextual relationships between known attacks; as such, we can use the discovered relationships to create prediction models that have the capability of discovering unknown attacks. We do not only rely on the observed anomalous behavior as in anomaly detection techniques but also on the combination of steps the attackers utilize from two or more known attacks to initiate an unknown attack.
- How significant is the contextual information available about the target domain in creating effective attack prediction models? Additionally, is it possible to effectively use the partial information about the target domain, such as information about the current configuration on the targeted host, to decrease the rate of false positives and achieve a high true positive rate? We, therefore, focus on using proper context modeling techniques to fuse and use such information to improve the detection rate of attacks and decrease false positives. For instance, these days a substantial amount of security-based contextual information is organized in ontologies or taxonomies. Extracting such a background knowledge and fusing it in the intrusion detection systems is one of the objectives of this work. Since we are utilizing multiple categories of contextual information, this not an easy task. The major challenge is to utilize the best combination of contextual models to use such information in a coherent, unified, and context-sensitive attack prediction process.

1.3 Approach Summary

The approach followed in this work focuses on utilizing the five elements of contextual information—activity, individuality, time, location, and relations—to improve the detection rate of cyber-attacks while decreasing the false alarms. We formulate the problem of cyber-attack detection so that it benefits from meta-content represented through several knowledge-based prediction models. Our prediction models are created as part of one framework using information about one or more of the contextual aspects discussed earlier. The produced models are implemented in a layered manner on top of machine learning-based intrusion techniques, and their effectiveness is measured. Since our framework is supposed to work on top of IDSs, the low-level, event-based analysis performed by such systems is still needed. By contrast, a major advantage of the prediction models we created is to analyze situations rather than individual events.

The relation aspect of context is modeled as graphs on which we identify contextual relationship between different types of suspicious and benign activities extracted from labeled intrusion detection datasets. Semantic reasoning is performed on these graphs to create semantic link networks (SLNs) with nodes representing cyber-attacks or benign activities and edges representing the strength (weight) of semantic relationships between them. The stronger the relationship between nodes, the higher the possibility they co-occur under a particular context. Consequently, identifying one suspicious node can help to proactively avoid another. The initial relationships between nodes are formed in terms of similarity between activities that cause them, and/or their source, target and time of occurrence. Therefore, SLNs are created by utilizing activity, time, or location contextual features or combinations of them. The semantic inference is performed to identify the most feasible relationships between nodes. Since SLNs are created solely based on data extracted from intrusion detection datasets, one needs to adapt them using domain knowledge about nodes mainly to improve the quality of the resulting semantic relationships. Such knowledge has been extracted from taxonomies of attacks. An SLN takes an initial prediction produced by a particular machine learning-based detection technique. It then retrieves other relevant nodes using the pre-identified relationships.

The activity contextual information is utilized to create activity-based profiles which are feature-based attack profiles that we utilize to distinguish among known attacks, unknown attacks, and benign activities. Attack profiles are utilized to decrease uncertainty about the predictions produced by SLNs; thus, they are applied to filter out some predictions of SLNs. A novel technique is utilized to create attack profiles using the features extracted from suspicious network connections. This technique takes into account the contextual relationships between known attacks to generate their corresponding profiles. It is an entropy-based technique which projects future possible actions/activities of an adversary and the unknown paths the adversary may take.

The time and location contextual information is utilized to form time- and location-based relationships between nodes on SLNs. Consequently, we created two types of SLNs: the first one is created with time and location information, and the second is created with such information. The effectiveness of both forms of SLNs in terms of attack detection and false positive rates is then measured.

The individuality contextual information is utilized to create host profiles. Host profiles are used to check if the predictions produced by other prediction layers are relevant based on the current configuration on the target hosts. Some of these predictions can lead to real damage or unauthorized access to user data if the target system is not patched. Other predictions need to be filtered out if they are not relevant based on the current situation on the targeted hosts.

Validation: To validate our framework, we relied on two criteria, first to conceptualize the problem and thus validate that our solution is theoretically valid and second to experimentally validate the effectiveness of our framework. Theoretically, we show several theoretical properties in our framework using graph, probability, and information theories. We validate the viability of infusing contextual information in creating semantic links between cyber-attacks and the role of such information for detecting events that are close to each other in context; we then mathematically demonstrate that using conditional entropy to create attack profiles results in identifying known attacks. Using conditional entropy to create contextual relationships between known attacks, we also demonstrate the validity for using attack profiles to detect unknown attacks as variants of known attacks.

Experimentally, we performed an extensive validation on our framework using benchmark intrusion detection datasets. The purpose of our validation is to demonstrate the usefulness of contextual information in creating attack prediction models that can be applied in conjunction with the intrusion detection techniques.

Our approach has been evaluated mainly on two datasets. The first dataset consists of network connections; each connection consists of all activities performed during sessions of particular duration. The features of each connection include the data portion of network packets. Each connection is classified as a benign activity or an attack with specific type. The second dataset consists of IP flows which only aggregate traffic-based activities and their time and location without any content features. Each flow is classified as benign or suspicious of a specific type.

During the validation phase, numerous experiments are conducted to measure the effectiveness of the prediction models in our framework. The experiments focus on measuring the detection rate of known and unknown cyber-attacks when our prediction models are applied on top of other machine learning-based intrusion detection techniques.

Initially, our experiments emphasize on measuring the detection rate of known cyber-attacks using SLNs, attack profiles, and host profiles. We then compare our experimental results with several machine learning techniques.

Moreover, we focus on measuring the detection rate of unknown attacks using combinations of several prediction models in our framework. New connections/flows with unknown attack signatures are added during the validation phase to demonstrate the effectiveness of our framework in terms of identifying unknown attacks. We compare the prediction models we created to detect unknown attacks with several semi-supervised anomaly detection techniques.

Several metrics are utilized to measure the effectiveness of our prediction models, including precision, recall, true positive and false positive rates, and ROC curve. In addition, our framework has also been tested from other perspectives such as its efficiency when applied at runtime.

1.4 Contributions

The framework introduced in this work makes significant contributions to several research problems:

First, we produce a substantial enhancement in detecting cyber-attacks using contextual semantic relationships. As opposed to other statistical intrusion detection models, the creation and reasoning of semantically augmented network of related security attacks using contextual information leads to improvements in the detection rate of cyber-attacks.

Second, we improve the detection rate of unknown (zero-day) attacks by utilizing activity and relation aspects of contextual information to discover such attacks as variations of known attacks. A novel technique is utilized in creating attack profiles based on activity features extracted from network traffic. The technique to create attack profiles utilizes the contextual relationships between known attacks; therefore, it assists in predicting unknown attacks by projecting future possible actions/activities of an adversary and the paths the adversary might take.

Third, we decrease the false positive rate in detecting cyber-attacks by utilizing multiple categories of contextual information to create intelligent prediction models and apply them in a layered manner. Contextual semantic relationships are represented as SLNs that are used to expand the predictions of other intrusion detection techniques. Then activity and individuality contextual information is utilized to create context profiles that are applied as filters to the predictions of SLNs. Attack profiles are created by taking into account both the activity and relation contextual information, and they are used to restrict the predictions produced by SLNs to improve the precision of predicting real attacks. Host profiles decrease false positives by utilizing individuality contextual information. Host profiles provide an awareness of a situation on the targeted hosts up to the point the decision to generate an alert is about to be made. They consist of information such as the current configuration on the targeted host and the patches applied. Host profiles are applied on top of SLNS to remove some predictions generated by SLNs if they are not relevant to the targeted hosts.

Fourth, we automate the manual and daunting process of human decisions about the possible semantic relationships between security incidents by utilizing a hybrid approach that considers the contextual patterns in the data and the domain knowledge-based approach to automate the generation of these relationships. The quality of the produced relationships cannot be produced neither by simple data mining techniques nor by manual investigation performed by domain experts. The produced relationships assist in gaining better understanding of the possible impacts of events that target computer networks.

Fifth, we demonstrate the effectiveness of utilizing contextual information in increasing true positive (TP) and decreasing false positive (FP) rates when network data at different granularity level is analyzed using the prediction models in our framework. We show that contextual information can be still applied to identify cyber-attacks from IP flows which lack the data portion of network packets which is usually needed to discover attacks by the existing IDSs.

Sixth, we reduce the uncertainty in the process of discovering unknown attacks using a combined approach that utilizes attack profiles with semi-supervised anomaly detection techniques. We also create a semiautomatic tuning mechanism to adjust the relative importance a domain expert gives to TPs and FPs during the investigation for unknown attacks from network data.

Seventh, we empirically prove the feasibility of utilizing time and location aspects of context in creating better-quality semantic relationships which are used to detect activities that are part of multistep attacks. In flow-based detection mode in our framework, time and location aspects of context discovered several forms of semantic relationships among suspicious activities that cannot be discovered by utilizing similarity over other activity features.

Eighth, we improve the efficiency and effectiveness of the detection of unknown attacks using dimensionality reduction and linear data transformation techniques. The linear data transformation technique is applied in a way that preserves the contextual relationships between known attacks. In particular, attack profiles are re-created as linear discriminant functions and utilized in detecting unknown attacks in a reduced dimensional space.

Ninth, we validate the effectiveness and efficiency of our framework on benchmark intrusion datasets and demonstrate that it achieves better effectiveness compared to other data mining (signature and anomaly-based attack detection) techniques. The prediction models of our framework significantly enhance the intelligence of cybersecurity systems to gain more situation awareness so that they become less dependent on humans to take decisions about security incidents.

References

1. Jones, A.K., Sielken, R.S.: Computer system intrusion detection: a survey. Computer Science Technical Report: Computer science technical report, Department of Computer Science, University of Virginia (2000)
2. Powell, D., Stroud, R.: Malicious and accidental fault tolerance for internet applications conceptual model and architecture. Technical report series-University of Newcastle Upon Tyne Computing Science: Technical report series, University of Newcastle Upon Tyne Computing Science (2001)
3. Halme, L.R.: Ain't misbehaving a taxonomy of anti-intrusion techniques. Comput. Secur. **14** (7), 606 (1995)
4. Kruegel, C., Valeur, F., Vigna, G.: Intrusion detection and correlation: challenges and solutions, vol. 14, Advances in Information Security, Springer (2004)
5. Axelsson, S.: Intrusion detection systems: a survey and taxonomy. Chalmers Univ: Technical report (2000)
6. Debar, H., Dacier, M., Wespi, A.: Towards a taxonomy of intrusion detection systems. Comput. Network. **31**(8), 805–822 (1999)
7. Debar, H., Dacier, M., Wespi, A.: A revised taxonomy for intrusion detection systems. Ann. Telecommun. **55**(7), 361–378 (2000)
8. Roesch, M.: Snort intrusion detection system. http://www.snort.org. Accessed 22 June 2014

9. Shon, T., Moon, J.: A hybrid machine learning approach to network anomaly detection. Inform. Sci. **177**(18), 3799–3821 (2007)
10. Song, J., Takakura, H., Kwon, Y.: A generalized feature extraction scheme to detect 0-day attacks via IDS alerts. In: Proceedings of the International Symposium on Applications and the Internet, pp. 55–61, Turku, Finland (2008). 1442004: IEEE Computer Society. doi: 10.1109/saint.2008.85.
11. Guan, Y., Ghorbani, A.A., Belacel, N.: Y-means: a clustering method for intrusion detection. In: IEEE Canadian Conference on Electrical and Computer Engineering, vol. 2, pp. 1083–1086, Montreal, Canada (2003) doi:10.1109/CCECE.2003.1226084. Accessed 4–7 May 2003
12. Gross, T., Specht, M.: Awareness in context-aware information systems. In: Mensch & computer conference, vol. 1, pp. 173–182, Citeseer, Germany, (2001)
13. Zimmermann, A., Lorenz, A., Oppermann, R.: An operational definition of context. In: Proceedings of the 6th International and Interdisciplinary Conference on Modeling and Using Context (Context'07), pp. 558–571 Roskilde University, Denmark (2007)
14. AlEroud, A., Karabatis, G.: A system for cyber attack detection using contextual semantics. In: 7th International Conference on Knowledge Management in Organizations: Service and Cloud Computing, pp. 431–442. Salamanca, Spain (2012)
15. AlEroud, A., Karabatis, G.: A contextual anomaly detection approach to discover zero-day attacks. In: ASE International Conference on Cyber Security, pp. 40–45. Washington, DC, USA (2013a)
16. AlEroud, A., Karabatis, G.: Discovering unknown cyber attacks using contextual misuse and anomaly detection. ASE Sci. J. **1**(1), 106–120 (2013)
17. AlEroud, A., Karabatis, G.: Toward zero-day attack identification using linear data transformation techniques. In: IEEE 7th International Conference on Software Security and Reliability (SERE'13), pp. 159–168. Washington, DC (2013c). doi:10.1109/SERE.2013.16. Accessed 18–20 June 2013
18. AlEroud, A., Karabatis, G.: Context infusion in semantic link networks to detect cyber-attacks: a flow-based detection approach. In: Eighth IEEE International Conference on Semantic Computing (2014a) Newport Beach, CA, USA, IEEE
19. AlEroud, A., Karabatis, G.: Detecting zero-day attacks using contextual relations. In: Ninth International Knowledge Management in Organizations Conference, vol. 185, Springer, Santiago, Chile (2014b)
20. Aleroud, A., Karabatis, G., Sharma, P., He, P.: Context and semantics for detection of cyber attacks. Int. J. Inf. Comput. Secur. **6**(1), 63–92 (2014). doi:10.1504/ijics.2014.059791

Chapter 2
A Framework for Contextual Information Fusion to Detect Cyber-Attacks

Ahmed AlEroud and George Karabatis

Abstract The focus of this research is a novel contextual approach that will be used in detecting zero-day cyber-attacks, generating possible zero-day attack signatures, and automatically measuring their risk on specific software components. In general, zero-day attacks exploit a software vulnerability that has not been discovered, and it is called zero-day vulnerability. This work proposes an approach to identify both zero-day attacks (in real time) and also zero-day vulnerabilities by examining known software vulnerabilities.

The proposed work is an innovative approach, which automatically and efficiently extracts, processes, and takes advantage of contextual information to identify zero-day attacks and vulnerabilities. Contextual information (time, location, etc.) identifies the context that can be used to infer relations between entities, such as cyber-attacks. These relations are called contextual relations. We propose methods to generate zero-day attack signatures using graph-based contextual relations between (1) known attacks and (2) vulnerable software components. These are certainly hard problems to solve, and we doubt that incremental improvements in IDSs will result in a significant solution that drastically improves their effectiveness. Consequently, we propose a substantially different and novel approach: contextual relations, if used intelligently, can reduce the search space in IDSs so that zero-day attacks can be identified in realistic and practical amount of time. There are several reasons that led us to investigate the use of contextual relations to detect zero-day attacks. First, the traditional data mining and pattern recognition techniques lack the desirable effectiveness since they focus on analyzing the data without the use of context. To better identify suspicious activities, direct and indirect contextual paths need to be identified among these activities. These are usually identified manually by domain experts (e.g., identifying relations between cyber-attacks). However, it is quite daunting and challenging to identify all possible

A. AlEroud (✉)
Department of Computer Information Systems, Yarmouk University, Irbid 21163, Jordan
e-mail: Ahmed.aleroud@yu.edu.jo

G. Karabatis
Department of Information Systems, University of Maryland, Baltimore County (UMBC), 1000 Hilltop Circle, Baltimore, MD 21250, USA
e-mail: georgek@umbc.edu

I.M. Alsmadi et al. (eds.), *Information Fusion for Cyber-Security Analytics*, Studies in Computational Intelligence 691, DOI 10.1007/978-3-319-44257-0_2

relations via manual investigation. Second, there are several contextual relations that need to be identified among vulnerabilities to predict which ones can lead to zero-day attacks and the software modules they are located, thus, empowering us to generate possible signatures for these attacks.

2.1 Limitations of the Current Research

The current intrusion detection techniques, which utilize some context aspects to detect known and zero-day attacks, have significant limitations. These limitations fall in three major categories as follows:

1. *Lack of semantic relations when context is applied to detect attacks*
   The intrusion detection approaches, which have been discussed in Chap. 2, consider some contextual information in their operations. However, such approaches struggle when it comes to apply *semantic relations* with context. Utilizing semantic relations in a specific context is quite significant for predicting cyber-attacks. The attack graph approaches utilized by Noel et al. [1–4] consider the relationships between machines with vulnerabilities, where an attack graph of machines which have vulnerabilities is constructed before the events occur. An alert is raised at runtime if its corresponding events are mapped to adjacent vulnerabilities in the attack graph. While the attack graphs incorporate some contextual information such as the relationships between the vulnerabilities of machines, their major limitations are threefold.
   First, an attack graph requires to be traversed at runtime in order to create possible attack scenarios. However, an attack graph contains in general too many paths. Traversing all paths to discover possible attack scenarios makes the detection process not efficient at runtime.
   Second, the attack graph is only aware of known vulnerabilities; thus, it cannot be used to detect unknown or zero-day attacks.
   Third, updating such graphs, with new knowledge about the domain (e.g., information about machine vulnerabilities), in a real-time manner is not feasible. As new vulnerabilities are added, the paths, in these graphs between different vulnerabilities, need to be updated in order to add new attack scenarios. Due to these limitations, attack graph approaches might result in high false alarm rate, specifically if machines are recently patched against the exploits in attack graph. Attack scenarios need to be precomputed and well defined by incorporating semantic relations between contextually related attack scenarios. We believe that it is more feasible to pre-calculate the contextual semantic relationships between possible attacks by (1) finding the similarity between these attack features, (2) pre-calculating the possible paths that connect these attacks, and (3) automatically augmenting such relations using domain expertise.
   Few studies have investigated the issue of incorporating semantic information in the process of detecting cyber-attacks. Mathew et al. [5] address the third limitation of attack graphs. They present a strategy for a real-time situation

awareness of multistage cyber-attacks. They adopt a practical correlation approach that considers high-level multistage attack semantics of heterogeneous sensor events without delving into vulnerability level details. Attack models called guidance templates are used to guide this correlation approach and produce dynamic attack hypotheses that are relevant to an evolving situation. However, there are two limitations in this approach. *First*, it does not maintain any statistical models to discover relationships between attacks, such as the similarity between attack features. While defining guidance template is definitely useful to capture relations between attacks, identifying such relations in a manual manner is not effective to capture all possible relations between attacks. *Second*, this approach completely ignores the information about the target environment, such as information about host vulnerabilities. Thus, this approach would result in a higher false alarm rate, if the hosts were patched against these vulnerabilities, but the IDS was not aware about it.

2. *Overlooking the similarity with known attack contextual profiles to discover potential zero-day attacks*
   A zero-day attack is an attack, which utilizes the previously unknown vulnerability in a computer application. A zero-day attack occurs on day zero of awareness of the vulnerability that causes it. The developers have no knowledge about zero day to address and patch such vulnerability. Since approaches that utilize vulnerability databases fail to detect zero-day attacks, there have been few studies addressing zero-day attack detection problem using machine learning techniques such as unsupervised anomaly detection techniques [6, 7], support vector machines (SVM) [8], and clustering approaches [7, 9].
   The main assumptions in these approaches are twofold: first, to assume any instance (e.g., connection) *that does not conform to a well-defined notion of normal activity as a zero-day attack* and, second, to assume any instance *that does not match the known attack signatures as a normal activity*. However, there are two main shortcomings in these assumptions; for the first assumption, it is not always possible to generate well-defined normal profiles. For the second assumption, it ignores the fact that in general, the zero-day attacks have some degree of similarity with one or more known attack signatures. The previous approaches, however, look at the problem as semi-supervised anomaly detection one, focusing only on discovering zero-day attacks as anomalies that deviate from normal profiles. Since it is not easy to define normal profiles because there are too many of normal patterns, utilizing only semi-supervised anomaly detection techniques to detect zero-day attacks might lead to a high false positive rate. As reported by Leung [6], considering semi-supervised anomaly detection, such as one-class SVM (using the inner product technique), leads to more than 48 % of false positives (normal activities predicted as zero-day attacks).
   Figure 2.1 gives an example of the problem that existing approaches have in detecting zero-day attacks. The figure shows several normal profiles as one cluster. These profiles can be created using machine learning techniques, and they describe normal signatures; thus, it can be used to match incoming connections at runtime. The figure also shows another two clusters of known attack

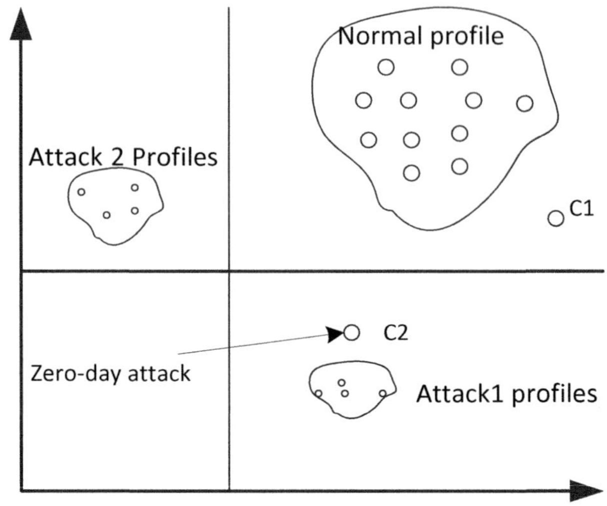

Fig. 2.1 The limitation of detecting zero-day attack using anomaly detection

profiles. For simplicity, we assume that a specific IDS has knowledge about these two attacks only. Each attack cluster consists of several profiles for the same attack.

Let us also assume that there are two network connections C_1 and C_2 that target a system which utilizes a semi-supervised anomaly-based ID technique to detect attacks. While the network connection C_1 is far from both attack 1 and attack 2 profiles, it may be predicted, by the semi-supervised anomaly techniques, as a zero-day attack because it is relatively far from normal profiles. However, since C_1 is not similar to any attack profiles, it is very possible that it is a new pattern of normal activities. Thus, the typical prediction, by the anomaly detection technique, would be most likely a false positive. On the other hand, if C_1 similarity with attack 1 and attack 2 is considered, it may be deemed as a normal activity because it is not similar to any attack profile.

Additionally if the similarity between connection C_2 features and both attack profiles is considered, C_2 may be declared as zero-day attack, since (1) it is far normal profiles, and thus it can be declared as an anomaly, and (2) it is also similar to attack 1 profiles. Thus, for C_2 connection we have two evidences that it could be a zero-day attack. Considering similarity with profiles of known attacks can be significant to avoid a large amount of false positives. Thus, we aim to combine both semi-supervised anomaly detection techniques and similarity with known attack profiles in order to detect zero-day attacks. We want to examine if such a combination would result in improving the zero-day attack detection rate, minimizing the false positive rate, and improving the computational efficiency in the detection process.

3. *Existing attack detection techniques are not computationally feasible*
   The network data, which has to be analyzed by IDS, contains too many numerical and categorical attributes. Thus, the dimensionality of this data is relatively high. When such data is analyzed at system runtime to detect possible intrusions, the computational time of the detection process is definitely high. In anomaly detection approaches, this problem is more noticeable. These techniques need to compare the incoming connections with too many normal profiles in order to discover zero-day intrusions, and thus these techniques are not computationally efficient.
   There are several approaches that have been applied to reduce the dimensionality of ID data, such as principal component analysis (PCA) [10–12], singular value decomposition (SVD) [10], local linear embedding (LLE) [13], and linear discriminant analysis (LDA) [14]. However, the existing approaches apply the dimensionality reduction process on the entire distribution of the dataset. Regardless of how effective these approaches are, the dimensionality reduction techniques have not been utilized to discover the local *context* where a particular attack occurs, that is, the dimensionality reduction techniques have not been performed, by considering only the instances of that attack in the ID data. The local context, such as the *activities* which identify the context in which a particular attack happens, is very important to investigate for zero-day attacks. As mentioned earlier, such attacks have a similarity with known attack profiles. Thus, the ability to accurately identify the local contexts of attacks (i.e., their context profiles) has two benefits in the reduced dimensional space. *First*, it minimizes the computational time of the detection process as the incoming connections need to be compared with few attack profiles described in the reduced dimensional space. *Second*, it can be used to detect zero-day attacks that occur in a similar context of particular known attack.

As a conclusion of our discussion, there is a need for an approach that provides the following functionalities to effectively detect known and zero-day attacks:

- To apply semantic relations with context in the process of intrusion detection
- To consider the characteristics of an environment (e.g., the patches on hosts) in detecting known and zero-day attacks
- To consider both anomaly detection and similarity with known attack context profiles in detecting zero-day attacks
- To apply dimensionality reduction and other transformation techniques, within a particular context, not on the entire distribution of data
- To consider other contextual aspects in data such as location and time in detecting known and zero-day attacks

In order to address the limitations discussed above, we propose a framework, which will be utilized to extract, model, and use contextual information in detecting both known and unknown attacks.

To address the first limitation, the proposed framework applies semantic networks of attacks and several types of context profile, which describe specific environment, to detect known attacks.

To address the second limitation, the proposed framework considers semi-supervised anomaly detection and attack profile similarity to detect zero-day attacks.

To address the third limitation, the proposed framework utilizes data transformations through linear discriminant analysis, which can efficiently handle numerical features in a reduced dimensional space.

In this chapter, we describe our proposed contextual framework to detect known and zero-day attacks. The proposed framework will be validated through a series of experiments, conducted on an intrusion detection dataset.

The chapter is organized as follows: In Sect. 2.1, we discuss the limitations of the existing research approaches, which utilize context in their operation and have been utilized to detect known and zero-day attacks. Next, in Sect. 2.2, we describe the components of the proposed contextual infusion framework, which will be used to detect known and zero-day attacks. The proposed framework aims to address the limitations of the existing approaches. The next section describes the components of the proposed framework in details.

2.2 A Framework for Contextual Information Fusion to Detect Known and Zero-Day Attacks

In this section, we present our framework, which consists of several contextual models that utilize different contextual information categories for the detection of known and zero-day attacks. Figure 2.2 shows a high-level architecture of our framework.

The process of creating, using, and evaluating this framework consists of three phases:

I. *Static phase*: during which ID data are being used to generate contextual models ahead of time that will be used to detect known and zero-day attack at runtime phase. The static phase consists of *data preprocessing*, *contextual information extraction*, and the *contextual information modeling* subphases as outlined below:

  - *Data preprocessing phase*: during this phase, several data preprocessing techniques are applied on ID data. Data preprocessing is needed to facilitate extracting contextual information, which will be used in creating contextual models (e.g., semantic networks) that have the capability to detect attacks.
  - *Contextual information extraction*: during this phase, several categories of contextual information (e.g., the similarity between attacks to identify

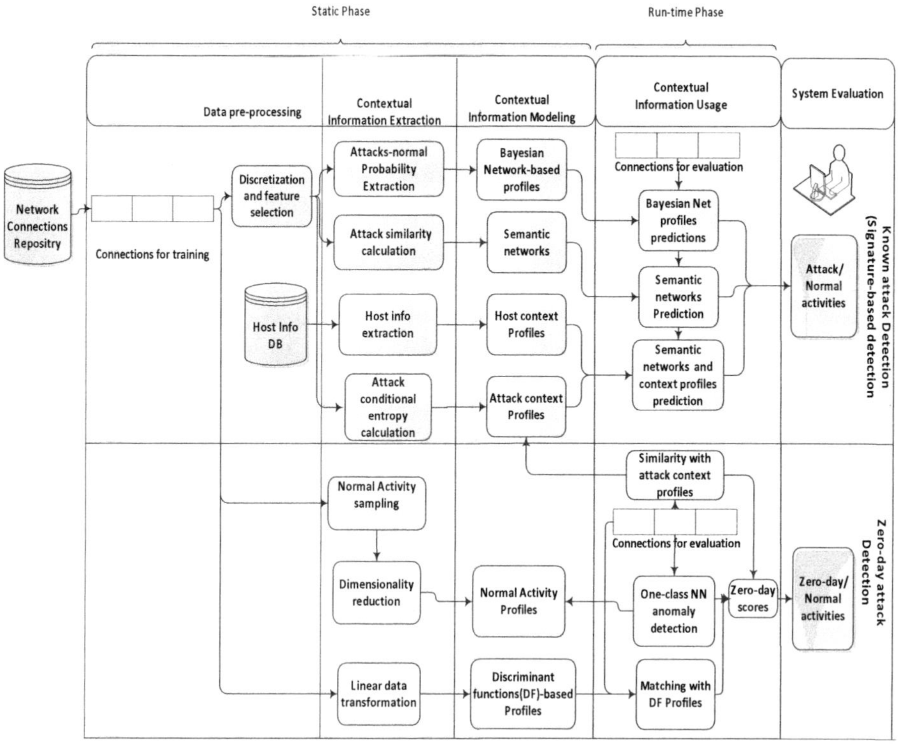

Fig. 2.2 The proposed framework for contextual information infusion to detect known and zero-day attacks

relation context) are extracted and identified. The extracted contextual information is then used to create contextual models as profiles.

- *Contextual information modeling*: during this phase, the contextual information extracted in the previous phase is modeled using several context modeling techniques. The contextual models are utilized at runtime to detect known and zero-day attack.

II. *Detection (runtime) phase*: during this phase, contextual model implementations are utilized to analyze an incoming connection and determine whether it is a known, a zero-day attack, or a normal activity. The runtime phase contains a contextual information usage subphase, which utilizes different types of profiles to discover attacks.

III. *System evaluation phase*: during this phase, a series of experiments have been conducted to validate the effectiveness and efficiency of the contextual models we utilized to detect known and zero-day attack. We use several measures during the evaluation phase, such as precision, recall, F-measure, true positive (TP) and false positive (FP) rates, detection accuracy (AC), and the ROC curves. The evaluation process is discussed in the experiment chapter.

It should be emphasized that the phases above have been applied to create contextual models to detect either known or zero-day attacks. The phases of creating these models are therefore customized based on the detection mode, which can be either *a known attack* (signature-based) detection mode or *zero-day attack* detection mode.

We next provide a high-level description of the ID data used as input, the detection modes, and phases utilized in creating the contextual models for each mode:

(a) *The network connection repository*: The system takes its input data from this repository, which serves as a database that stores data about connections that target computer networks. This data is collected by network sensors (e.g., IDSs), and it can be represented as features, which describe individual characteristics of the connection, such as connection *protocol*, connection *duration*, *services requested*, packet flow statistics, etc. The data is presented in a high-level format called *connection records*. Each connection consists of 41 features (38 numerical and 3 categorical). The connections are labeled as normal or attacks, where the attacks are categorized in one of the following four categories: user to root (U2R), remote to local (R2L), denial of service (DoS), and probe. The connections are divided into two parts:

   1. *The training part*: these connections have been utilized during the static phase to create the contextual models we utilize to detect known and zero-day attacks.
   2. *The evaluation part*: these are treated as incoming connections, which are used during runtime phase, to evaluate the rate of attack detection and also to measure how efficient the contextual models are.

(b) *System detection modes*
The framework is designed to detect attacks in two modes: the *known attack (signature-based) detection mode*, during which the known attacks are identified, and *the zero-day attack detection mode*, during which zero-day or unknown attacks are identified. We now provide a high-level description of both known and zero-day attack detection modes and the phases utilized in creating and using the contextual models for both modes:

   1. *The known attack (signature-based) detection mode*: The contextual models in this mode are used to detect known attacks, whose signatures can be detected within a specific context. The models in this mode are created through the following phases:
      I. *Data preprocessing*: During this phase, discretization is used to convert the numerical features into bins. Feature selection has been also utilized to identify features that best describe attacks and normal contexts hidden in training connections. The outcomes of feature selection and discretization processes are used to extract contextual information.
      II. *Contextual information extraction*: During this phase, contextual information is identified and extracted after data preprocessing is carried out

on training connections. The extracted information is used later to create contextual models that are eventually utilized during runtime to detect known attacks. During this phase, the following extraction procedures are carried out:

- *Attack and normal probability extraction*: The conditional and joint probabilities of attacks and normal activities are extracted, based on the occurrence of particular features (activities) in the training connections. The probability information extracted during this phase is used to model the *activity context*, which describes the conditions of occurrence of attacks and normal activities as Bayesian network-based context profiles.
- *Attack similarity calculation*: Several similarity coefficients are utilized to calculate the pairwise similarity between different types of attacks extracted from training connections. The similarity between attacks is used to model the *relation context* information as *semantic networks*.
- *Attack conditional entropy calculation*: The training connection records are used to calculate the conditional entropy of attacks given particular features. The features, which result in low conditional entropy values, are used to model the *activity context* as *attack context profiles*. These profiles are used to describe the preconditions of attack occurrence. Such profiles are used at runtime to predict the occurrence of different types of attacks.
- *Host information extraction*: Information about the environment of hosts, such as the recent patches, operating systems, applications, application versions, etc., is extracted from the host info DB. Automatic vulnerability scanners might be aware of collecting information about the computing environment. The information about hosts and their environment is used to model the *individuality* context information about hosts as *host profiles*.

III. *Contextual information modeling*: During this phase, the contextual information extracted during contextual information extraction phase is modeled using the following types of profiles:

- *Bayesian network-based profiles*: The extracted attacks and normal probability information are used to create these types of profiles. These profiles consist of different attack types, normal activities, connection feature values, and the probability of specific attack/normal activity given these feature values.
- *Semantic networks*: The attack similarity values calculated earlier are used to create semantic networks, which are graphs with nodes representing attacks and edges representing relationships between them. Domain knowledge about attacks is also added to these networks to further improve the quality of semantic relations between

attacks. The semantic relationships among attacks are used to expand predictions at runtime.

- *Attack context profiles*: The conditional entropy of attacks given feature values is used to create attack context profiles. These profiles consist of several known attack signatures. Each profile consists of an attack name and the connection features which minimize the uncertainty about the occurrence of that attack (i.e., the features which minimize conditional entropy of that attack). These profiles are used at runtime to detect attacks when the signatures described in their context profiles are triggered.
- *Host context profiles*: The information extracted about hosts is used to create host context profiles. These profiles incorporate contextual information about hosts and their environment and are used to filter out potential attacks that are out of host's current context.

IV. *Contextual information usage*: During this phase, the contextual models created during the previous phase are utilized to analyze runtime connection and examine if they are possible known attacks or normal activities. The types of predictions, which are carried out by contextual models at runtime, are:

- *Bayesian network profile predictions*: This type of prediction is performed as follows: the runtime connection features are analyzed. Then, the probabilities of each attack and the normal activity are calculated. The prediction with the maximum probability is considered the correct prediction. This prediction can also be passed to semantic networks, which will be used to expand the prediction made by Bayesian network profiles.
- *Semantic network predictions*: The initial prediction made by Bayesian network is selected as a starting node to search in semantic network for additional relevant attacks. Based on the strengths of that node's relation with other nodes in the semantic network, some of these nodes are added to the initial prediction. This expansion process helps in predicting these attacks ahead of time, especially if the initial attack is a step in a multistage attack.
- *Semantic network and context profile predictions*: Expanding the initial predictions may be helpful to include some semantically related attacks to the initial one; nevertheless, this expansion would result in adding some attacks which are not relevant to the current context. Thus, context profiles are used to filter out (discard) some semantic network predictions, which are not relevant to the current context. The process of discarding the nonrelevant semantic network predictions is done based on the *activity* context, that is, the flow of activities (feature values) in the corresponding incoming connection, on which the predictions are made. The feature values of these connections might not match the context profiles of attacks predicted

by semantic networks. Therefore, these predictions need to be filtered out using attack context profiles. Host context profiles are used to filter out the predictions, which are not relevant to the current status of host targeted by the connection. Each host is supposed to have a specific context. For instance, the host could be patched against specific vulnerability if such vulnerability is related to a particular attack that has been predicted by a semantic network; this attack needs to be discarded since it is not relevant to that host's *individuality* context.

2. *The zero-day attack detection mode*: In this mode, several contextual models are used to detect zero-day attacks given an incoming connection, which does not conform to a well-defined notion of normal activity or a known attack context profile. In this detection mode of our system, the connections used at runtime are not passed to Bayesian networks or to semantic networks because these models are specific to detect known attacks. The only part we reuse, from known attack detection models to discover zero-day attacks, is attack context profiles. The contextual models in the zero-day attack detection mode are created and used through the following phases:

   I. *Contextual information extraction*: During this phase, contextual information about normal and attack connections is identified and used to create specific contextual models that can be used to discover zero-day attacks. The following extraction procedures are carried out during this phase:

      - *Normal activity sampling*: This process is carried out to select representative normal training connections, which are used as part of an anomaly-based zero-day attack detection technique. These connections are represented as normal activity profiles, and they are used to examine runtime connection records for possible zero-day attacks.
      - *Linear data transformation*: The linear data transformation process utilizes linear discriminant analysis (LDA) technique on the training connections to create several discriminant functions that linearly separate normal activities from attack activities. The extracted linear discriminant functions are used to create several profiles, which are used at runtime, to estimate the probability of zero-day attack given the incoming connection record features as inputs.

   II. *Contextual information modeling*: During this phase, the contextual information extracted during contextual information extraction phase is modeled using the following types of profiles:

      - *Normal activity profiles*: The representative normal connections are stored as normal activity profiles. These profiles are used at runtime to detect zero-day attacks using a semi-supervised anomaly detection technique. A dimensionality reduction process has been carried out on these profiles to improve the efficiency of the detection process. These normal profiles are then stored in a reduced dimensional space, by utilizing only the most significant eigenvectors to describe their characteristics.

- *Discriminant function-based profiles*: Each of these profiles consists of discriminant functions, which linearly describe attacks and normal activity patterns. The purpose of generating discriminant function-based profiles is to use them at runtime to estimate the probability of zero-day attack, given an incoming connection that does not match known attack context profiles.

III. *Contextual information usage*: During this phase, the profiles which are created during contextual information modeling phase are utilized at runtime to detect possible zero-day attacks. The suspicious incoming connections, which have no matching known attack context profiles, are processed using one or more of the techniques outlined below:

- *Similarity with known attack context profiles*: During which, the incoming connection features are matched with several known attack context profile features to calculate the similarity between connection record and these known attack profiles. The output of such matching is the maximum similarity value. The higher this value is, the greater the possibilities that the incoming connection record could be a zero-day attack, while we do not claim that all attacks detected using similarity with attack profiles are zero-day attacks as some of them might be known attacks. However, we focus mainly on the success rate of our approach in detecting zero-day attacks which are unknown to our framework components as we show in our experiments.
- *One-class NN anomaly detection using normal activity profiles*: The deviation of incoming connection features from normal activity profiles is calculated. An anomaly score is assigned to each connection, where the high anomaly score indicates that the incoming connection could be a zero-day attack.
- *Estimated probability calculation discriminant function (DF)-based profiles*: The connection records are passed to several linear discriminant functions to estimate the probability of attack labels given the features of connections. Once the estimated probabilities are calculated using the corresponding discriminant functions, the highest estimated probability value is selected and used as a zero-day score. This score is used to declare a possible zero-day attack.

We next describe in detail the process of creating contextual models and use their implementation to detect attacks in known and zero-day attack detection modes.

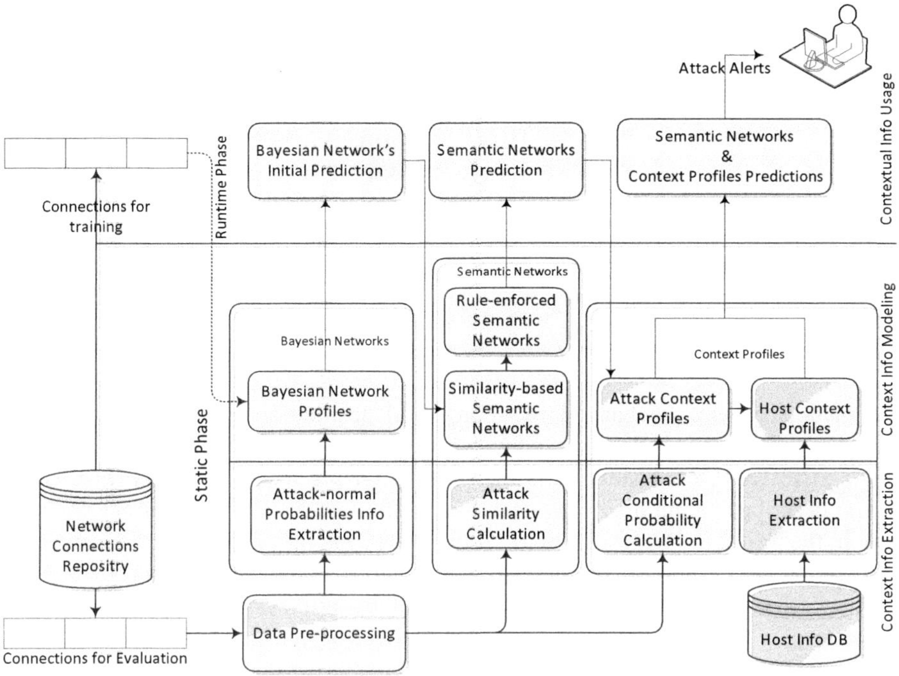

Fig. 2.3 The misuse detection components

2.2.1 *The Known Attack (Signature-Based) Detection Mode*

In this section, we describe the details of the contextual model we create and use during this mode.

Figure 2.3 illustrates the major components and processes, which are created and used during this mode. The *static phase* consists of several pre-calculated contextual models that are being used at runtime. Let us describe in detail the input data utilized and the contextual model which are created.

2.2.1.1 Network Connection Repository

This is the standard NSL-KDD database [15, 16] which is used to create and evaluate the effectiveness of contextual model implementations in terms of known attack detection rate in this mode. The database consists of connection records each of which represents a time window of two seconds of interaction between specific source and target hosts. Each record in the dataset is labeled with the name of an actual attack or "normal" if there is no attack.

Table 2.1 shows a sample of the connections with some connection features. For instance, the *duration* describes the duration of the connection; *src_bytes* and

Table 2.1 A sample of connection records with selected features

Connection_id	Duration	Protocol	Service	Flag	src_ bytes	dst_ bytes	Count	Label
82755	0	TCP	IRC	REJ	0	0	477	Satan
82970	0	TCP	bgp	REJ	0	0	324	Satan
101057	0	TCP	Courier	SH	0	0	1	Nmap
220670	0	TCP	Domain	SH	0	0	1	Nmap
9	0	UDP	domain_u	SF	29	0	2	Normal
76	0	UDP	domain_u	SF	44	115	1	Normal
13632	5	ICMP	eco_i	SF	18	0	1	Ipsweep
13638	0	ICMP	eco_i	SF	18	0	1	Ipsweep

dst_bytes are bytes exchanged by source and destination and vice versa. *Flag* describes the status flag of the connection. *Protocol* is the connection protocol (e.g., TCP, UDP, etc.). *Service* is the destination service (e.g., telnet, Ftp, etc.). *Count* describes the number of connections to the same host as the current connection in the past 2 s. The last column, *label*, represents the actual attack, if any, this connection led to. Some connections are normal activities. We use this attribute to evaluate the correctness of attacks predicted by our system. The data in this repository is divided into two parts, the connections, which have been used to create the contextual models to detect known attacks (or the training connections), and the connections which have been used to evaluate these contextual models in detecting these attacks (the evaluation connections). Some preprocessing steps are required to create the contextual models. Next, we describe these data preprocessing steps.

2.2.1.2 Data Preprocessing

The data preprocessing in this mode, as shown in Fig. 2.4, consists of two processes, *discretization* and *feature selection*. Most of the features in the dataset are numerical. Continuity in these feature values makes it more challenging to apply data mining concepts to the dataset. We discretized the continuous features in the training connections. Bins are automatically created based on the label values present in the dataset using a supervised binning approach.

The dataset has 41 columns (including the label). This means that each label is described by 40 features. In order to minimize the computational complexity, we carried out feature selection using the correlation-based feature selection method proposed by Hall et al. [17] on this dataset to yield the highly ranked features shown in Table 2.2. Previous work done on this dataset utilizes most of these features in their experiments. The preprocessed data is used to create several contextual models to detect zero-day attacks. Let us start with one of these models, which are Bayesian network-based profiles.

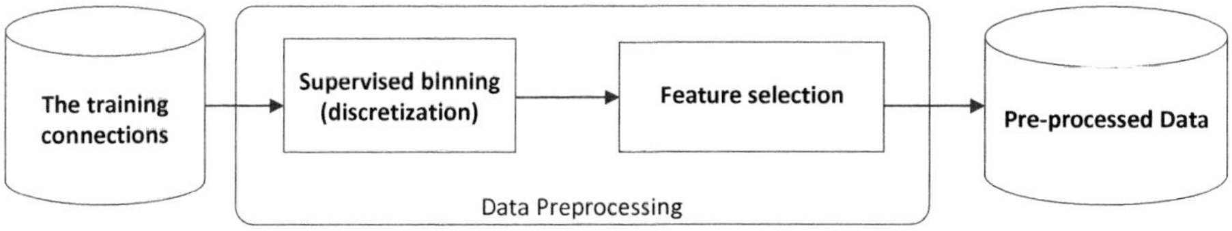

Fig. 2.4 The data preprocessing steps

Table 2.2 Correlation-based feature selection method on NSL-KDD dataset

Feature	Description
Duration	Duration of the connection
Land	1 if connection if from/to the same host/port, 0 otherwise
Protocol	Connection protocol (e.g., TCP, UDP)
same_srv_rate	% of the connections to the same service
Flag	Status flag of the service
dst_bytes	Bytes sent from destination to the source
Count	Number of connections to the same host as the current connection in past two seconds
diff_srv_rate	% of connections to different services
dst_host_diff_srv_rate	% of different services on the current host
Hot	Number of "hot" indicators
dst_host_same_src_port_rate	% of connections to the current host having the same src port
wrong_fragment	Number of wrong fragments
src_bytes	Bytes sent from source to destination
Service	Destination service (e.g., telnet, Ftp)

2.2.1.3 Using Bayesian Networks (BN) for Known Attack Detection

Bayesian networks are directed acyclic graphs that have been used to infer causal relationships based on conditional probability of a node given its parent. We use a simplified version of Bayesian network called simple Bayesian network, which consists of one parent node and many children leaf nodes. The parent node represents the label to be predicted, which can be a specific attack or a normal activity; the child nodes represent some connection features which are the observations that will be used to calculate attack probability. The processes of creating and using Bayesian networks to detect attacks in known attack detection mode consist of the following phases:

Contextual information extraction (*attack and normal probabilities info extraction*): Information about probabilities of attacks and normal activities, given specific observation in the dataset, are calculated. In BN, each prediction can be an attack or normal activity, and it represents an unobserved parent node (i.e., the label) to be predicted. Initially, the training connections are used to calculate the probability of each feature value given a specific attack or a normal activity. Once the conditional probabilities of features given specific attacks or normal patterns are generated, they can be stored in Bayesian network-based profiles.

Table 2.3 The structure of Bayesian network-based context profiles

Attack name	Feature name	Lower bound	Upper bound	Categorical feature value	Probability
Rootkit	src_bytes	2347.51	5678.765		0.01639
Portsweep	src_bytes	2347.51	5678.765		0.00149
Neptune	src_bytes	2347.51	5678.765		0.00756
Normal	src_bytes	2347.51	5678.765		0.00500
Smurf	src_bytes	2347.51	5678.765		0.00136
Smurf	Protocol			TCP	0.96825
Loadmodule	Protocol			TCP	0.90476
Perl	Protocol			TCP	0.77777
Guess_Password	Protocol			TCP	0.98165
…..	…..	…..	…..	…..	…..

Contextual information modeling (Bayesian network-based profiles): The probabilities extracted in the previous phase are modeled as Bayesian network-based profiles. Table 2.3 shows the structure of Bayesian network-based profiles. Each profile represents the *activity context* of a specific attack. We also created a separate profile for normal activities. The attack name represents the name of the attack for which the profile is created. The feature name represents the name of the feature used in creating Bayesian network-based profiles. In the case of numerical values, lower bound represents the smallest value of the bin for numerical features that have been discretized. The upper bound represents the highest value of the bin for the numerical features which have been discretized. The categorical feature value represents the value of features which are nominal. It should be noted that the Bayesian network-based attack context profiles consist of all selected feature values, along with the probability of success of an attack over the given feature value; thus, Table 2.3 shows only a few features in each profile.

Contextual information usage (*Bayesian network profile prediction at runtime*): The BN-based profiles are used to examine runtime connections to find the probabilities of different attacks/normal activities. The combined probability of any label a_i, which can be either an attack or normal activity, is calculated as follows:

$$P(a_i|F) = (P(f_1|a_i) \cdot P(f_2|a_i) \ldots, P(f_n|a_i) \cdot P(a_i))/P(F) \tag{2.1}$$

where $P(a_i|F)$ is the probability of label a_i given the connection feature vector F, which consists of connection features $\{f_1, \ldots, f_n\}$, under the assumption that the features are conditionally independent. The probability $P(f_i|a_i)$ reveals how often each feature value f_i occurs with a_i in the dataset; this value is retrieved from the corresponding Bayesian network-based profile. The basis for retrieval is the feature values possessed by this connection record. For numerical value attributes, the corresponding interval is looked upon in the Bayesian network-based profiles, and the probability associated with this feature interval is recorded. For categorical features, it is matched with the corresponding value in BN-based profiles, and the associated conditional probability is recorded. $P(a_i)$ is the prior probability of a_i occurrence, whose value is extracted from training connections. Finally $P(F)$ is the probability of occurrence of the feature vector F. The system calculates the probability of each label (attack, normal activity), given the connection record features, to identify the label with maximum probability, which is passed as an initial prediction to the semantic network.

2.2.1.4 Using Semantic Networks (SN) for Detection of Known Attacks

A semantic network or net is a graphic notation for representing knowledge in patterns of interconnected nodes and arcs. Computer implementations of semantic networks were first developed for artificial intelligence and machine translation, but earlier versions have long been used in philosophy, psychology, and linguistics. Sowa gives a descriptive outline on the types and use of semantic networks in different disciplines [18, 19]. We used semantic networks to model the *relation* aspect of context. We created semantic networks as graphs where nodes represent attacks and edges model semantic relationships between such attacks. We used semantic networks to infer relevant attacks that are semantically related to the initial prediction produced by Bayesian networks, utilizing relationships which cannot be captured by simple classification techniques. A semantic network is created at the static phase, and each one of its nodes represents one of the 22 attacks in the dataset. Normal activity is also treated as a node in the semantic network. To avoid the intricate task of manually constructing and maintaining the semantic network, we adopt an automatic approach which consists of two phases. In the first phase, we created a similarity-based semantic network (i.e., the first mode network). In the second phase, we created a rule-enforced semantic network (i.e., the second mode network). The similarity-based or first mode network is a graph which identifies the degree of relevancy among attacks based on similarity values to connect such attacks using edges. The second mode network modifies the first mode and adjusts it by adding domain expertise; let us explain first how similarities between attack features are utilized to extract the contextual relationship among them to construct the first mode semantic networks.

Contextual information extraction (*attack similarity calculation*): Let $A = \{a_1, \ldots, a_n\}$ be the set of network attacks, where each $a_p \in A$ is associated with a set of characterizing features forming an *attack feature vector* $V_{a_p} = \left\langle f_{a_{p1}} \ldots, f_{a_{pm}} \right\rangle$ where $f_{a_{pj}}$ is the normalized frequency of feature j occurring with attack a_p. The feature values may have numerical and/or categorical values; hence, prior to creating an attack feature vector, we apply binning to convert the numerical features into bins. This is necessary since some similarity measures utilized in creating semantic networks require categorical values to calculate similarity among attacks. A sample of attack feature vectors is shown in Table 2.4. We determine the similarity between attacks as follows: first, we generate a single *universal feature vector* V (the first column in Table 2.4), which is the union of all

Table 2.4 Universal and attack feature vectors

	Attack feature vector		
Universal feature vector	Guess_Password	Imap	...
Dst_host_srv_serro_1	0.924528302	0. 83333333	
Dst_host_srv_serro_2	0.037735849	0.043333333	
Dst_host_srv_serro_3	0.38867925	0.5	
Dst_host_srv_serro_4	0	0.166666667	
Dst_host_srv_serro_5	0. 18867925	0.166666667	
Duration_1	1	1	
Duration_2	0	0	
Duration_3	0	0	
Duration_4	0	0	
REJ	0	0	
Rsto	0.849056604	0	
RstosO	0	0	
...			

features extracted from attack feature vectors. The universal feature vector V is used to calculate the similarity between attack feature vectors $V_{a_1}, \ldots, V_{a_n}$. We utilize several *similarity coefficients* to calculate the similarity among attacks. These coefficients serve as an effective tool for measuring the similarity among vectors. Lewis and Janeja [20] surveyed 35 different coefficients that can be used to find the similarity between vectors V_{ap} and V_{aq}. The inputs to similarity coefficient measures utilized in this work are binary vectors of zeros and ones; thus, we convert each attack feature vector V_{a_i} into a binary vector of zeros and ones by applying the absolute cutoff data transformation technique [21] using 0.5 as a cutoff point. We found experimentally that smaller cutoff points result in high similarity between attacks that belong to dissimilar categories.

To avoid the intricate task of manually constructing and maintaining the SLNs, we adopt an automatic approach which consists of two phases similarly to the process followed in Karabatis et al. [22]. In the first phase, we create similarity-based SLNs (first mode networks). In the second phase, we create rule-enforced SLNs (second mode networks). The similarity-based or the first mode SLN is a graph which identifies the degree of relevancy between nodes based on direct or indirect similarity relationship among them. The second mode SLN modifies the degree of relevancy in the first mode by applying domain knowledge in creating the relationships between nodes.

Domain-specific example: Let us consider an initial SLN (see Fig. 2.5) to explain how to generate relevance scores among its nodes. The nodes in the SLN represent some attacks in the DARPA dataset [23] which contains suspicious and benign connections. Suppose that the objective is to calculate the relevance score $rs'_{(Warezclient \rightarrow Ftp\,write)}$ between the nodes representing the Warezclient and Ftp_Write attacks.

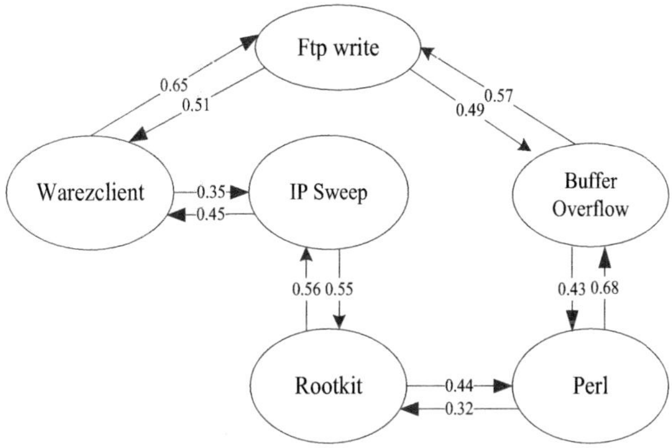

Fig. 2.5 An initial SLN example for some attacks in DARPA dataset

Table 2.5 DARPA dataset attack taxonomy

Attack category	Attack name
Denial of service (DoS)	Smurf, Neptune, Back, Teardrop, Pod, Land
Remote to local (R2L)	Warezclient, Guess_Password, Warezmaster, Imap, Ftp_Write, Multihop, Phf, Spy
User to root (U2R)	Buffer_Overflow, Rootkit, Loadmodule, Perl
Probe	Satan, Ipsweep, Portsweep, Nmap

Figure 2.5 shows the six attacks in the SLN and the transition probabilities between them. The relevance score $rs_{(Warezclient \to Ftp\, write)}$ should be the maximum relevance score over all possible paths connecting Ftp_Write to Warezclient. After two reasoning steps, the resulting $rs = 0.5$, however, the direct path $t_{1\,(Warezclient \to Ftp\ write)}$ with length = 1 results in $rs = 0.65$; thus, the final relevance score $rs_{(warezclient \to Ftp\, write)}$ is the maximum of these two *rss* which is 0.65.

Now let us adjust the *rs* from Warezclient to Ftp_Write to generate a second mode SLN. We will use the DARPA dataset attack taxonomy to perform this adjustment. This taxonomy classifies some cyber-attacks in four categories as shown in Table 2.5. It has been created by security specialists who state most of the attacks which belong to the same category have stronger semantic relationships between them, compared to those that belong to different categories.

Since Ftp_Write and Warezclient attacks belong to the same category (i.e., $\beta_{(n_i,\, n_j)} - rs_{(n_i,\, n_j)} > 0$), the new adjusted relevance score $rs'_{(warezclient \to Ftp\, write)}$ is $0.65 + (|1 - 0.65| \times 0.66) \approx 0.88$. It is noteworthy to mention that domain knowledge stored in taxonomies is an important aspect in adjusting the values of *rs*. However, creating SLNs using taxonomies only is not sufficient to discover all relationships between nodes. Therefore, the *SBSLNs* are still needed, in particular to discover relationships that cannot be revealed using taxonomies (e.g., nodes in different categories). SLNs receive as input (starting node) the most probable node that corresponds to a specific connection as predicted by the BN. This "initial predicted node" serves as the starting node in searching for other semantically related nodes based on a user-defined threshold (tr' threshold).

2.2.1.5 Attack Profiles (APs) to Discard Inaccurate Predictions

The SLNs include benign activity nodes that might be included in the predictions made to a specific connection based on the value of rs' threshold. Therefore, we may have a scenario where the set of predictions made to a specific connection includes both attacks and benign activities. It is then necessary to apply context-based filtering to avoid such scenario and filter out nonrelevant predictions. Therefore, we created attack profiles as a second prediction model; they consist of preconditions represented as features to trigger the occurrence of specific attacks, and they work as filters on the predictions of SLNs to discard a predicted attack n_i when the features of an incoming connection do not match the profile of attack n_i. When the connection is a benign activity, *APs* are supposed to remove all attack predictions made to such a connection. When the connection is an attack, the *APs* must remove all nodes that correspond to benign activities from the prediction list. Another purpose of *APs* is to identify attacks that are relevant to a particular context and to filter out some predictions made by SLNs that are "out of context," thus reducing the amount of inaccurate predictions.

This work proposes a novel technique in which we utilize conditional entropy to create *APs*. The technique assumes that there could be several attacks that co-occur in similar contexts, and therefore it is fairly significant to predict them together. In addition, the proposed technique preserves the contextual relationship produced by SLNs. When *APs* are applied to the predictions produced by SLNs, they remove only the nonrelevant attacks. Lastly, the technique we proposed to create *APs* does not only help in filtering nonrelevant predictions but also identifying unknown attacks as we shall elaborate shortly. *APs* are initially created by measuring the conditional entropy of attacks based on the occurrence of each feature observed in a labeled dataset that consists of TCP connections. The *APs* denoted by AP_{n_i} for each attack n_i are created using the features that decrease the conditional entropy, thus decreasing the uncertainty about the occurrence of n_i. Conditional entropy in this context is the amount of information needed to infer the degree of uncertainty about n_i based on the occurrence of feature f. The conditional entropy for each single attack n_i on the condition of occurrence of f is

$$H(n_i|f) = -\sum_{j=1}^{v} P\left(n_i,f_j\right)\log_2 P\left(n_i|f_j\right) \tag{2.2}$$

where $P\left(n_i,f_j\right)$ is the joint probability of n_i and f_j (one possible value of f) and $P\left(n_i|f_j\right)$ is the conditional probability of n_i given f_j. In the context of attack prediction from TCP connections, we utilize the features of labeled connections to find the conditional entropy for each attack n_i on the condition of occurrence of different values of feature f.

The sum of conditional entropy for n_i conditioned on all values of f is a measure of the local conditional entropy. Local conditional entropy measures the contribution of a particular feature to the occurrence of a particular attack. The lower the value of local conditional entropy of attack n_i given f, the better the feature f can predict the occurrence of attack n_i. Since SLNs predict several related attacks, our objective is to preserve the relation aspect of context, namely, the majority of the predictions produced by SLNs. We create *APs* by considering the fact that there could be several attacks co-occurring in similar contexts (e.g., both attacks have a goal to get root access). To be aware of such relationships between attacks, features that predict related attacks are considered in creating *APs*. If one has a partial knowledge about the target domain (e.g., the group of attacks that might co-occur together from a taxonomy of attacks or a domain expert knowledge), the conditional entropy for each group of related attacks (given such taxonomy of attacks) when conditioned on f can be calculated. This value indicates the global entropy which measures the importance of a feature for predicting the occurrence of each group of related attacks. The lower the value of global conditional entropy, the better the discrimination between different contexts. Let G_{n_i} denote the set of features which gave the lowest global conditional entropy values with the set $N' = \{n_1, \ldots, n_k\}$ of related attacks (e.g., attacks that belong to similar category). Let L_{n_i} denotes the set of features which gave the lowest local conditional entropy values when conditioned on a specific attack $n_i | n_i \in N'$. The features in G_{n_i} and L_{n_i} convey better quality information; consequently, they are used in creating *AP*. The resulting *APs* are supposed to identify the combination of features that lead to one or more related attacks. Therefore some *APs* can be similar so that the corresponding attacks can be triggered together. The set of attacks which are semantically relevant is not supposed to be discarded by attack profiles. Accordingly, *APs* for semantically related attacks need to have similar content, i.e., an overlap between *APs* of related attacks is expected.

Domain-specific example: In this example, we demonstrate the process of creating attack profiles for the attacks in the DARPA intrusion detection dataset. Based on the domain expert categorization of the attacks shown in Table 2.5, the features in the set G_{n_i} with low global entropy values are formulated for each attack category as shown in Table 2.6. The number of features for each attack category is chosen according to a domain expert analysis conducted by Gupta et al. [24] who selected the most important features for each category based on attack semantics. Similarly, we select a new set of features L_{n_i} for each attack n_i. Table 2.7 shows some related attacks in the R2L attack category and the set of features $(G_{n_i} \cap L_{n_i})$ for each attack (e.g., *service*, *number of shells*, *number of files accessed*, *flags*, *is hot login*, *logged in* for Warezclient attack).

The value column ([V]) lists some values of these features. The occurrence of each feature with the corresponding attack is expressed as a conditional probability

Table 2.6 Important features in Gn_i for attacks in DARPA data/per category

Feature name	Attack category Gn_{Probe}	Gn_{DoS}	Gn_{R2L}	Gn_{U2R}
Duration	√	√	√	
protocol_type	√	√		
Service	√	√	√	√
Flag	√	√	√	
src_bytes	√	√	√	
wrong_fragment			√	
Hot	√			
num_failed_logins			√	√
logged_in			√	
num_compromised			√	
root_shell				√
num_root			√	√
num_file_creations			√	√
num_shells			√	√
num_access_files			√	√
is_hot_login			√	√
is_guest_login			√	
Count				
dst_host_same_srv_rate		√		
dst_host_serror_rate		√		
dst_host_srv_serror_rate		√		
dst_host_rerror_rate		√		
Number of features	*6*	*9*	*14*	*8*

Table 2.7 Sample attacks in R2L category and the corresponding features $(G_{n_i} \cap L_{n_i})$

Feature	Service		Number of shells		Number of files accessed		Flag		Is hot login		Logged in	
attack	[V]	[P]	[V]	[P]	[V]	[P]	[V]	[P]	[V]	[P]	[V]	[P]
Warezclient	Ftp_data	0.91	0	1	0	1	SF	1	0	0.96	1	0.98
	Ftp	0.09	1	0	1	0	–	0	1	0.04	0	0.02
Ftp_Write			0	1			SF	1	0	0.95	1	0.90
			1	0			–	0	1	0.05	0	0.10
Imap	Imap4	1	0	1	0	1			0	0.98	0	0.92
	–	0	1	0	1	0			1	0.02	1	0.08

in the column [P]. As shown in the table, the shaded feature values have been selected to create the attack profile entries for the corresponding attack. For instance, the final *AP* for Warezclient and Ftp_Write attacks is expressed as a set of feature value pairs:

$$\lceil Service_{is(Ftp_{data})}, No\_of\_Shell(0), No\_of\_files\_accessed(0), flag('SF'), is\_hot\_login(0)$$
$$logged\_in(1) \rightarrow warezclient\rceil$$
$$\lceil No\_of\_Shell(0), flag('SF'), is\_hot\_login(0), logged\_in(1) \rightarrow ftp\_write\rceil$$

Based on the similarity of their profiles, Warezclient and Ftp_Write are contextually related. By looking at their profiles, the set of features which are selected to create AP_{ftp_write} is a subset of features used to create $AP_{warezclient}$; consequently, there is a high probability that these two attacks are initiated under similar circumstances. APs are used to discard some predictions made by SLNs. The features of a connection c (for which a set of predictions N' are produced by SLNs) are matched with attack profiles. When the features of c do not match the profile $AP_{n_i} \middle| n_i \in N'$, the attack n_i is discarded from N'. As such, the final objective is to keep attacks that are relevant and contextually related. For instance, if the set of final predictions contains *Warezclient* attack as a prediction, there is a high probability that the same set contains *Ftp_Write* as a prediction as well. Incoming connections passing through SLNs but do not match any APs are deemed as benign activities.

2.2.1.6 Host Profiles (HPs) to Discard Inaccurate Predictions

The attack profiles do not convey any contextual features about the targeted hosts such as the patches applied against some attacks; thus, the domain experts must make assumptions about the missing information (e.g., the patches applied to the targeted) which may lead to incorrect predictions, namely, false positives. This situation needs to be addressed by gathering contextual features about the network hosts to filter out potential attacks that are not relevant, based on the current state on the target host. The information gathered represents the individuality features about the target hosts. This information can be collected using a patch management system or host monitoring tools. In our framework such information is represented as host profiles.

A host profile (*HP*) for a particular host h_d consists of logic facts about that host; each fact belongs to a specific type t. Example fact types are patches applied in response to the discovered vulnerabilities, applications on such host, its operating system, services running, etc. HPs are utilized in a process called individuality-based filtering through which these profiles are applied to filter out nonrelevant predictions produced by SLNs and APs.

As part of host-based filtering, our framework is aware of the existing facts about known attacks. An attack n_i is said to be out of context of host h_d, when all facts about such an attack that have a particular type t contradict with all host facts of the same type t. In other words, there is at least one requirement of the attack that is not satisfied on the targeted host. Once a context mismatch (described in Chap. 3) is detected, the attack is considered as nonrelevant and it is filtered out. Context mismatch is detected as follows: first, all facts $f_{1(n_i)}..f_{m(n_i)}$ of type t about a specific predicted attack n_i are retrieved. The context mismatch between the facts about a specific attack n_i and those of host h_d occurs when for all facts of type t about the attack n_i, there is no relevant fact of the same type on the host h_d (e.g., the target host is patched against all known vulnerabilities that cause the attack). If the attack n_i is out of context of the host h_d, it needs to be discarded from the predictions. It is noteworthy to mention that the filtering extent of *HPs* depends mainly on the amount of information collected about the targeted hosts; therefore, there is no assumption about the completeness of such profiles. Individuality-based information is considered complete if all details about hosts are profiled in real-time manner; however, there are several reasons that lead to incomplete *HPs*. First, not all information is documented about each host, since this might lead to more overhead during the detection process. Second, even a small organization with a single server can expect to spend time reviewing a handful of critical patches per month. Therefore, if information about particular host is not available (e.g., patches against a specific attack n_i), our framework considers that h_d is vulnerable to the attack n_i, and therefore a context match is triggered if such an attack is in the prediction list. We create a synthetic individuality-based setting to generate *HPs* and use them to discard nonrelevant attacks. The details are discussed in the experiment chapter. The following attack scenario illustrates the applicability of *H Ps* on top of SLNs and *AP*.

Attack Scenario 3.2: The Perl attack is a user to root (U2R) type of attack where attackers set the user ID to root in a Perl script and create a root shell. Usually the intention of an attacker is to perform other types of L2R such as DoS and buffer overflow attacks, specifically, on systems with the Apache server installed. One form of this attack allows local attackers to launch a symlink using an e-command to overwrite arbitrary files on the targeted hosts that have *Red Hat Linux* operating system. This attack is initiated when the attacker exploits the -e option vulnerability with reference *CVE-1999-1386*. If the same system has an Apache server, the “sioux” vulnerability (*CVE-1999-1199*) allows remote attackers to establish a denial of service back attack using GET requests containing a bulky number of “\” characters. Suppose that the target host h_d has *Red Hat Linux version 5.0 operating system* and *an Apache server version 1.3.1* installed on it. Suppose that there is no information available about patches against Perl on the target host. Let connection $c \subseteq h_s \times h_d$ be established between an attacker host h_s and a target host h_d at a particular time interval. The following facts are derived about the target host h_d at the time of connection: *hasOS* (*redhat linux5.0*) $\bigwedge$ *hasApplication*(*Apache server 1.3.1*) $\bigwedge$ *Patched*(*CVE-1999-1199*) $\bigwedge$ *Patched*(*CVE-1999-0513*). Additionally, the following facts are known about the Perl attack (*existOS*(*Redhat linux5.0*) and *CausedByVulnerability*(*CVE-1999-1386*)), the DoS Back attack (*exist Application* (*Apache server 1.3.1*) $\bigwedge$ *CausedByVulnerability*(*CVE-1999-1199*)), and the DoS Smurf attack (*CausedByVulnerability* (*CVE-1999-0513*)). Smurf attack is a method by which an attacker can send a moderate amount of traffic and cause a virtual explosion of traffic at the target host. Since we do not have any facts on the

Table 2.8 SLNs and context profile prediction for connection on host h_d

Initial prediction	Predictions of SLNs	rs' $Rootkit \rightarrow n_j$	AP_{n_i} match	HP_{h_d} match	Possible attack?
Rootkit	Buffer_Overflow	0.80	Y	Y	Y
	Back	0.56	Y	N	N
	Perl	0.74	Y	Y	Y
	Smurf	0.45	N	N	N

victim host h_d about the patches applied against vulnerability *CVE-1999-1386*, h_d is considered vulnerable to Perl attack.

Table 2.8 shows the predictions that correspond to the connection c after the initial prediction is made (column 1); other relevant nodes are retrieved by SLN (column 2), and after attack/host profiles are applied to filter out the nonrelevant predictions (columns 4 and 5). The SLNs predicted several relevant attacks based on 0.4 rs' threshold. Such attacks are deemed relevant to "Rootkit" (i.e., the starting node in SLN) which is a U2R attack.

Out of these, only two attacks (Buffer_Overflow and Perl) are kept since their corresponding *APs* match the features of connection c. Back attack is filtered out since h_d is patched against this attack. Smurf attack is filtered out since its *AP* does not match the features of connection c. Buffer_Overflow attack is kept in the final prediction list since there are no sufficient evidences based on information in the profile of h_d that patches have been applied against this attack. The next chapter will discuss the creation of our prediction models using time and location features in a network environment with network flows.

2.3 Experiments and Analysis

In this set of experiments, we present the results of the experiments that have been conducted to test the effectiveness of our framework.

2.3.1 Experiments on Discovering Known Attacks in TCP Connections

In this section we describe the results of several experiments that measure the effectiveness of the prediction models when applied to identify known attacks from TCP connections. SLNs and context profiles are applied in a layered manner to identify semantically related attacks that occur in similar contexts. We first provide some preliminary details about the implementation of our prediction models, the setting under which the experiments are conducted, the datasets utilized for evaluation, the evaluation metrics used, and the types of experiments performed.

Implementation, dataset, and experiment settings: We developed a prototype system which includes the implementation of our prediction models using an Oracle database. Several data mining tools are used to perform the preprocessing steps in our approaches. In particular, we used Weka [25] and KNIME [26] data mining tools to perform preprocessing tasks such as feature selection along with extracting the probabilities of attacks and benign activities to create the BN-based classifier. We utilized PL/SQL to implement several other preprocessing steps such as the creation of feature vectors and similarity coefficients. For this set of experiments, we used the DARPA intrusion detection dataset [27], which is considered as a benchmark on which different methodologies can be evaluated. The DARPA dataset was created based on a simulation of a military network consisting of three "target" machines running various operating systems and services. Another three machines were then used to spoof different IP addresses to generate network traffic. Finally, there was a sniffer that recorded all network traffic using the TCP dump format. The dataset includes a wide variety of intrusions simulated in a military network environment. It also contains normal traffic. The TCP dump format of this dataset is summarized into connection sessions. Specifically, a connection is a sequence of TCP packets starting and ending at some well-defined times. A review that has been published in 2009 by Chih et al. [28] shows that the majority of intrusion detection approaches utilized this dataset as a benchmark to validate their methods. Due to privacy reasons, the same review demonstrates that very few research approaches use nonpublic datasets to validate their intrusion detection methods.

In this set of experiments, we utilize a pre-identified subset of 976,067 connections from this data, out of which 544,000 connections are used during the training phase to create our prediction models and 432,067 connections are used during the evaluation (runtime) phase. The training data contains 22 types of attacks, and each connection in the training date is labeled as either an attack or a benign activity. Connections that contain attacks belong to one of the four attack categories: DoS, R2L, U2R, or probing. Denial of service (DoS) attacks occur when the attacker tries to prevent legitimate users from using a service. Remote to local (R2L) attacks occur when the attacker does not have an account on the victim machine and tries to gain access. User to root (U2R) attacks occur when the attacker has local access to the victim machine and tries to gain a super user privileges. Probe attacks occur when the attacker tries to gain information about the target host. The connections in the evaluation part are also labeled as benign activities or attacks. The distribution of connections utilized during the training and evaluation phases is shown in Table 2.9.

The data preprocessing steps consist of two processes, discretization (binning) and feature selection. Most of the features in this dataset are numerical. Continuity in these feature values makes it harder to apply some data mining techniques. Therefore, the continuous features are automatically discretized using equal width binning approach [25]. The dataset has 41 features (including the label). This means that each connection is described by 40 features. In order to decrease the computations during evaluation time, we carried out feature selection using the

Table 2.9 Connections used in experiments from DARPA intrusion detection dataset

Category	Connections to create the prediction models	Connections to evaluate the prediction models
Benign	147,250	110,620
Probe	4,112	4,171
DoS	391,458	300,853
R2L	1130	16,354
U2R	50	69
Total	*544,000*	*432,067*

correlation-based feature selection method proposed by Hall [17] to yield the highly ranked features. The preprocessed data is used to create the prediction models described earlier. A BN classifier is created by extracting information to generate the probability of each attack and benign activity given the features extracted from the dataset. In BN, each prediction can be an attack or a benign activity, and it represents an unobserved parent node (i.e., the label) to be predicted at runtime. Once the conditional probabilities of nodes based on the occurrence of specific features are generated, they are stored as profiles and used to predict attacks/benign activities based on matching features of incoming connections.

Similarity-based SLNs (*SBSLNs*) are created based on the similarity information extracted from the dataset. Each node in the *SBSLN* represents one of the 22 attacks found in the dataset. Benign activity is also treated as a node in the SLNs. Our experiments were performed on a server with Intel Pentium D Dual Core 3.4 GHz CPU with 8 GB RAM running 64-bit Windows. Several types of experiments are conducted as part of this set of our experiments.

Precision, recall, and F-measure are used as evaluation metrics, as defined below:

$$P = \frac{TP}{TP + FP} \tag{2.3}$$

$$R = \frac{TP}{TP + FN} \tag{2.4}$$

$$F = \frac{\left(1 + \beta^2\right) \times PR \times DR}{\beta^2 \times (PR + DR)} \beta^2 = 1 \tag{2.5}$$

TP, *FP*, and *FN* represent the true positives, false positives, and false negatives, respectively. A *TP* occurs when a specific incoming connection is correctly recognized by the prediction model as an attack. TPs for a connection labeled as an attack are expected to be the actual attack (the label) and the attacks that are contextually related to it (e.g., both attacks targeted Ftp application). The latter are identified based on many real-world attack scenarios described in the common vulnerability exposure (CVE) database [29]. A *FP* occurs when a specific connection under evaluation is incorrectly recognized as an attack. A *FN* occurs when a specific incoming connection is incorrectly recognized by the system as a benign activity, but in reality it is an actual attack.

2.3.2 Applying SLNs and Context Profiles (CPs)

In this set of experiments, we measure the effectiveness of context profiles (both attack profiles (*APs*) and host profiles (*HPs*)) when they are applied as filters to remove potential incorrect predictions produced by SLNs. Some nodes retrieved by SLNs can be false positives, that is, the connection itself is a benign activity but predicted as an attack. This scenario occurs when the set of predictions that correspond to a specific connection contains both attacks and benign predictions. In addition, a prediction is incorrect when the set of predictions generated by SLNs contains an attack that is not relevant based on the profiles of the targeted hosts. One main issue in this experiment is the creation of *HPs*. Different hosts have different sets of patches, application types, operating systems, etc. Host vulnerabilities and the required patches can be discovered using vulnerability management tools. Additionally, various tools can be used to create organizational maps. However, this task is beyond the scope of this work. Instead, we created a testbed environment that resembles a realistic organization based on information extracted from the connections in the dataset we experimented on. There are some metadata features in each connection that indicate facts or evidences that we utilized to create host profiles. We utilize the connection features (service, flags, and protocol type) to create host profiles (HPs) given the constraint that no knowledge is available about the type (label) of such a connection. *First*, we randomly assign each target host h_d an identifier which is its IP address. *Second*, we extract the metadata features from connections destined for host h_d, and we identify all possible attacks that may contain these features. The possible attacks are selected from each attack category to make sure that the host profile that corresponds to h_d contains facts about attacks in all categories. *Third*, we utilize the common vulnerability exposure (CVE) database to query about all possible configurations that pertain to these attacks. The collected data is used to generate attack facts and host facts. The host facts include patches against some attacks in each attack category. Given that our framework has no knowledge about the type of an incoming connection, the *HPs* might discard any prediction generated by SLNs, including some relevant predictions. Note that the majority of the remaining predictions, after APs are applied, are expected to be relevant based on context. Additionally, few predictions are expected to be irrelevant but have not been discarded by *APs*. Since the probability of removing a relevant or a nonrelevant prediction by *HPs* is the same, and there are more relevant predictions at this point, there is a higher probability not to discard them by *HPs* (given that such profiles contain partial information about the targeted hosts). The advantage of using this method is to create hypothetical yet realistic host contexts that can be used to filter out some predictions based on what is known about the targeted hosts. For this reason we used *HPs* as a complementary layer to validate our detection approach. We created 100 different *HPs* based on the data collected using the protocol, service, and flag features. Given an incoming connection that targets h_d, *HPs* discard attack predictions made by SLNs if the required patches have already been applied to that host. The attack predictions about which

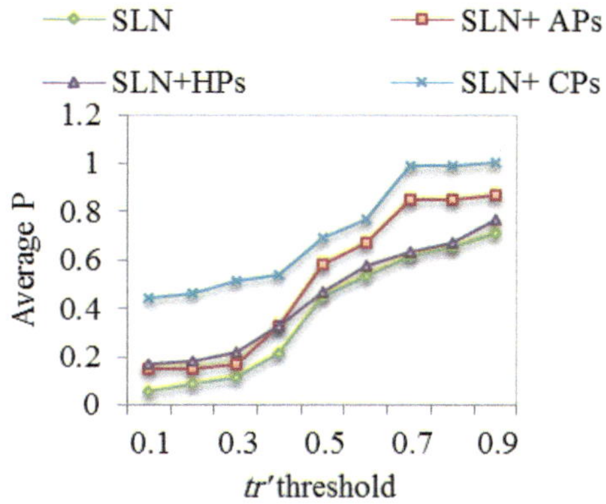

Fig. 2.6 Average precision for 2nd mode SLNs and context profiles

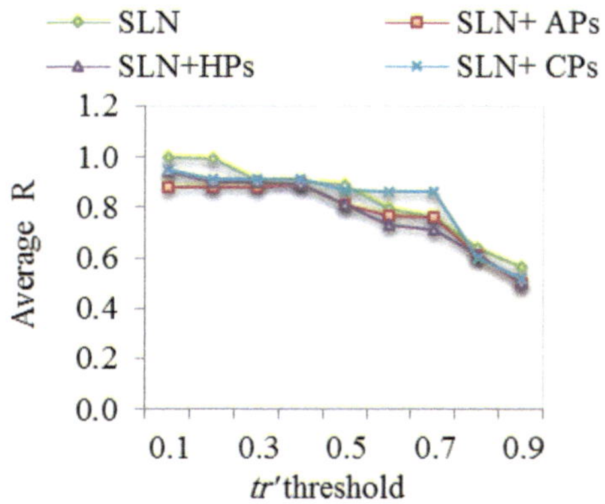

Fig. 2.7 Average recall for 2nd mode SLNs and context profiles

the host profile has no knowledge are kept in the final prediction list, and they are considered true positives. The best performing 2nd mode SLNs created using Anderberg (AD) similarity coefficient are used in this experiment with one or both types of context profiles. In particular, we evaluate (1) SLNs without any context profiles, (2) *APs* when applied as filters on the predictions of SLNs (SLN+ APs), (3) *HPs* when applied to the predictions of SLNs (SLN+ *HPs*), and (4) *APs* and *HPs* when applied in a compound fashion to the predictions of SLNs (SLNs+ *CPs*). We added the following constraint when *HPs* are applied on top of SLNs and *APs*: if the incoming connection does not match any *AP*, the connection is deemed as a benign activity and we skip the *HPs*. Finally, the predictions of SLNs, which still remain after *APs* and *HPs* are applied, are the ones that characterize the incoming connection; as a result the target host h_d is deemed vulnerable to them.

Figures 2.6 and 2.7 show the average precision and recall results in this set of experiments. Figure 2.6 illustrates that utilizing *APs* on top of SLNs improves precision. *APs* are quite efficient in differentiating between the context of suspicious and benign activities. The improvement achieved when *APs* are applied to the predictions made by SLNs is explained by the reduction achieved in false positive rate. As noticed from the results shown in Fig. 2.6, the *HPs* without *APs* are not very effective when applied to the predictions made by SLNs, resulting in an average precision of 0.63. This is interpreted by predicting some benign connections as attacks (i.e., false positives). The reasons for this result are twofold: (1) these connections were incorrectly predicted as attacks by SLNs, and (2) the context of the targeted hosts also matches some of these predictions. In the end, these incorrect predictions made by SLNs are not discarded by *HPs*. This situation explains the benefits of utilizing *AP* profiles to recognize false positives before *HPs* are applied. When all layers are used together, the SLNs and both context profiles (shown as SLNs+ CPs line in figures), the precision is almost 0.98 at the 0.7 *tr*' threshold compared to 0.62 when SLNs are used without any context profiles and 0.63 when only *HPs* are applied on top of SLNs.

Figure 2.7 shows the average recall in this experiment. SLNs without any profiles have a slightly higher recall value at small *tr*' threshold; however, the difference is not very high compared to using SLNs with context profiles. The SLNs have an average recall of 1 at 0.1 and 0.2 thresholds. As observed, the recall of SLNs and other context profiles starts to decline after 0.7 level of *tr*' threshold. This decline indicates that the SLNs start missing some relevant predictions beyond the 0.7 threshold (they become too selective). The recall for SLNs is 0.76 at 0.7 threshold compared to 0.71 when SLNs are used with *HPs* and 0.86 when both context profiles are utilized on top of SLNs. The relatively higher recall of CPs at higher threshold values (0.5–0.8) can be interpreted by the layered filtering mode (applying first *APs*, followed by *HPs*), making them more effective in handling connections that represent benign activities versus SLNs without any context profiles. Overall, the layered filtering mode achieves good precision and recall values. Figure 2.8 shows the results of F-measure values, when context profiles are used on top of SLNs. The best F-measure value when *APs* are applied on top of *SLNs* is 0.8, and it is achieved at 0.7 *tr*' threshold. By contrast, this value is approximately 0.68 when SLNs are used without any context profiles. Overall, when both context profiles are applied in a layered manner on top of SLN, the *F*-measure value is approximately 0.92, and it is obtained at the 0.7 *tr*' threshold. As observed in

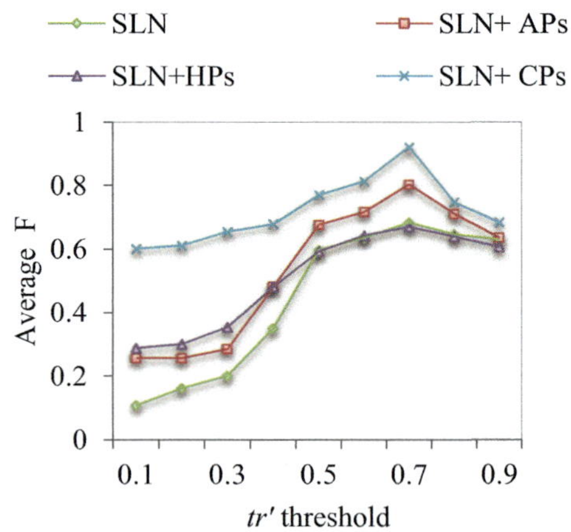

Fig. 2.8 Average F-measure for 2nd mode SLNs and context profiles

Fig. 2.8, the F-value obtained by applying the *HPs* on the results of SLNs is about 0.66 at 0.7 threshold which is lower than when *APs* is applied. *In summary, the HPs show some limitations in handling benign connections when no APs are used. When both context profiles are used together, a higher precision is achieved resulting in higher F-measure values.*

2.3.3 Factors Affecting the Effectiveness of APs

The effectiveness of *APs* which are utilized to filter out nonrelevant predictions might be affected by several factors. *First*, the effectiveness of *APs* might be affected by the distribution of attacks and benign activities in the training data used to create them. In particular, the proportion of attacks to benign activities in the training data used to create *APs* might affect the capability of the resulting profiles to discriminate between the context of cyber-attacks and benign activities at runtime. *Second*, the effectiveness of *APs* might be affected by the number of features utilized in creating them. This section demonstrates the results of two experiments that we conducted to compare the effectiveness of *APs* when changing these two factors. In the first experiment, two training subsets with different distribution of attacks to benign activity are utilized to create *APs*. The resulting *APs* from each subset are then applied to the predictions of SLNs, and their effectiveness is compared. In the second experiment, we focus on varying the number of features to create *APs*, and we then measure the effect of such changes on the effectiveness of these profiles. Let us give more details about the creation of *APs* before demonstrating the settings of both experiments. In general, our objective during the creation of *APs* is to decrease the overlap between profiles that correspond to attacks which occur in different contexts. An observation by Gupta et al. [24] on the same dataset reveals that attacks which belong to different categories occur under different preconditions. Thus, they presumably must have different attack profiles. As a second observation, we notice some attacks belonging to the same category but having different features. Thus, although some attacks belong to the same category, they do not necessarily occur under identical circumstances.

The methodology we utilize to create *APs* using the DARPA intrusion detection dataset takes these two observations into consideration. Besides the features that have high weights (low global entropy) in each attack category, we focus on features that characterize the context in which a specific attack occurs (i.e., features that have low entropy with each attack).

The first observation has been partially considered by Gupta et al. [24] where the authors utilize a local (per attack category) feature selection method which iteratively constructs feature conjunctions to increase the conditional log-likelihood when added to a conditional random field attack prediction model. The direction in Gupta et al. work is to select a specific set of features for each attack category to create an attack prediction model that differentiates between different contexts. Similarly, for the creation of *APs*, we select the features from each attack category

Table 2.10 A subset S_2 of connections selected from DARPA dataset

Attack category	Attack type	Number of connections
Remote to local (R2L)	dict, dict_simple, Ftp_Write, guest, Multihop, Phf, Spy, warez, Warezclient, Warezmaster	2723
Denial of service (DoS)	Land, syslog, Teardrop	1124
User to root (U2R)	Eject, eject-fail, ffb, ffb_clear, format_fail, format_clear, format, imap, load_clear, load_clear, Loadmodule, perl_clear, perlmagic, Rootkit	81
Benign	–	174,873

so that the resulting *APs* of attacks in different categories are dissimilar. Prior to running the first experiment, we utilize a new subset S_2 of connections from the DARPA intrusion detection data in addition to the main subset S_1 utilized earlier in our previous experiments. The majority of connections in S_2 are benign activities and a few attack connections that belong to the three categories shown in Table 2.10. This subset has been extracted from the TCP dump format of DARPA dataset by Perona et al. [30], and it consists of attacks in the categories *DoS*, *R2L*, and *U2R*. There are no flooding (probe) attacks in this subset. The connections in S_2 are divided into training and evaluation parts. The training part is used to create *AD*-based SLNs and *APs*. To create *APs*, the connections which contain attacks are divided into disjoint categories $(c_1 = Probe, c_2 = DoS, c_3 = R2L, c_4 = U2R$ for the connections in the first subset S_1) and ($c_1 = DoS, c_2 = R2L, c_3 = U2R$ for the second subset S_2). For each category c_i, we created the set $Gc_i = \{f_1, \ldots, f_m\}$ which contains the features that give the lowest global conditional entropy given the attacks in that category. For each category, we selected about the same number of features to those selected by Gupta et al. [24] (see Sect. 4.5). The features which contribute to the occurrence of a specific attack n_i are also considered in the creation of attack profiles. Therefore, we selected a new set of features $L_{n_i} = \{f_1, \ldots, f_n\}$ which have the lowest local conditional entropy for a specific attack n_i. We then start creating AP_{n_i} for each attack using the features in $Gc_i \cap Ln_i$. For the first experiment, the incoming connections selected from S_1 and S_2 are initially processed by a BN classifier, and then they are passed to SLNs to retrieve additional relevant predictions and *APs* to filter out the nonrelevant ones.

The values of precision, recall, and F-measures are reported in Figs. 2.9, 2.10, and 2.11. The figures clearly illustrate better effectiveness of *APs* when more attack connections are used to create the profiles (SLNs+APs_S_1). The average precision, recall, and *F*-values are better when the subset S_1 is used in this experiment. The reason we observe relatively lower precision and recall results with S_2 is related to the strength of relationships between benign and attack nodes in the SLNs created using fewer attack connections (using the subset S_2). In these networks, attacks and benign activity nodes are not well separated. While the 2nd mode SLNs are expected to lower the probability of this problem by adjusting relationships between nodes, some false positives are still expected resulting in lower precision values.

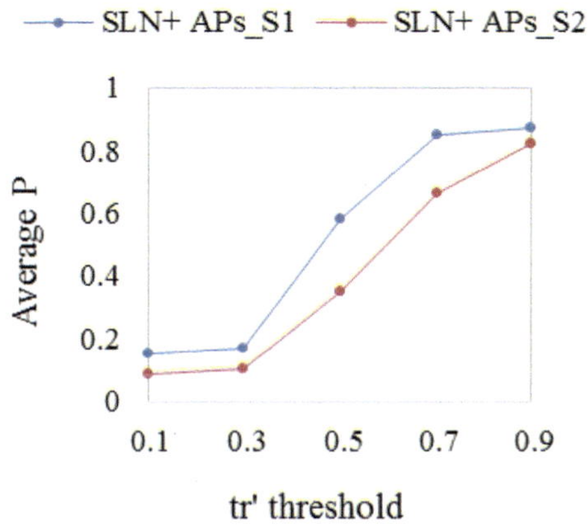

Fig. 2.9 Average precision for AD-based 2nd mode SLNs and APs using S1 and S2

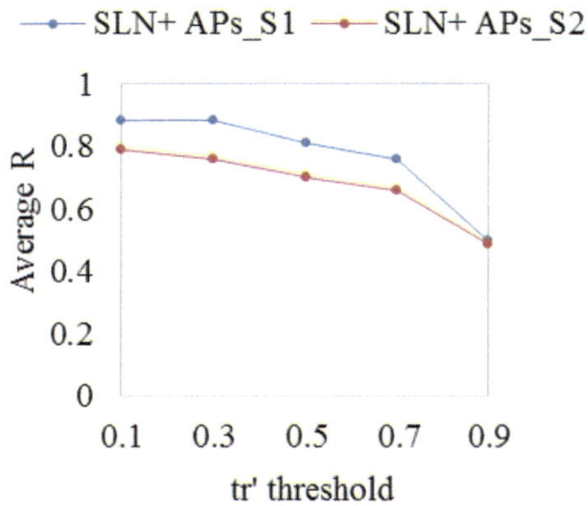

Fig. 2.10 Average recall for AD-based 2nd mode SLNs and APs using S1 and S2

The primary reason is related, however, to the quality of features utilized in creating *APs*. When fewer connections from each attack category are utilized in creating *APs*, the boundaries between contexts become less. With this limitation, the selected features become less discriminatory across different contexts, thus, affecting the overall detection rate. The second experiment measures the effect of the number of features selected for each attack (the number of features in the set L_{n_i}) to create its profile on F-values. We experimented with a range of 4–14 features. The experiment is conducted on both subsets of connections S_1 and S_2. Figure 2.12 shows the effectiveness of SLNs and *APs* in terms of F-values when changing the number of features to create *APs*. It can be observed that utilizing six to eight features from the set L_{n_i} produces relatively better results. For computational efficiency, we only utilize the six highly ranked features from L_{n_i} to create *APs* for attacks in S_1 and seven features to create *APs* for attacks in S_2. Several profiles are found similar when the connections in the subsets S_1 and S_2 are used to create *APs*. In the end, each attack is identified using one profile.

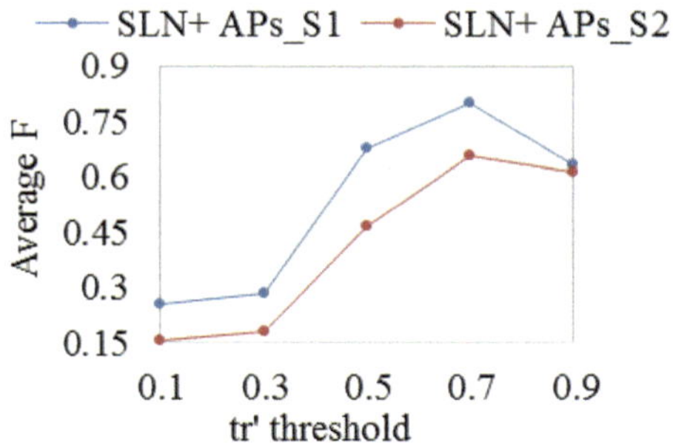

Fig. 2.11 Average F-measure for AD-based 2nd mode SLNs and APs using S1 and S2

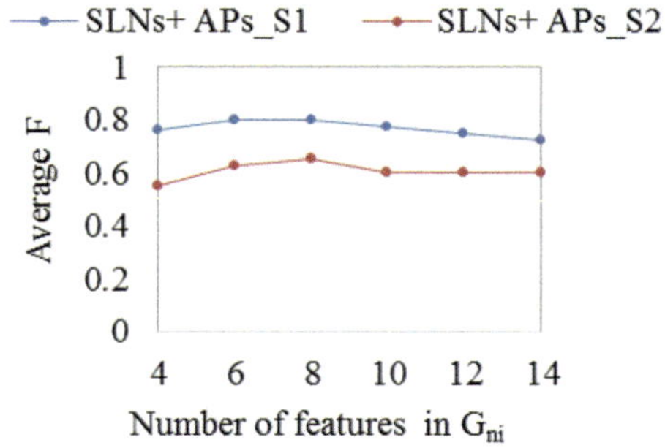

Fig. 2.12 The effects of number of features used in creating APs on F-measure values

References

1. Noel, S., Jajodia, S.: Understanding complex network attack graphs through clustered adjacency matrices. In 21st Annual Computer Security Applications Conference AZ, USA, 5—9 December 2005, pp. 10 pp.–169. doi:10.1109/CSAC.2005.58 (2005)
2. Noel, S., Robertson, E., Jajodia, S.: Correlating intrusion events and building attack scenarios through attack graph distances. In: 20th Annual Computer Security Applications Conference (CSAC'04), Tucson, AZ, USA, pp. 350–359. doi:10.1109/CSAC.2004.11 (2004)
3. Noel, S., Sushil, J., O'Berry, B., Jacobs, M.: Efficient minimum-cost network hardening via exploit dependency graphs. In: Proceedings. 19th Annual Computer Security Applications Conference, Orlando, FL USA, 8–12 December 2003, pp. 86-95. doi:10.1109/CSAC.2003.1254313 (2003)
4. Ritchey, R., O'Berry, B., Noel, S.: Representing TCP/IP connectivity for topological analysis of network security. In: Proceedings. 18th Annual Computer Security Applications Conference, Las Vegas, Nevada, pp. 25–31. doi:10.1109/CSAC.2002.1176275 (2002)
5. Mathew, S., Upadhyaya, S., Sudit, M., Stotz, A.: Situation awareness of multistage cyber attacks by semantic event fusion. In: Military Communications Conference, 2010—milcom 2010, San Jose, CA, 31 October 2010–3 November 2010, pp. 1286–1291. doi:10.1109/MILCOM.2010.5680121 (2010)
6. Leung, K., Leckie, C.: Unsupervised anomaly detection in network intrusion detection using clusters. In Proceedings of the Twenty-eighth Australasian conference on Computer Science, Newcastle, NSW, Australia. Australian Computer Society, Inc., pp. 333–342 (2005)
7. Portnoy, L.: Intrusion detection with unlabeled data using clustering. Data Mining Lab, Department of Computer Science, Columbia University (2001)
8. Song, J., Takakura, H., Kwon, Y. A generalized feature extraction scheme to detect 0-day attacks via IDS alerts. In: Proceedings of the 2008 International Symposium on Applications

and the Internet, Turku, Finland, pp. 55–61. 1442004: IEEE Computer Society. doi:10.1109/saint.2008.85 (2008)
9. Hendry, G.R., Yang, S.J.: Intrusion signature creation via clustering anomalies. In: Proc. of SPIE Bellingham, WA, vol. 6973, pp. 69730C–69731 (2008)
10. Kuchimanchi, G.K., Phoha, V.V., Balagani, K.S., Gaddam, S.R. Dimension reduction using feature extraction methods for Real-time misuse detection systems. In: Proceedings of the Fifth Annual IEEE SMC Information Assurance Workshop, West Point, New York. IEEE, pp. 195–202 (2004)
11. Liu, G., Yi, Z., Yang, S.: A hierarchical intrusion detection model based on the PCA neural networks. Neurocomputing **70**(7–9), 1561–1568 (2007). doi:10.1016/j.neucom.2006.10.146
12. Siraj, M.M., Maarof, M.A., Hashim, S.Z.M.: Intelligent clustering with PCA and unsupervised learning algorithm in intrusion alert correlation. In: Fifth International Conference on Information Assurance and Security (IAS '09). Xi'an, China, 18–20 August 2009, vol. 1, pp. 679–682. doi:10.1109/IAS.2009.261 (2009)
13. Zheng, K., Qian, X., Wang, P.: Dimension reduction in intrusion detection using manifold learning. In: International Conference on Computational Intelligence and Security (CIS'09). Beijing, China, vol. 2, pp. 464–468. IEEE (2009)
14. Li, X.-B.: A scalable decision tree system and its application in pattern recognition and intrusion detection. Decis. Support Syst. **41**(1), 112–130 (2005). doi:10.1016/j.dss.2004.06.016
15. Tavallaee, M., Bagheri, E., Wei, L., Ghorbani, A.A. A detailed analysis of the KDD CUP 99 data set. In IEEE Symposium on Computational Intelligence for Security and Defense Applications (CISDA'09), Ottawa, ON, 8–10 July 2009, pp. 1–6. doi:10.1109/CISDA.2009.5356528 (2009)
16. Tavallaee, M., Bagheri, E., Lu, W., Ghorbani, A.: NSL-KDD Dataset. http://iscx.ca/NSL-KDD/ (2009)
17. Hall, M.A.: Correlation-Based Feature Selection for Machine Learning. The University of Waikato (1999)
18. Sowa, J.F.: Principles of Semantic Networks. Morgan Kaufmann Pub., San Mateo, CA (1991)
19. Sowa, J.F.: Semantic networks. Encyclopedia of Cognitive Science (2006)
20. Lewis, D.M., Janeja, V.P.: An empirical evaluation of similarity coefficients for binary valued data. Int. J. Data Warehous. Min. **7**(2), 44–66 (2011)
21. Pensa, R.G., Leschi, C., Besson, J., Boulicaut, J.F.: Assessment of discretization techniques for relevant pattern discovery from gene expression data. In: Proceedings ACM BIOKDD, vol. 4, pp. 24–30 (2004)
22. Karabatis, G., Chen, Z., Janeja, V.P., Lobo, T., Advani, M., Lindvall, M., et al.: Using semantic networks and context in search for relevant software engineering artifacts. J. Data Semant. **5880**(1), 74–104 (2009). doi:10.1007/978-3-642-10562-3_3
23. Lippmann, R. MIT Lincoln Laboratory KDD Attack Taxonomy. http://www.ll.mit.edu/mission/communications/cyber/CSTcorpora/ideval/docs/. Accessed 05 June 2014 (2014)
24. Gupta, K.K., Nath, B., Kotagiri, R.: Layered approach using conditional random fields for intrusion detection. IEEE Trans. Dependable Secure Comput. **7**(1), 35–49 (2010)
25. Frank, E., Smith, T., Witten, I.: Weka A machine learning software. Machine Learning Group at the University of Waikato. http://www.cs.waikato.ac.nz/ml/weka/ (2014)
26. Wiswedel, B., Ohl, P., Gabriel, T.: KNIME: Kontaz Information Miner. http://www.knime.org/ (2014)
27. Granchelli, D.: DARPA intrusion detection datasets. Massachusetts Institute of Technology (MIT): MIT Lincoln Labratory (1999)
28. Tsai, C.F., Hsu, Y.F., Lin, C.Y., Lin, W.Y.: Intrusion detection by machine learning: a review. Expert Systems with Applications **36**(10), 11994–12000 (2009)
29. Lawler, S., Meunier, P. Common vulnerabilities and exposures. http://cve.mitre.org/. Accessed 07 October 2014 (2012)
30. Perona, I., Gurrutxaga, I., Arbelaitz, O., Martín, J.I., Muguerza, J., Pérez, J.M.: Service-independent payload analysis to improve intrusion detection in network traffic. In: Proceedings of the 7th Australasian Data Mining Conference, SA, Australia, pp. 171–178. Australian Computer Society, Inc. (2008)

Chapter 3
Detecting Unknown Attacks Using Context Similarity

Ahmed AlEroud and George Karabatis

Abstract The security risk of unknown attacks has been considered something that security specialists cannot measure. Contrary to hardware faults, software flaws cannot be easily identified. While it is not easy to identify unknown vulnerabilities in software systems, unknown attacks can be detected at the network level as variations of known attacks. This chapter introduces novel techniques that mainly focus on utilizing attack profiles created earlier to identify unknown attacks as a variation of known attacks.

3.1 Importance of Contextual Relations to Identify Unknown Attacks

Figure 3.1 illustrates some shortcomings of most existing unknown attack detection techniques. The figure shows a benign activity profile as a set of normal activities (e.g., features of a network connection). Such a profile can be created using machine learning techniques that utilize the features of prior benign activities.

The features of incoming network connections (e.g., C_1) are compared with the features stored in such a profile to detect anomalies. The figure also shows another two profiles of known attacks; each one consists of several known features. For simplicity, we assume that a specific IDS has knowledge about these two attacks only. Let us also assume that there are two network connections C_1, C_2 that target a particular host. While the network connection C_1 is far from the profiles of attacks 1 and 2 (if a distance-based measure is used), it may be predicted by a semi-supervised anomaly technique as an unknown attack because it is relatively far from the profile of benign activity as well. Since C_1 consists of activities that are not similar to any attack profile, it is possible that it is a new pattern of benign activities;

A. AlEroud (✉)
Department of Computer Information Systems, Yarmouk University, Irbid 21163, Jordan
e-mail: ahmed.aleroud@yu.edu.jo

G. Karabatis
Department of Information Systems, University of Maryland, Baltimore County (UMBC), 1000 Hilltop Circle, Baltimore, MD 21250, USA
e-mail: georgek@umbc.edu

I.M. Alsmadi et al. (eds.), *Information Fusion for Cyber-Security Analytics*, Studies in Computational Intelligence 691, DOI 10.1007/978-3-319-44257-0_3

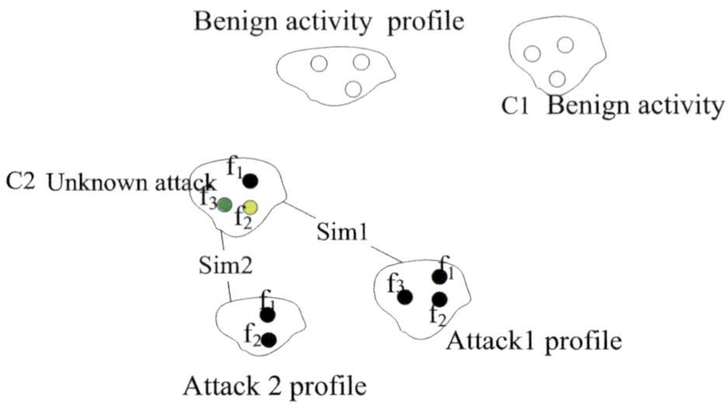

Fig. 3.1 The limitations of anomaly detection techniques to detect unknown attack

nevertheless, there is a high probability that C_1 will be classified as an anomaly by a supervised anomaly detection technique since it is not similar to the profile of benign activity. Such a prediction is considered a false positive. Instead, if the similarity between C_1 and both attack profiles is calculated, it may be deemed as a benign activity since it is not similar to any of these two profiles. As opposed to C_1, if the similarity between the features of connection C_2 and both attack profiles is considered, C_2 may be declared as an unknown attack since (1) it is far from the benign activity profile, and thus can be declared as an anomaly, and (2) it is partially similar (in terms of its features) to attack 1 and 2 profiles. Thus, for connection C_2, we have evidence that it could be an unknown attack.

The previous scenario points to the importance of using a hybrid (signature-anomaly-based) approach to detect unknown attacks. However, such an approach still has some limitations and is not guaranteed to detect the majority of unknown attacks. One major issue in utilizing similarity with attack profiles to detect unknown attacks is the assumption that these profiles are contextually independent. In general, there could be several types of contextual relationships between the attacks for which such profiles are created. These relationships include but not limited to causal relationships, sequential relationships, hierarchical relationships, etc. Moreover, the final objective of related attacks is usually similar. One question is whether identifying such relationships is important to detect unknown attacks. Let us consider again the scenario shown in Fig. 3.1. Suppose that there is a contextual relationship between attack 1 and attack 2. Such a relationship if identified "in a systematic way" would definitely affect the features that would be selected to create the profiles of both attacks. Based on the strength of the contextual relationship between the two attacks, these features can be selected so that one profile complements the other or, at a minimum, both profiles have some similarity. As shown in Fig. 3.1, and due to the assumed contextual relationships between attack 1 and attack 2, the set of features used in creating the profile of attack 2 is a subset of the features used in creating the profile of attack 1. Consequently, the profile of attack 2 contains a subset of activities that initiate attack 1. If all possible contextual relationships between attack 1 and other known attacks are considered, several similar (sub)-profiles of the profile of attack 1 will be created. To this end, an attacker exploiting a zero-day vulnerability to initiate an unknown attack modifies the activities that lead to attack 1 to initiate a zero-day attack. Usually, the scope of modification the attacker made on the signature of attack 1 to initiate his unknown attempt might not be identified easily by an existing security mechanism such as an IDS.

Based on the scope of such a modification, the similarity between the activities initiated by an attacker and the profile of attack 1 might be very low due to attacker's intention to mislead the IDS so that the zero-day attack attempt can succeed. Therefore, it is highly possible that a signature-based IDS will classify the attacker attempt as a benign activity if only the similarity with the profile of attack 1 is utilized. However, if the similarity with other sub-profiles of attack 1 is considered, there is a higher probability that a subset of attacker activities matches one of these profiles. Since such profiles contain fewer features that can lead to attack 1, their contribution to its occurrence is not very high. They mainly capture attacker sub-steps to initiate a zero-day attempt as a variation of attack 1. Consequently, if such a variation is captured, there is a higher probability to detect attacker attempt.

Based on Fig. 3.1 above, suppose that C_2 contains a sequence of features (f_1,f_2,f_3). Suppose that the similarity between the features of such a connection and the features in attack profiles is used as a metric to identify unknown attacks. If such a similarity is calculated, the similarity value "sim_2" (between the features of C_2 and the profile of attack 2—having one feature mismatch) will be higher than the similarity value "sim_1" (between the features of C_2 with the activities in the profile of attack 1—having two feature mismatches). Increased feature mismatches between C_2 and profile 1 lower the probability of detecting C_2 as an unknown attack. Such a mismatch is less between C_2 and the features in the profile of attack 2. While the profile of attack 2 has fewer features than those in the profile of attack 1, both attacks are still related in regard to the context in which they occur. This observation has a significant implication: If the contextual relationships between attack 1 and other attacks are utilized in creating attack profiles, the context in which attack 1 occurs can still be partially captured even if the attacker modifies the signature of attack 1. The utilization of contextual relationships leads to one or more sub-profiles of attack 1 with fewer number of features. These profiles are supposed to be less sensitive to changes the attackers make on the sequence of activities that lead to attack 1 to initiate an unknown attack attempt. Compared to matching with the profile of attack 1, matching the features of incoming connections with these sub-profiles results in higher similarity values on which one relies to declare unknown attacks based on specific certainty level. The use of contextual relationship helps in creating several variants of the same profile. This can be a significant factor to avoid a large amount of false positives and negatives. In summary, we want to examine if utilizing contextual relationships in creating attack profiles would result in improving the detection rate of unknown attacks and decreasing false positives. The detection of unknown attacks by analyzing network traffic has been mainly addressed using some machine learning approaches such as support vector machines (SVMs) [1, 2], clustering [3, 4], and signature generation [5]. None of these techniques focus on utilizing contextual relationships between known attacks to generate new attack signatures to successively identify unknown attacks. We utilize information entropy and linear data transformation techniques to generate feature-based and linear function-based attack profiles and use them to identify unknown attacks. The proposed techniques systematically utilize contextual relationships between known attacks to generate attack profiles that capture the combinations of activities attackers exploit to initiate unknown attacks. Our approach utilizes similarity with these profiles to identify unknown attacks. A semi-supervised anomaly detection technique is utilized on top of these profiles to decrease the uncertainty in the process of identifying unknown attacks. The following attack scenario demonstrates the importance of context similarity to identify unknown attacks.

Attack Scenario 3.1:
Suppose that a host h_1 provides a secure shell (ssh) service. An attacker on an external host h_0 initiating ssh connection exploits several vulnerabilities in ssh service to gain root privileges on host h_1 (e.g., by exploiting vulnerability CVE-2007-5616 on the target server). CVE-2007-2063 is another vulnerability which if exploited can cause arbitrary processes to be stopped. It may enable an attacker to create files with insecure permissions. Several mitigation strategies can be applied to avoid such an attack, specifically, configuring ssh server to use nonstandard ports and restricting access to ssh servers. Suppose that an attacker is using the following sequence of activities to gain root privileges on h_1:

Attack 1: Step1:
`ssh conn` $h_0 \rightarrow h_1$ by exploiting CVE-2007-5616

Step2:
`ssh conn` $h_0 \rightarrow h_1$ by exploiting CVE-2007-2063

Since these steps are known to IDSs, they might be detected. Additionally, they might be nonrelevant if the target host is patched against the corresponding vulnerabilities. Now suppose that the ssh service running on host h_1 has a zero-day vulnerability. If the system is vulnerable to CVE-2007-2063, an attacker will be able to exploit both vulnerabilities (the zero-day vulnerability and CVE-2007-2063) to gain root access on the target host h_1 as follows:

Attack 2: Step1
`ssh conn` $h_0 \rightarrow h_1$ by exploiting the zero-day vulnerability

Step2
`ssh conn` $h_0 \rightarrow h_1$ by exploiting CVE-2007-2063

Therefore, attack 2 becomes a zero-day (unknown) attack that cannot be discovered directly by a signature-based security mechanism (e.g., IDS). However, based on our example, there is an overlap in the contextual preconditions under which attacks 1 and 2 occur. Although the two attacks are different, the final objective is the same, that is, to gain root access. Therefore, the zero-day attack 2 can be considered as a modified sequence or a variant of the known attack 1, and it is possible to discover attack 2 if a mechanism exists to detect zero-day attacks as variations of known attacks.

This chapter is organized as follows. Section 3.2 describes the usage of attack profiles to identify unknown attacks. In Sect. 3.3 we utilize a semi-supervised anomaly detection technique to improve the effectiveness of attack profiles to detect unknown attacks. In Sect. 3.4 attack profiles are re-created and used as linear discriminant functions to detect unknown attacks.

3.2 Using APs to Detect Unknown Attacks

One major advantage of the technique we previously utilized to create attack profiles is the fusion of contextual relationships between known attacks to create their corresponding profiles. Consequently, the attack profiles for related attacks are created as potential combinations of steps that can lead to a major attack n_i with specific objective (e.g., gain access to root). In general, when an attacker attempts to modify the signature of that attack by exploiting a zero-day vulnerability, the new signature $n_i^{'}$ might have some common characteristics or an overlap with one or more features in the corresponding profiles of n_i. Unknown attacks can be discovered based on such a high overlap between the features of $n_i^{'}$ and the profiles of n_i. As a general rule, the connections which correspond to unknown attacks will be recognized as benign activities according to the existing settings in our framework. Therefore, examining some connections to recognize unknown attacks occurs after such connections are examined by the classification model m, the SLNs, and the attack profiles and inaccurately classified as benign activities. The majority of predictions that correspond to these connections are false negatives which are unknown attacks but predicted as benign activities. As a result, our focus is to create a prediction model that has the capability to recognize such connections as unknown attacks. Therefore, we utilize partial similarity with attack profiles as a measure to detect unknown attacks. A high similarity between the features of a connection c and the features of a specific attack profile AP_{n_i} is an indicator of a possible unknown attack pattern in such a connection.

The new attack patterns may be similar but not identical to the attack n_i. This clue is used to discover potential attacks that are unknown to the prediction models utilized earlier. We capture such a similarity using Anderberg (AD) similarity measure. Initially, the similarity values $\mathrm{SIM}(\mathrm{AP}_{n_i}, c),\ldots, \mathrm{SIM}\left(\mathrm{AP}_{n_p}, c\right)$ between each profile AP_{n_i} of a known attack and the incoming connection c are calculated. The similarity is calculated between the features of the profile AP_{n_i} which can be represented as a binary feature vector v_{n_i} and the corresponding features of the incoming connection record c which can be represented as another binary vector v_c. Calculating profile similarity score $\mathrm{SIM}(\mathrm{AP}_{n_i}, c)$ between vectors v_{n_i} and v_c is carried out by finding the number of matches and mismatches between feature vectors (as described in Sect. 4.3). The maximum similarity score MaxS from the set $\left\{\mathrm{SIM}(\mathrm{AP}_{n_i}, c),\ldots, \mathrm{SIM}\left(\mathrm{AP}_{n_p}, c\right)\right.$ is selected to determine if the incoming connection c is an unknown attack (i.e., a zero-day attack). If such a score is very low, it might indicate that the connection is possibly a benign activity. Instead, if it is relatively high, it indicates a potential unknown attack. A user-defined similarity threshold Z_{PS} is used to decide whether the incoming connection is an unknown attack or a benign activity.

Algorithm 3.1 summarizes the steps needed to examine connections to identify possible unknown (zero-day) using profile similarity. Lines (1–9) summarize the process of calculating the similarity between c_j and attack profiles as well as selecting the MaxS (the closest attack profile). Lines (10–16) summarize the steps needed to label c_j as an unknown attack or a benign activity based on the values of Max$S$ and the Z_{PS} threshold. The complexity in the process of discovering unknown attacks is $O(N \times P)$ where N is the number of connections under analysis and P is the number of attack profiles.

Algorithm 3.1 Similarity with APs to detect unknown attacks

Input: a set of connections conn $= \{c_1, \ldots, c_n\}$
APs $= \{\text{AP}_{n_i}, \ldots, \text{AP}_{n_p}\}$
a profile similarity threshold Z_{PS}
Output: $Rc_j \in \{\text{Zday}_{\text{A}}, \text{bengin}\}$
1. **Begin**
2. Max$S = 0$
3. **For** $j = 1$ to n
4. **For** $i = 1$ to p do
5. **Find** SIM(AP_{n_i}, c_j)
6. **If** SIM(AP_{n_i}, c_j) > Max$S$ **then**
7. Max$S$ = SIM(AP_{n_i}, c_j)
8. **End if**
9. **End For**
10. **IF** Max$S > Z_{\text{PS}}$
11. $Rc_j = \text{Zday}_{\text{A}}$
12. **Else**
13. Rc_j = bengin
14. **End if**
15. **Retrieve** Rc_j
16. **End for**
17. **End**

3.3 Using One-Class Nearest Neighbor (1CNN) on Top of APs to Detect Unknown Attacks

The connections which partially match attack profiles might expose unknown attacks that do not conform to a well-defined notation of the benign behavior; however, the use of profile similarity score (Max$S$) as the only indicator to discover unknown attacks may not be sufficient to discover all false positives. If the boundary of the context in which the majority of benign activities occur is well defined and c is found similar to the connections observed in that context, then the connection c predicted as an unknown attack is possibly a benign activity; therefore, a scheme to decrease the uncertainty about the predictions produced by attack profiles is needed. Primarily, the context of benign activities needs to be considered in the process of predicting unknown attacks in order to decrease the number of false positives. While it is challenging to determine the context boundaries for benign activities, anomaly detection techniques can be used to identify such context. Semi-supervised anomaly detection techniques have been utilized to identify the context of benign activities when data that contains labels for benign activities is available [6]. Such techniques create a model that identifies the context boundaries for benign activities and use it to identify anomalies. Semi-supervised anomaly detection techniques presume that benign activities usually occur in dense regions, while anomalies occur far from their closest neighbors in such regions. Semi-supervised anomaly detection techniques utilize density measures to estimate the anomalous behavior for each data instance. Data instances which are located in a neighborhood with low density can be declared as anomalies. Conversely, instances which are located in a dense neighborhood are declared as benign activities. We propose to use one of these techniques—the *one-class nearest neighbor* (1CNN)—to decrease the uncertainty in the process predicting unknown attacks. The 1CNN anomaly detection algorithm [7] models the contextual preconditions that characterize benign activities as normal activity profiles (NPs) that can be represented as raw features or projected in a subspace. 1CNN has been initially proposed to detect anomalies that might expose suspicious activities. Generally, nearest neighbor-based anomaly detection techniques require distance or similarity functions (such as Euclidean distance) to be defined between features). The (1CNN) method accepts a specific connection c as an unknown attack (i.e., rejects it as benign activity) when its local density is less than the local density of its nearest neighbor—selected from the normal activity profiles (NPs). Usually, the first nearest neighbor is used for local density estimation. The following is the acceptance criteria that are used to calculate the anomaly score AS for an incoming connection c and compare it to a user-defined threshold Z_{AS}:

$$\mathrm{AS}(c) = \frac{\left\| c - \mathrm{NN}^{\mathrm{NP}}(c) \right\|}{\left\| \mathrm{NN}^{\mathrm{NP}}(c) - \left(\mathrm{NN}^{\mathrm{NP}}\left(\mathrm{NN}^{\mathrm{NP}}(c)\right)\right) \right\|} > Z_{\mathrm{AS}} \tag{3.1}$$

The equation above is used to compare the distance from a connection c to its nearest neighbor $\mathrm{NN}^{\mathrm{NP}}(c)$ selected from the set of normal activity profile NPs; let us call this distance d_1. The distance from this nearest neighbor $\mathrm{NN}^{\mathrm{NP}}(c)$ to its nearest neighbor $\mathrm{NN}^{\mathrm{NP}}(\mathrm{NN}^{\mathrm{NP}}(c))$ is called d_2. The connection c is accepted as an unknown (zero-day) attack based on the ratio between d_1 and d_2 as shown on Fig. 3.2.

Since NPs are supposed to be preselected, the nearest neighbor for each profile can be identified prior to prediction time. Thus, the calculation of the expression $(\mathrm{NN}^{\mathrm{NP}}(\mathrm{NN}^{\mathrm{NP}}(c)))$ at run-time can be avoided.

The 1CNN algorithm has several predefined thresholds such as the acceptance threshold Z_{AS} (default = 1) and the number of nearest neighbors k to consider (default = 1). The parameter Z_{AS} can be changed to smaller or larger values. Additionally, increasing the number of neighbors k decreases the local sensitivity of the method and therefore reduces its sensitivity to noise. We show the effect of tuning these threshold values in the experiments chapter.

To identify unknown attacks using 1CNN, we mainly focus on the scenario in which 1CNN is applied on top of attack profiles to decrease the uncertainty in the process of detecting unknown attacks. Let us describe this scenario further: Once

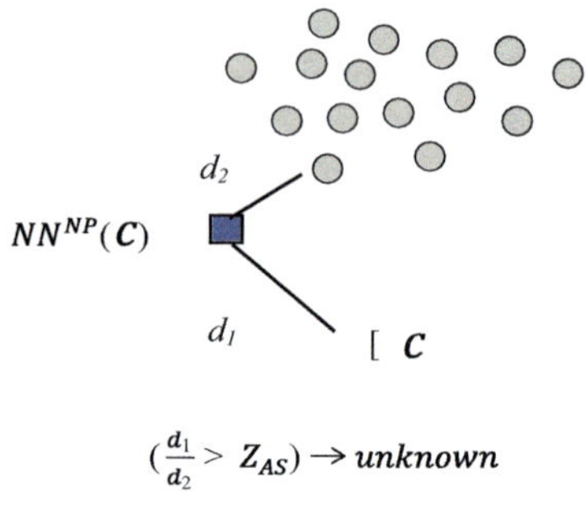

Fig. 3.2 The description of one-class nearest neighbor (1CNN)

the incoming connection partially matches some attack profiles, the Max*S* is calculated to find the closest attack profile to the connection c. If the Max*S* score is less than the threshold Z_{PS}, the connection is deemed as benign (see Algorithm 3.1). Usually, if the Max*S* score which corresponds to connection c is low (e.g., ≈ 0), it may be more reasonable to deem such a connection as a benign activity, and therefore it needs not to be processed by the 1CNN which requires more density-based computation. On the other hand, if the Max*S* is higher than Z_{PS}, the connection needs to be further processed using 1CNN anomaly detection technique. The anomaly score AS(c) is used to classify the connection as an unknown attack or a benign activity. If AS(c) $> Z_{AS}$, the connection is accepted as an unknown attack; otherwise it will be recognized as a benign activity.

The complexity of discovering unknown attacks using 1CNN is $O(N \times P \times D)$, where N is the number of connections under analysis, P is the number of NPs, and D is the number of dimensions for each NPs.

3.4 APs as Discriminant Function Profiles (DFPs) to Detect Unknown Attacks

The effectiveness of the feature-value model of the attack profiles APs can be affected by the type and number of features used to create such profiles. The majority of features in network traffic data are numerical. Due to discretization of numerical features, the feature-value model used to create APs can lead to information loss which has negative effects on prediction accuracy of attacks. Overall, there is a high possibility of feature mismatches between incoming connections and APs which usually lead to incorrect predictions. In summary, the representation model of attack profiles might have an effect on the detection rate of unknown attacks. To decrease the effect of feature discretization, we re-create attack profiles using numerical features in a transformed subspace. There are several data transformation techniques; however, we need a technique that can discriminate between suspicious and benign activities in the transformed space. A transformation technique called linear discriminant analysis (LDA) serves this objective. LDA has the advantage of creating prediction models on datasets with too many numerical features [8]. Even more, linear discriminant analysis preserves the linear relationships between variables to maximize the discrimination between classes. Thus, the correlation among them would be unchanged after such transformation. Discriminant analysis techniques take the advantage of aggregate statistics (e.g., the mean of numerical features and their covariance in the original space) to produce a few sets of discriminant functions that describe different types of nodes (e.g., attack and benign activity) in a transformed space. As shown in [9], LDA utilizes a single discrimination function to classify new instances in binary classification problems (i.e., when the number of categories is two). Since the benign activities and attack types cannot be easily identified using a single classification function, the original LDA classification technique is not very effective for intrusion detection tasks. To handle this limitation, a recent approach is proposed to use LDA in conjunction with other techniques, such as decision trees (DT) to produce multiple classification rules [9]; however, such an approach focuses on identifying known attacks. We applied LDA through a data transformation process to re-create attack profiles as linear functions that can be used to identify unknown attacks. Each function defines one or more types of attacks and is created using pre-identified sets of connections. Each set of connections used in creating DFPs consists of a sample of benign connections and all connections pertaining to attack types that have similar attack profiles. In particular, LDA is utilized in the following manner to re-create the attack profiles as DFs:

1. Bi-class sampling: This step is carried out by selecting many sets of connections; each one contains two types of connection benign activities and attacks. The purpose of creating these sets of connections is to define linear discriminant functions for each type of attacks. Each connection set C contains n_1 connections pertaining to attacks with similar APs. Additionally, it consists of n_2 connections of specific benign activity type.
2. Pooled covariance matrix creation: The covariance matrices are created to extract sufficient yet representative statistics about the correlation between features. In order to define the correlations between the features for each set of connections selected in step 1, a matrix called pooled covariance matrix is required. Using the number of connections n_1, n_2 in each set of connections C, the pooled covariance matrix is created. To illustrate this, let us assume that the data contains only two numerical features f_1,f_2. The number of bytes transmitted from source to destination is one example of numerical features. Let r_1,r_2 denote covariance matrices for benign activities and attack connections in C. The pooled covariance matrix R is then defined by

$$R = \frac{(n_1 - 1)r_1 + (n_2 - 1)r_2}{(n_1 + n_2 - 2)} \tag{3.2}$$

3. Discriminant function definition: Given $\overline{c}_1$,$\overline{c}_2$ that denote the mean vectors of n_1, n_2 connections in C and the pooled covariance matrix R, we define two linear

discriminant functions (since each connection set contains two types of connections: benign and attack). Each discriminant function consists of two linear coefficients (based on our scenario with two features) and a function constant. The constant for each function is calculated as follows:

$$\text{temp1} = (\overline{c}_1)R^{-1} \tag{3.3}$$

$$\text{temp2} = (\overline{c}_2)R^{-1} \tag{3.4}$$

$$\text{cons}_1 = -0.5.\ \text{temp}_1.\ \overline{c}_2{}^T + log(n_1/n) \tag{3.5}$$

$$\text{cons}_2 = -0.5.\ \text{temp}_2.\ \overline{c}_2{}^T + log(n_2/n) \tag{3.6}$$

Where R^{-1} is the inverse of the pooled covariance matrix calculated in Eq. (6.2) and log(n_i/n) is the prior probability of each class (i.e., benign and attack classes). −0.5 is a predetermined number (based on Anderson rule [9]) that is used to assign a connection c to class one (i.e., benign) if

$$(\overline{c}_1 - \overline{c}_2)^T R^{-1} c > \tfrac{1}{2}(\overline{c}_1 - \overline{c}_2)^T R^{-1} (\overline{c}_1 + \overline{c}_2)^T + \log\left({}^{n_2}/_{n_1}\right)$$

or to class two (attack) otherwise. The results are stored in vectors W_1 and W_2 as follows:

- W_1 is the discriminant function coefficients for benign connections in C and a function constant [$\text{cons}_1\text{temp}_1$].
- W_2 is the discriminant function coefficients for attack connections in C and a function constant [$\text{cons}_2\text{temp}_2$].

The result of applying the previous steps is linear functions that are represented as profiles. Let each of these profiles be denoted by DFP_{n_i}. Each profile describes the contextual preconditions that lead to a specific attack n_i as a linear function. Such functions are used to estimate the possibility of unknown attacks given the features of incoming connection. For any connection c with numerical features, e.g., (f_1, f_2) delivered to DFP_{n_i}, a linear score LS(c,DFP_{n_i}) is calculated.

The linear score is supposed to indicate the probability of a connection being an unknown attack. The linear score is not a normalized probability and may even take negative values. To obtain an estimated normalized probability of unknown attacks (zero-day), we apply softmax transformation [10] on linear scores to generate these probabilities. The outcome of the previous step is several estimated probabilities of attacks (one from each discriminant function); the one with the maximum value Max_{EP} is selected. The selected value is compared to a tunable user-defined zero-day estimated probability threshold Z_{EP}. The Z_{EP} is used to declare the incoming connection as a new unknown attack if its estimated probability Max_{EP} value > Z_{EP}; otherwise, it is declared as a benign activity. Algorithm 3.2 summarizes the steps performed to identify unknown attacks from a set of incoming connections $c_1, \ldots, c_n$. Lines (1–10) summarize the steps needed to calculate the Max_{EP} of each connection. Lines (11–18) summarize the steps needed to decide if c_j is an unknown attack when Max_{EP} is greater than the user-defined threshold Z_{EP}. The complexity s of discovering unknown attacks using the algorithm above is $O(N \times P \times D)$ where N is the number of connections under analysis, P is the number of DFPs and D is the number of coefficients in each function. The following example explains the process of detection of unknown attacks using linear discriminant functions.

Algorithm 3.2 DFPs profiles to detect unknown attacks

Input: A set of connections conn $= \{c_1, \ldots, c_n\}$
DFPs $= \{\text{DFP}_{n_i}, \ldots, \text{DFP}_{n_p}\}$
Z_{EP} a zero-day estimated probability threshold
Output: $Rc_J \in \{\text{Zday}_\text{A}, \text{bengin}\}$
Begin
$\text{Max}_{\text{EP}} = 0$
For $j = 1$ to n **do**
For $i = 1$ to p **do**
Find $\text{LS}(c_j, \text{DFP}_{n_i})$
$\text{EP}(c_j, \text{DFP}_{n_i}) = \textit{softmax transformation}\,(\text{LS}(c_j, \text{DFP}_{n_i}))$
If $\text{EP}(c_j, \text{DFP}_{n_i}) > \text{Max}_{\text{EP}}$ **then**
$\text{Max}_{\text{EP}} = \text{EP}(c_j, \text{DFP}_{n_i})$
End if
End for
If $\text{Max}_{\text{EP}} > Z_{\text{EP}}$ **then**
$Rc_j = \text{Zday}_\text{A}$
Else
$Rc_j = \text{benign}$
End if
Retrieve Rcj
End for
End

Domain-Specific Example: Table 3.1 shows a sample of five connections.

The first two are unknown attacks (in this context, the signature of these two attacks is assumed to be absent from the training data), and the last three are benign activities. Suppose that these connections have been predicted as benign by other prediction models (e.g., SLNs + APs).

The linear discrimination space in this example consists of the following pre-calculated pooled covariance matrix R for a particular set of connections that consists of many instances of an attack n_i and benign activities:

$$R = \begin{bmatrix} 3.2539 & -0.2235 \\ -0.2235 & 0.6015 \end{bmatrix}$$

Let also assume that the data in our example consists of two numerical features duration and source bytes. The discriminant function coefficient matrix W is shown below (the first column contains function constants; the second and third are the coefficients). For simplicity, we consider only one set of discriminant functions that

Table 3.1 Unknown attacks and benign connections with labels

C_{ID}	Duration	Source bytes	SLNs + APs prediction	Actual label
C_1	302	700	Benign	Unknown sql attack
C_2	151	1587	Benign	Unknown xterm attack
C_3	20	900	Benign	Benign
C_4	16	1500	Benign	Benign
C_5	200	800	Benign	Benign

identifies the profile of a specific attack n_i and a benign activity type (rows 1 and 2 in W):

$$W = \begin{bmatrix} -18.82 & 0.0676 & 0.0152 \\ -7.443 & 0.0352 & 0.010 \end{bmatrix}$$

Using R and W, the linear score LS of the attack n_i and the benign activity for the connections in Table 3.1 are calculated as follows. First is formulating the connection matrix P using the connections in Table 3.1. The matrix P consists of the duration and source byte columns and another column with "1" entry. This column needs to be added to P and it is multiplied by function constants. It is very important to note that the addition of such a column has no effect on the estimated probability values, since the values in this column are the ones multiplied by the function constants. Second is multiplying P by W^T to find the linear scores (LSs) for the attack n_i and benign activity. Lastly, the estimated probabilities EP for the attack n_i (Zero$_d$EP) and benign activity (Normal EP) are calculated by applying softmax transformation on LSs. The final EP values for the connections are as follows:

$$\mathrm{EP} = \begin{bmatrix} \text{Connection} & \text{Zero}_d\,\text{EP} & \text{Normal EP} \\ C_1 & 0.0884 & 0.116 \\ C_2 & 0.849 & 0.151 \\ C_3 & 0.002 & 0.998 \\ C_4 & 0.043 & 0.957 \\ C_5 & 0.318 & 0.681 \end{bmatrix}$$

Assuming that the Z_{EP} threshold is 0.6, then C_1 and C_2 are declared as unknown attacks based on the corresponding EP values, while C_3, C_4, C_5 are labeled as benign activities. More details on tuning the zero-day threshold Z_{EP} are discussed in the experiment chapter.

3.5 Experiments on Prediction Models to Discover Unknown Attacks

The technique utilized to create APs identifies relationships between attacks at the feature level based on context. Finding such relationships leads to several variants of the same attack. The objective of the experiments illustrated in this section is to examine if such variants can assist in predicting the steps performed by adversaries to initiate unknown attacks by measuring the effectiveness of APs and discriminant function profiles (DFPs) in detecting unknown attacks as variants of known attacks. More experiments are also conducted to compare the effectiveness of APs and 1CNN anomaly detection techniques in terms of their capability to detect unknown attacks.

This set of experiments is conducted on DARPA intrusion detection dataset which contains 14 types of unknown attacks in its testing part that do not exist in the training dataset. Table 3.2 shows details on these attacks. Most of the connections that contain such types of unknown attacks are predicted by the prediction models in our framework as benign activities (since they are unknown). Generally, the classification models we utilized earlier to discover known attacks do not have the ability to predict unknown attacks since they do not have the corresponding signatures. Since the prediction models we utilized earlier predict correctly most of the benign connections in the testing data (i.e., they have a high TN rate), we added new benign connections as shown in Table 3.2 to test if the APs we used to detect unknown attacks would correctly identify these connections as benign activities. The same set of the training connections—we utilized to create prediction models to identify known attack—is also utilized to create the prediction models of unknown attacks.

About 10 % of the benign connections in the training data are used to generate normal profiles (NPs). The latter are needed by the 1CNN anomaly detection technique. Three experiments are conducted to measure the detection rate of unknown attacks by analyzing incoming connections that partially match APs. In the first experiment, we measure the effectiveness of each prediction model individually. In the second experiment, we utilize the 1CNN anomaly detection technique on top of APs to examine if we can lower the FP rate by combining both

Table 3.2 Unknown types of attacks in DARPA intrusion detection dataset

Category	Unknown attack type	Number of connections
Denial of service (DoS)	Mailbomb, Processtable, Udpstorm	5115
Remote to local (R2L)	Httptunnel, named, Snmpgetattack, Snmpguess, Worm, Xlock, Xsnoop0	10,776
User to root (U2R)	Sqlattack, Xterm	8
Probe	Mscan, Saint	516
Benign	–	20,233
Total		36,648

prediction models in a layered manner. The third experiment focuses on the run-time efficiency for the prediction models of unknown attacks. The last part of this section demonstrates a comparison between the prediction models for unknown attacks in our framework versus other existing techniques.

According to several previous works, TP and FP rates are the major metrics that are commonly used to evaluate intrusion detection techniques [11, 12]. Therefore, we measure the effectiveness of unknown attack detection techniques in terms of both TP and FP rates. The true-positive (TP) rate refers to the probability that the prediction model produces an unknown attack when there is an actual attack. The false-positive (FP) rate is the probability that the prediction model produces an unknown attack when there is no attack. We also utilize the ROC curve (receiver operating characteristics) to compare between the prediction models in terms of both TP and FP rates. Since the experiments in the previous sections focus on the detection rate of known attacks, the main focus of the subsequent set of experiments is the detection rate of unknown attacks.

3.5.1 Using 1CNN, APs, and DFPs to Detect Unknown Attacks

The first experiment is conducted using (1) one-class nearest neighbor (1CNN) anomaly detection technique to detect unknown attacks as anomalies, (2) attack profiles (APs) to identify unknown attacks using profile similarity, and (3) discriminant function profiles (DFPs) to estimate zero-day attack probability. We utilize the anomaly score AS to discover potential unknown attacks in connections that do not fully match any APs. In the setting of this experiment, we used the Z_{AS} threshold values in the range from 0.5 to 2 to measure the changes in the TP and FP rates. Originally, the algorithm uses the value 1 as a default threshold. We ran this experiment using the top six dimensions of the NPs (normal activity profiles) extracted in a dimensionality reduction process performed using singular value decomposition (SVD). Similarly, the connections under evaluation are transformed to the same subspace through a projection process.

Figure 3.3 shows the results of this part of our experiment. Using the anomaly score AS as a threshold to identify unknown attacks, we observe high TP and FP rates at low values of Z_{AS} threshold. At higher values of Z_{AS} threshold, the FP rate has been diminished. One of the best combinations of TP and FP rates, 0.8 and 0.17, respectively, is attained at 1.7 level of Z_{AS}. Since the relative importance of TP and FP is to be identified by domain experts, we utilize two parameters $\alpha, \beta | 0 < \alpha, \beta < 1$ to find the best threshold value that achieves the domain expert expectation. The parameters α and β express the relative importance the domain expert assigns to TP and FP. If the domain expert is concerned more about detecting unknown attacks while expecting some false positives, β needs to be greater than α.

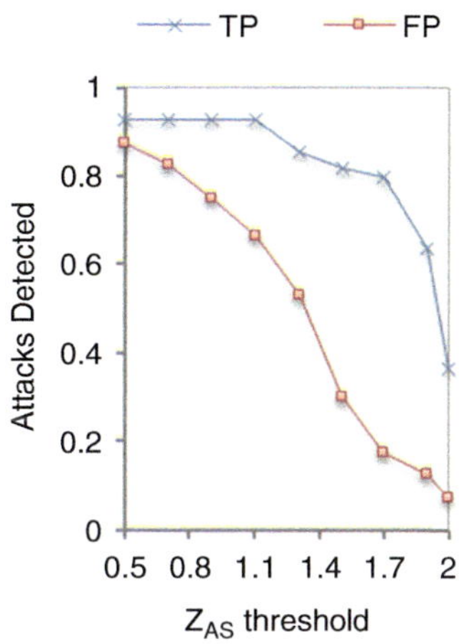

Fig. 3.3 TP and FP rates for detecting unknown attacks using 1CNN

Table 3.3 Optimizing the detection rate of unknown attacks (1CNN)

Z_{AS}	α	β	$\alpha.\ (1–FP) + \beta.\ TP$
0.5	0.3	0.7	0.686
0.7	0.3	0.7	0.701
0.9	0.3	0.7	0.724
1.1	0.3	0.7	0.749
1.3	0.3	0.7	0.739
1.5	0.3	0.7	0.782
1.7	**0.3**	**0.7**	**0.807**
1.9	0.3	0.7	0.707
2	0.3	0.7	0.532

Table 3.3 shows a scenario in which a domain expert gives higher weight (0.7) to β. Based on the assigned weights, 1.7 is the threshold value that leads to the highest rate of detecting unknown attacks. The corresponding TP and FP rates (from Fig. 3.3) are 0.80 and 0.17, respectively. If the domain expert expects to see less false positives (e.g., $\alpha = 0.6$, $\beta = 0.4$), the value of the best threshold will go higher (1.9). As a second step in this experiment, we measure the detection rate of unknown attacks using similarity with APs which are created as feature-based profiles to discover potential unknown attacks. This part of the experiments is carried out by varying the values of profile similarity threshold Z_{PS} from 0.1 to 0.9. For any connection C, if its profile similarity score is greater than Z_{PS}, it is labeled as an unknown attack; otherwise the connection is labeled as benign. Figure 3.4 illustrates the results of this experiment. The TP rate is higher at small values of Z_{PS}; when Z_{PS} is 0.1, the TP rate is about 0.98. The FP rate is also high at low values of Z_{PS}. As the values of Z_{PS} increase, a decrease in the values of FPs is observed, though we start missing some true positives (i.e., attacks) as well. Similarity with APs as a prediction metric for unknown attacks achieves satisfactory TP rates at small values of Z_{PS}. At 0.8 Z_{PS} threshold, the TP rate is

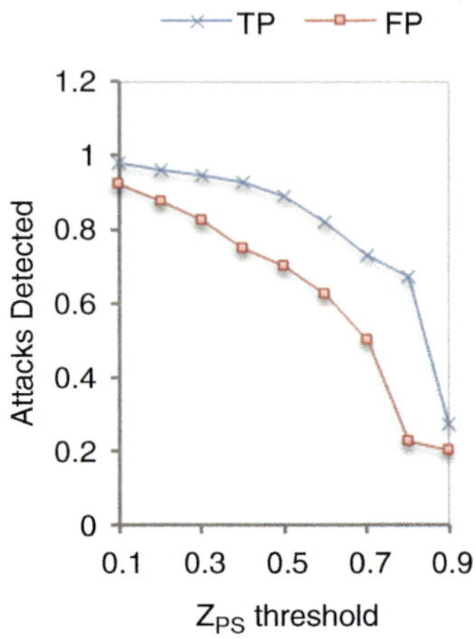

Fig. 3.4 TP and FP rates for detecting unknown attacks using similarity with APs

approximately 0.67, and the FP rate is 0.22. The results of this experiment show that the TP rate is lower when only the similarity score with APs is utilized to discover unknown attacks, in particular, at high values of Z_{PS} threshold. The main reason is related to numerical features utilized in creating APs. Unknown attacks have several unique characteristics, and matching the features of incoming connections with the discretized numerical features in APs becomes an ineffective approach for similarity calculation. This leads to a high mismatch rate between the features of incoming connections and the features in APs, resulting in a low similarity between the incoming connections that include unknown attacks and the APs. Therefore, we re-create APs as discriminant function profiles (DFPs) and examine their effect on detecting unknown attacks. For each set of attacks with similar APs, their corresponding DFPs are created using the numerical features in the dataset. The highly ranked ten features used to create DFPs are selected using correlation-based feature selection technique proposed by Hall et al. [13]. The high EP (see Chap. 6) value calculated on a specific incoming connection indicates that such a connection is similar but not identical to a specific known attack. Therefore, such a connection has not been discovered as a known attack using APs. Connections which do not match any APs are passed to the DFPs. Once the connections are processed by DFPs, the maximum estimated probability EP calculated for each connection is compared to a user-tunable threshold (Z_{EP}). Alerts about unknown attacks are raised if the estimated probability for a specific connection is greater than Z_{EP}.

Figure 3.5 shows the results of this experiment. The values of Z_{EP} are in the range 0.1–0.9. When the Z_{EP} threshold has small values (<0.6), both TP and FP rates are high; however, at higher values of Z_{EP} threshold, the FP rates are diminished. For instance, at 0.8 value of Z_{EP}, the FP rate is approximately 0.1 and the TP rate is 0.83. We also observe that few true positives are missed when Z_{EP} is greater than 0.5. We compare between the prediction models utilized to identify unknown attacks in terms of both TP and FP. We used ROC curve to perform this comparison as shown on Fig. 3.6.

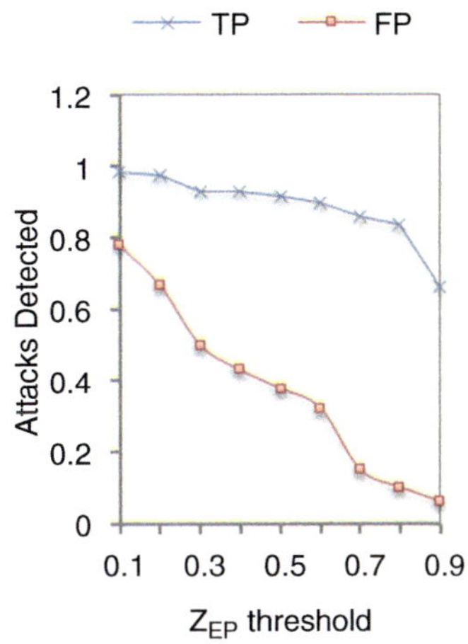

Fig. 3.5 TP and FP rates for detecting unknown attacks using DFPs

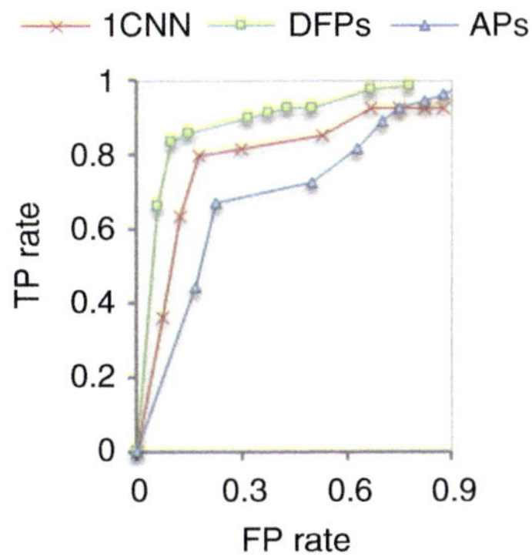

Fig. 3.6 ROC curve for the prediction models used to detect unknown attacks

The effectiveness of each prediction model is shown on a line that connects several discrete operating points. These points correspond to specific values of thresholds Z_{PS}, Z_{AS}, Z_{EP} utilized earlier. The figure indicates that the DFPs achieve better detection rate in terms of both TP and FP rates, and they can be explained by two reasons:

First, DFPs show better handling of the numerical features in network data compared to other prediction models.

Second, while DFPs are created in a transformed space, they are based on the similarities and relationships discovered between APs. As such we still need APs initially to create DFPs. The latter are however more effective than APs in terms of projecting possible actions/activities of attackers and the paths they might take as linear functions.

3.5.2 Using 1CNN on Top of APs and DFPs to Detect Unknown Attacks

The second experiment focuses on measuring the effect of utilizing 1CNN on top of APs and DFPs to identify unknown attacks. This experiment is conducted in two phases as follows:

Applying 1CNN on Top of APs. We selected a profile similarity threshold Z_{PS} based on which the connections predicted as benign activities by APs are further processed by 1CNN anomaly detection technique. The connection for which the similarity score with APs is less than Z_{PS} threshold is deemed as benign activity, and it is not processed further by 1CNN. The connection for which the similarity with APs is greater than Z_{PS} is processed further using 1CNN. Such a connection is declared as an unknown attack if its anomaly score AS is also greater than a specific value of the Z_{AS} threshold. We selected 0.5 as a profile similarity (Z_{PS}) threshold. We also selected five Z_{AS} operational points (the Z_{AS} values [1.3, 1.5, 1.7, 1.9, and 2]) to conduct this experiment. We then compare the predictions made by 1CNN only, with the predictions made when 1CNN is applied on top of APS. It should be noted that the purpose of this phase in our experiment is not to select the combinations of Z_{AS} and Z_{PS} thresholds that give the highest FP and TP rates, but our purpose is to examine if anomaly detection techniques such as 1CNN, when applied to refine the predictions made by APs, decrease the FPs generated when utilizing profile similarity score

Applying 1CNN on Top of DFPs. We selected specific value for Z_{EP} threshold based on which the connections are further processed by 1CNN. Each connection for which the EP value is less than Z_{EP} is deemed as benign connection, and it is not further processed by 1CNN. The connection for which the EP value calculated using discriminant functions is greater than Z_{EP} threshold is processed further using 1CNN. Such a connection is declared as an unknown attack if its anomaly score AS is also greater than a specific Z_{AS} threshold. We selected 0.5 as (Z_{EP}) cutoff point to further process connections using 1CNN. We also selected five Z_{AS} operational points (the Z_{AS} values [1.3, 1.5, 1.7, 1.9, and 2]) to run this experiment. We then compared the predictions made by 1CNN only with the predictions produced when 1CNN is applied on top of DFPs.

Figure 3.7 shows the results of this experiment. Each line connects several discrete operating points. These points represent the threshold values of Z_{AS}. The figure shows that utilizing 1CNN on top of attack profiles results in relatively lower FP rate and higher TP rate, specifically when 1CNN is applied on top of DFPS. Applying 1CNN on top of other prediction models leads to less uncertainty about the predictions made by DFPs and APs. Thus, there is no need to analyze these connections using 1CNN. *To conclude, the experiments demonstrate that applying an approach which combines prediction models that captures contextual relationships between attacks with an appropriate anomaly detection technique leads to lower FP and higher TP rate in the process of detecting unknown attacks.*

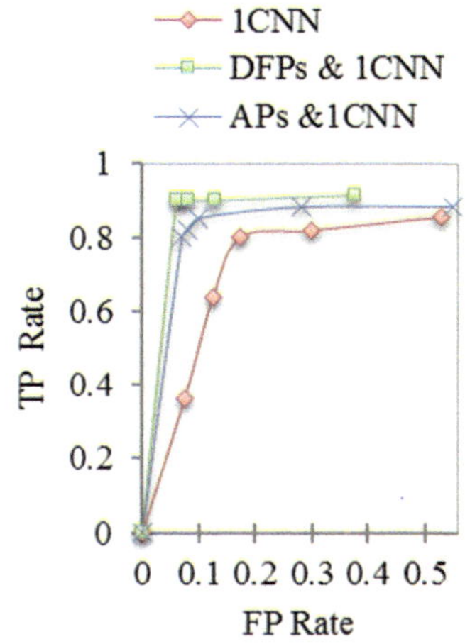

Fig. 3.7 ROC curve for detecting unknown attacks using 1CNN and APs, 1CNN and DFPs

3.5.3 Time-Based Efficiency for Identifying Unknown Attacks

We conducted a third experiment to measure the time efficiency of the prediction models utilized to discover unknown attacks in terms of time needed to examine incoming connections at run-time. The elapsed time in seconds during the evaluation (testing) phase is measured and used as a metric for comparison under the following settings:

- Changing the number of dimensions used to calculate the anomaly score (AS), the profile similarity score (SIM), and the estimated probability (EP). The incoming connections are transformed to a reduced *k*-dimensional space where *k* refers to the number of dimensions of NPs used by 1CNN, the number of linear coefficients of DFPs, or the number of features used to calculate similarity with APs.
- Using small set of benign connections to create NPs that are used by 1CNN anomaly detection technique to discover unknown attacks. The larger the number of benign connections to create NPs, the more the computational time the 1CNN requires to calculate the anomaly score. To make our comparison consistent among the prediction models of unknown attacks, the number of the NPs that we selected to run this experiment is small and equals the number of attack profiles in APs and DFPs.

Figure 3.8 shows the results of this comparison. The time it takes to examine incoming connection that does not match APs is 47 s when ten dimensions are utilized to identify unknown attacks using similarity with APs. In contrast, the figure shows that the time the 1CNN anomaly detection algorithm takes is approximately 60 s to process incoming connections. This computational difference is due to the overhead needed to find the two nearest neighbors. Now, let us consider

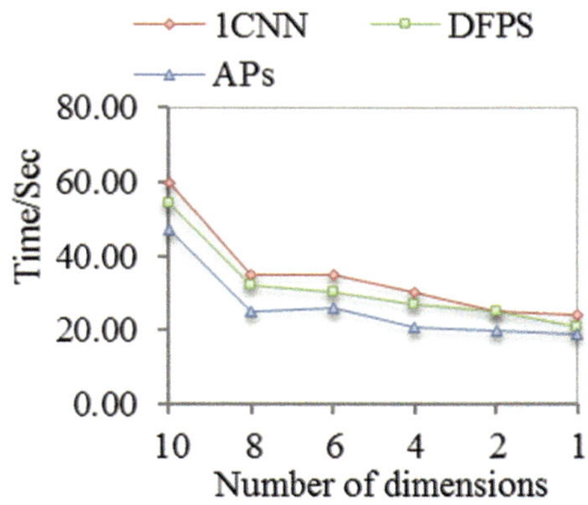

Fig. 3.8 Time-based efficiency of unknown attack detection techniques

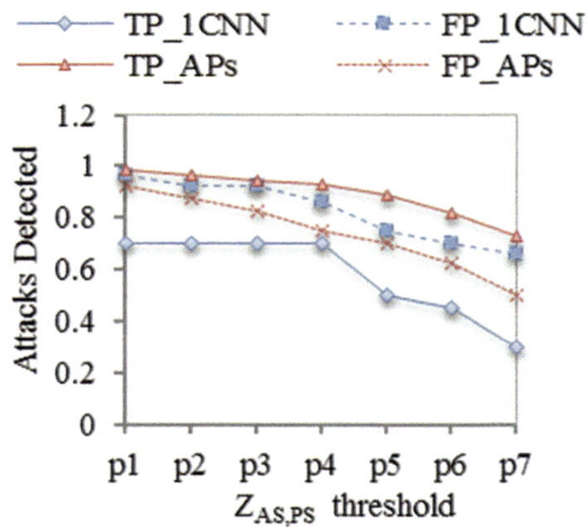

Fig. 3.9 TP and FP rates for detecting unknown attacks using APs and 1CNN (with few NPs)

Fig. 3.9 on the right. The figure compares the effectiveness of APs and 1CNN when the number of NPs utilized to detect anomalies using 1CNN equals the number of APs, that is, when very few *NPs* are utilized to detect anomalies using 1CNN. The figure shows that the effectiveness (measured using TP and FP rates) of APs when seven Z_{PS} operation points in the range (0.1–0.7) are used is better than that of 1CNN using seven Z_{AS} operation points in the range 0.5–1.7. This part of our experiments indicates that the effectiveness of 1CNN depends on the number of benign patterns used to identify unknown attacks as anomalies. Overall, density-based anomaly detection techniques (such as 1CNN) attain good detection rate of unknown attacks when they are trained using sufficient patterns of benign activities; however, this requires a large number of normal activity signatures which increase the computation time. In contrast, one can notice that by capturing contextual relationships between known attacks, such as in the case of APs, fewer patterns are needed to identify unknown attacks with relatively less computational time.

3.5.4 A Comparison with Existing Unknown Attack Detection Techniques

We compared the prediction models that we created to detect unknown attacks with those of existing ones. In particular, we compared several anomaly detection approaches such as one-class SVM [1] (see Sect. 2.5.1.4) and payload-based anomaly detector that has been utilized by Bolzoni et al. [14]. The latter models the normal application payload of network traffic in an unsupervised fashion. During the training phase, a profile byte frequency distribution and their standard deviation of the application payload flowing to a single host and port are calculated; then Mahalanobis distance is used during the detection phase to calculate the similarity of new data against the precomputed profiles. The detector compares this measure against a threshold and generates an alert when the distance of the new input exceeds this threshold. Since we performed dimensionality reduction during the detection process, we compared our prediction models with the PCASOM (principal component analysis and self-organizing map) approach [15]. PCASOM is a modified unsupervised neural network learning algorithm to perform PCA and use the resulting model to detect unknown attacks as anomalies.

We also compared our prediction models with other local-density-based anomaly detection techniques such as density-based local outliers which have been utilized in Lazarevic et al. [16] to assign to each data example a degree of being outlier. This degree is called the local outlier factor (LOF) a data example. The technique computes local reachability density of data connection c as inverse of the average reachability distance based on the minimum number of data example nearest neighbors of data connection c.

We also compared the prediction models in our framework with other probabilistic techniques such as Bayes estimator [17] and Gaussian field [18]. The Bayes estimator technique mainly estimates the prior and posterior probabilities of unknown attacks based on pre-identified knowledge about known attacks. The authors construct a naive Bayes classifier to classify the instances into normal instances, known attacks, and unknown attacks. The authors state that the advantage of pseudo-Bayes estimators is that no knowledge about new attacks is needed since the estimated prior and posterior probabilities of new attacks are derived from the information about normal instances and known attacks. In the Gaussian field approach, the problem is formulated on a graph, where the mean of the field is characterized in terms of harmonic functions. It can be viewed as having a quadratic loss function with infinity weight, so that the labeled data are clamped (fixed at given label values).

Table 3.4 summarizes the results of our comparison. All of the approaches under comparison conducted their experiments using the same dataset. Our comparison with these approaches is conducted using threshold values 0.8 for Z_{EP}, 0.8 for Z_{PS}, and 1.7 for Z_{AS}. Some of the existing approaches achieve good true-positive rates such as Bayes estimators and PCASOM. Bayes estimators also show a low false alarm rate. Compared to other approaches, the one-class SVM is not very effective

Table 3.4 A comparison with existing unknown attack detection techniques

Approach	TP	FP
One-class SVM (inner product) [1]	0.58	0.48
Anomalous payload-based IDS [14]	0.58	0.09
PCASOM [15]	0.77	–
Local outlier factor [16]	0.66	–
Bayes estimators [17]	0.72	0.07
Gaussian field [18]	0.53	–
DFPs, Z_{EP} threshold = 0.8	0.83	0.09
APs, Z_{PS} = 0.8	0.67	0.22
1CNN, Z_{AS} threshold = 1.7	0.8	0.17
DFPs + ICNN	**0.89**	**0.06**
APs + ICNN	0.85	0.10

as it leads to almost 0.48 of false alarms. The best TP is attained by DFPs with TP rate of 0.83. When DFPs are combined with 1CNN anomaly detection algorithm, the FP rate becomes less than 0.07. The comparison we performed in this section indicates that DFPs when combined with 1CNN are quite effective in detecting unknown attacks, particularly, compared to other classifiers such as neural networks (NN).

References

1. Shon, T., Moon, J.: A hybrid machine learning approach to network anomaly detection. Information Sci. **177**(18), 3799–3821 (2007)
2. Song, J., Takakura, H., Kwon, Y.: A generalized feature extraction scheme to detect 0-day attacks via IDS alerts. In: Proceedings of the international symposium on applications and the internet, Urku, Finland, pp. 55–61. 1442004: IEEE Computer Society (2008). doi:10.1109/saint.2008.85
3. Hendry, G.R., Yang, S.J.: Intrusion signature creation via clustering anomalies. In: Proceedings of SPIE, Bellingham, WA, pp. 69730C–69731 (2008)
4. Portnoy, L.: Intrusion Detection with Unlabeled Data Using Clustering. Data Mining Lab-Department of Computer Science, Columbia University, Technical report (2001)
5. Li, Z., Sanghi, M., Chen, Y., Kao, M.-Y., Chavez, B.: HAMSA: fast signature generation for zero-day polymorphic worms with provable attack resilience. In: IEEE symposium on security and privacy, pp. 32–47. IEEE (2006)
6. Chandola, V., Banerjee, A., Kumar, V.: Anomaly detection: a survey. ACM Comput. Surv. **41** (3), 1–58 (2009). doi:10.1145/1541880.1541882
7. Tax, D.M., Duin, R.P.: Data description in subspaces. In: 15th international conference on pattern recognition, pp. 672–675. IEEE (2000)
8. Mika, S., Ratsch, G., Weston, J., Scholkopf, B., Mullers, K.R. Fisher discriminant analysis with kernels. In: Proceedings of the IEEE signal processing society workshop, Madison, WI, pp. 41–48 (1999). doi:10.1109/NNSP.1999.788121
9. Li, X.-B.: A scalable decision tree system and its application in pattern recognition and intrusion detection. Decis. Support Syst. **41**(1), 112–130 (2005). doi:10.1016/j.dss.2004.06.016
10. Tuerk, A.: Implicit softmax transforms for dimensionality reduction. In: IEEE international conference on acoustics, speech and signal processing (ICASSP'08), Las Vegas, NV, pp. 1973–1976. IEEE (2008)

11. Gu, G., Fogla, P., Dagon, D., Lee, W., Skorić, B.: Measuring intrusion detection capability: an information theoretic approach. In: Proceedings of the ACM symposium on information, computer and communications security, Taipei, Taiwan, pp. 90–101. ACM (2006)
12. Proffitt, T.: How Can You Build and Leverage SNORT IDS Metrics to Reduce Risk? The SANS (SysAdmin, Audit, Networking, and Security) Institute, Boston, MA (2013)
13. Hall, M.A.: Correlation-Based Feature Selection for Machine Learning. The University of Waikato, Hamilton (1999)
14. Bolzoni, D., Etalle, S., Hartel, P.: Poseidon: a 2-tier anomaly-based network intrusion detection system. In: Fourth IEEE international workshop on information assurance (IWIA'06), London, pp. 146–156. IEEE (2006)
15. Liu, G., Yi, Z.: Intrusion detection using PCA-SOM neural networks. In: Third international symposium on neural networks, Chengdu, China, pp. 240–245 (2006)
16. Lazarevic, A., Ertoz, L., Kumar, V., Ozgur, A., Srivastava, J.: A comparative study of anomaly detection schemes in network intrusion detection. In: Proceedings of the third SIAM international conference on data mining, vol. 3, San Francisco, CA, USA, pp. 25–36. Society for Industrial & Applied (2003)
17. Barbara, D., Wu, N., Jajodia, S.: Detecting novel network intrusions using Bayes estimators. In: First SIAM conference on data mining, Chicago IL, pp. 1–17, Citeseer (2001)
18. Chuanliang, C., Yunchao, G., Yingjie, T.: Semi-supervised learning methods for network intrusion detection. In: IEEE international conference on systems, man and cybernetics (SMC'08), Seoul, Korea, 12–15 Oct 2008, pp. 2603–2608 (2008). doi:10.1109/ICSMC.2008.4811688

Chapter 4
Unwanted Traffic Detection and Control Based on Trust Management

Zheng Yan, Raimo Kantola, Lifang Zhang, and Yutan Ma

Abstract Networks such as the Internet, mobile cellular networks, and self-organized ad hoc networks have dramatically changed our daily life and brought tremendous benefits to us. However, they are also bogged down by unwanted traffic, which is malicious, harmful, or unexpected for its receivers. In order to control the unwanted traffic over the networks, especially the mobile Internet, we propose a generic scheme named TruCon for unwanted traffic detection and control based on trust management in this chapter. It can control unwanted traffic from its source to destinations in a personalized manner according to trust evaluation at a global trust operator, traffic and behavior analysis at hosts, and traffic observation at network service providers. Thus, the proposed scheme can conduct unwanted traffic detection and control by integrating distributed and centralized functions and supporting both defensive and offensive approaches of unwanted traffic control. We successfully applied the scheme to control SMS spam and unwanted contents in pervasive social networking and implemented it under the infrastructure of software-defined networking (SDN). System implementation and evaluation showed that the scheme is effective with regard to accuracy and efficiency for intrusion detection and unwanted traffic control. It is also robust against a number of internal misleading system attacks, such as hide evidence attack, bad-mouthing attack, and on-off attack, playing in conjunction with traffic intrusions. Meanwhile,

---

Z. Yan (✉)
State Key Laboratory on Integrated Services Networks, School of Cyber Engineering, Xidian University, Xi'an, China

Department of Communications and Networking, School of Electrical Engineering, Aalto University, Espoo, Finland
e-mail: zhengyan.pz@gmail.com; zyan@xidian.edu.cn

R. Kantola • L. Zhang
Department of Communications and Networking, School of Electrical Engineering, Aalto University, Espoo, Finland
e-mail: raimo.kantola@aalto.fi; lifang.zhang@aalto.fi

Y. Ma
State Key Laboratory on Integrated Services Networks, School of Cyber Engineering, Xidian University, Xi'an, China
e-mail: xiaomadream@126.com

I.M. Alsmadi et al. (eds.), *Information Fusion for Cyber-Security Analytics*, Studies in Computational Intelligence 691, DOI 10.1007/978-3-319-44257-0_4

the scheme can provide personalized unwanted traffic control based on unwanted traffic detection behaviors.

4.1 Introduction

Networks such as the Internet, mobile cellular networks, and self-organized ad hoc networks have dramatically changed our daily life and brought tremendous benefits to us. The Internet is the backbone of remote communications, networking, and computing. It carries a vast range of information resources and services, bringing unprecedented convenience to our daily life. Mobile cellular networks coupled with local connection technologies connect billions of mobile devices to the Internet and thus enable them to access the services and applications of the Internet, such as mobile email, short message service (SMS), mobile commerce, multimedia communications, and mobile social networking. Mobile cellular networks have therefore become a global platform for the provision of various applications and services. Mobile ad hoc network (MANET) can provide us exceptional social experience by satisfying our instant social networking needs, especially for familiar strangers in vicinity. Thus, MANET has a good prospect of becoming a practical platform for pervasive social networking and computing. In short, the Internet, mobile cellular networks, and self-organized ad hoc networks are indispensable in our modern life.

However, when we enjoy great convenience brought by networks, we also receive unwanted, unexpected, and even malicious contents. Normally, such unwanted contents are distributed by botnets and include malware, viruses, spam, intrusions, and unsolicited commercial advertisements. They could intrude user devices, occupy device memory, and irritate the user. Moreover, they burden network service providers by adding extra load into the network, which greatly increases the possibility of normal traffic congestion. Most importantly, the receivers have to pay for such unwanted contents in the form of wasting time, investing into and operating spam filtering, installing firewalls, virus scanning, malware and intrusion detection, and cleaning up after infection. Thus, developing a systematic method to control the unwanted contents in the networks has become a crucial task that brooks no delay.

Controlling unwanted traffic is difficult due to many technical and social reasons. First, unwanted traffic is a subjective opinion of a user in that some traffic might be viewed as unwanted by one user, while another treats it as useful. Second, the subjective notion of unwanted traffic and various types of Internet traffic make it difficult to develop a generic solution. Third, any unwanted traffic control (UTC) system could be the target of hackers. This fact requests that the designed system should be robust against various attacks. On the other hand, we note that security issues are difficult for ordinary users to comprehend, which leads to low security awareness. This implies that it is preferred to have an automatic and intelligent solution personalized for each user and with minimum involvement of the user. Trust management-based mechanism represents one of the effective methods to detect and control unwanted traffic through trust and reputation evaluation. Thus, a number of trust and reputation mechanisms have been proposed in the literature to control spam [1–8], SPIM (i.e., instant messaging spam [9]), SPIT (Spam over Internet Telephony [10]), and web page spam [11–14]. However, the aforementioned work lacks a generic, trustworthy, and personalized solution that can overcome all the above challenges.

In this chapter, we propose a generic scheme named TruCon for unwanted traffic detection and control based on trust management [44]. It can control unwanted traffic from its source to destinations in a personalized manner according to trust evaluation at a global trust operator (GTO), traffic and behavior analysis at hosts, and traffic observation at network service providers (NSPs). The trust of an entity contains two parts: the global trust that indicates the probability and nature of unwanted traffic sourced from the entity and the detection trust that specifies the previous detection performance of the entity. The detection trust is introduced herein to indicate the quality and credibility of a detection report. The motivation to analyze it in our system is to fight against unqualified reports, such as reports from intruded system entities, reports being intentionally framed, and reports detected with broken and poor detection tools. Our system hinders such unqualified reports by scaling up detection reports from an entity with a high detection trust value while scaling down detection reports from entities with a low detection trust value when calculating the global trust. In this way, we make sure that the global trust is calculated mainly from trusted reports and thus improve our system effectiveness. We assume that system effectiveness and feasibility follows from three characteristics: accuracy, efficiency, and robustness. Accuracy means that the (global) trust value of an entity must reflect the share of unwanted traffic that is sent by hosts subscribed or belonging to the entity. A 100 % accuracy implies that there are no false positives. Efficiency means that malicious senders are spotted quickly and that the share of false negatives is as low as it can be. Robustness means that the system will continue performing its task accurately and efficiently under any feasible attack strategies of internal misleading entities that do not conform to the unwanted traffic monitoring and reporting process in the system.

Concretely, we evaluate the global trust of each involved system entity in order to figure out if the traffic from it should be controlled for a receiver. The system entity can be a host, a corporate network, or an NSP. The evaluation is based on both unwanted traffic detection reports sent from the hosts and traffic monitoring and checking at NSPs. The system controls unwanted traffic with a combination of distributed and centralized functions. The host itself is capable of blocking traffic targeting on it based on local traffic and behavior analysis. We define that a counterapproach to unwanted traffic is defensive if it focuses on protecting hosts and networks from unwanted traffic intrusion using traffic and content analysis and blocking it based on local knowledge. The approach is offensive if it seeks to make the business of unwanted traffic less successful and profitable, for example, by punishing misleading or indifferent behaviors and encouraging good behaviors. Our proposed system filters unwanted traffic at each host in a defensive way and automatically controls traffic from a distrusted source in an offensive manner through trust management. In order to overcome potential attacks, we apply both the global trust and detection trust to validate the reports from the hosts and SPs and make use of reporting correlation to provide personalized UTC. Our design aims to provide a generic solution for various types of unwanted traffic over the networks, which is efficient in controlling unwanted traffic based on personal preferences and robust against various system attacks.

We successfully applied the scheme to control SMS spam and unwanted contents in pervasive social networking (PSN) and implemented it under the infrastructure of software-defined networking (SDN). System implementation and evaluation showed that the scheme is effective with regard to accuracy and efficiency for intrusion detection and unwanted traffic control. It is also robust against a number of internal misleading system attacks, such as hide evidence attack, bad-mouthing attack, and on-off attack, playing in conjunction with traffic intrusions. Meanwhile, the scheme can provide personalized unwanted traffic control based on unwanted traffic detection behaviors. In addition, the system acceptance by the network entities (NSPs and hosts) is further investigated based on game theory.

The rest of the paper is organized as follows. Section 4.2 gives a brief overview of related work. Section 4.3 introduces the system and threat models and our design goals. Then we describe the implementation and performance evaluation of applying TruCon to detect and control SMS spam in Sect. 4.4. Section 4.5 demonstrates implementation and performance analysis of TruCon in the context of PSN, followed by the implementation of TruCon based on SDN infrastructure in Sect. 4.6. Finally, conclusion is summarized in the last section.

4.2 Related Work Overview

4.2.1 Unwanted Traffic Detection Technology

A number of anti-spam techniques and applications have been proposed in the literature. Examples of existing solutions are whitelists/blacklists, header/content checks and rule-based filtering (e.g., SpamAssassin [15]), Bayesian analysis (e.g., SpamBayes [16]), sender authentication (e.g., Sender Policy Framework [17], Yahoo DomainKeys [18], etc.), challenge/response, Blackhole listing (e.g., SORBS, Kelkea MAPS), and distributed checksums (e.g., [19]). A problem of whitelists and blacklists is that they leave a sizable set of senders in the middle of the spectrum that are not classified. We also note that spam filters, intrusion detection systems, and firewalls are defensive tools, by defending against malware and intrusion attacks. They are good at collecting the evidence of suspect behaviors. An offensive tool would set the goal of winning the war against the shady and criminal ecosystem by making it or at least as many of its businesses as possible unprofitable. Firewalls may reside in routers and separate boxes and on hosts. A network-based firewall for battery-powered mobile devices can be efficient, while on the device a firewall is not practical; attacks or unwanted traffic could deplete the battery. Thus, a network-based offensive solution is preferred for the mobile Internet.

Meanwhile, a personalized solution for UTC is expected in practice. Obviously, whether the traffic is perceived as unwanted is subjective. Some traffic, classified as unwanted by a host, could be expected by others. A network-based or cloud-based solution controlled by a policy dynamically personalized for different individual hosts looks like an attractive option. The reports of various detection tools executed at a host can play as valuable inputs to the personalized control of unwanted traffic.

4.2.2 Unwanted Traffic Control Through Trust Management

Trust management has been applied to control unwanted traffic. It concerns evaluating, establishing, controlling, enhancing, and ensuring trust [20, 21]. In terms of UTC, trust management aims to control or filter traffic automatically based on the trust relationship between a traffic source and its receiver. A number of solutions were proposed to control unwanted traffic via trust and reputation mechanisms. Most of them target on email spam.

Most existing spam control solutions cannot provide personalized traffic control and overcome internal misleading attacks on the proposed system. A distributed architecture and protocol for establishing and maintaining trust between mail servers was proposed in [1]. The architecture is a closed-loop control system that can be used to adaptively improve spam filtering by automatically using trust information to tune the threshold of such filters. A layered trust management framework was proposed in order to help email receivers eliminate their unwitting trust and provide them with accountability support [5]. In Zhang et al. [6], the IPGroupRep clusters the senders into different groups based on their IP addresses and computes the reputation value of each group according to the feedback of email receivers. The reputation value can be used to indicate whether an incoming message is spam or not. However, the above solutions cannot provide personalized control and overcome attacks raised by malicious feedback.

Other spam control solutions adopt a different system structure or mechanisms from our solution, although some features are similar to ours. MailTrust filters out dishonest feedbacks to obtain an accurate trust value of each mail server [8]. The credibility-based reputation generation is similar to the detection trust in our solution. But MailTrust is a distributed reputation system, while ours is based on GTO, but also defends against distributed intrusions. A predictive approach based on static statistical analysis on the behavior of email senders was proposed in Tang et al. [3]. But this solution cannot be applied into an UTC system at runtime, like ours. It is not efficient to control fast spreading botnets. A multilevel reputation-based greylisting solution was proposed to improve the efficiency of traditional greylisting anti-spam methods by significantly reducing the transfer delay of messages caused by the additional greylisting level [2]. Comparing with the above work, the trust evaluation in our solution is not only based on the detection reports from hosts, the traffic, and behavior analysis at hosts but also the monitored behavior of unwanted traffic sources at NSP. Particularly, the control can be personalized according to the correlation of past detection reporting behaviors.

Highly related to our work, a reporter-based reputation system for spam filtering was proposed to filter spam [4]. The system uses the reports of highly reputable reporters for spam removal, while in our solution the traffic control is fine-grained by counting all collected reports with the detection trust as a discount and in a personalized manner. This work did not evaluate the system performance under various attacks, such as bad-mouthing attack. It did not discuss its applicability on other types of unwanted traffic. Further study and analysis are needed to make a generic solution that is applicable and effective to control various types of unwanted traffic and robust against system attacks.

A number of solutions attempted to overcome web page spam [7, 11–14], VoIP spam calls (SPIT [10]), and spam of instant messaging (SPIM [9]). These solutions are only applicable for a specific type of spam, not generic and suitable for controlling other types. But these methods can play as specific detection tools for detecting different types of unwanted traffic in our proposed solution. Literature still lacks a generic solution, which is efficient, accurate, and robust to control various unwanted traffic in a personalized manner.

4.2.3 SMS Spam Control

There are mainly two types of techniques currently applied to filter SMS spam: black- and whitelist and text classification [22]. Black- and whitelist allows mobile users to make a blacklist and a whitelist that contain the phone numbers or keywords used by an SMS spam filtering system. Thereby, every SMS coming from those phone numbers or containing the keywords in the blacklist will be put into a spam folder. One way in this approach is to filter SMS spam by comparing the words in SMS messages with the keywords saved in the blacklist. This technique has been practically deployed in mobile phones, e.g., an SMS Spam Manager running in Nokia Symbian phones and a Spam SMS Blocker running in Android phones. However, this technique does not perform very well in many situations because it depends on the keywords listed in the blacklist. Obviously, accuracy of SMS spam filtering and control is an issue. Some useful SMS messages could be filtered, and some real SMS spam cannot be controlled if the corresponding keywords are not listed in the blacklist. On the other hand, text classification distinguishes SMS spam from other messages based on message content. It can be applied at a mobile cellular operator to block SMS spam or on a smartphone to filter SMS spam. Text classification relies on the patterns of SMS spam. Some examples of patterns are word occurrences, length, and frequency of words in messages. Text classification uses pattern recognition algorithms such as naïve Bayes, support vector machine (SVM), artificial neural network (ANN), decision tree, k-nearest neighbor (kNN), and hidden Markov model (HMM) for SMS spam filtering. Deng and Peng proposed a method to filter Chinese SMS spam with a naïve Bayesian classification algorithm by introducing such attributes as the length of the SMS and rules found by statistics into an attribute set [23].

Although there are a number of existing anti-SMS-spam solutions as described above, they may not be very effective in practice. This is due to the nature of SMS messages. For example, the SMS messages have significantly less characters than emails. A standard SMS message can only contain 160 characters. So people tend to use nonstandard words in their messages, such as "how r u (how are you)" and "asap (as soon as possible)," which make it difficult for the abovementioned methods based on words or semantic contents of messages to accurately filter SMS spam. However, a trust management-based solution, such as TruCon, can overcome such a shortcoming because user-subjective feedback plays an essential role in the trust evaluation on SMS sources. In addition, the anti-SMS-spam toolkit installed in the mobile devices (as described above) can greatly help TruCon automatically generate detection reports.

4.2.4 Pervasive Social Networking and Its Unwanted Content Control

In both academia and industry, social activities supported by MANET were widely explored. Stanford MobiSocial Group developed Junction, a mobile ad hoc and multiparty platform for MANET applications [24]. ETHz Systems Group developed a pervasive social communication platform, named AdSocial [25]. In a floating content system, contents are only shared within an anchor zone in a best-effort manner [26]. In industry, quite a number of companies, such as Microsoft, Nokia, and Intel, conducted research in the area of PSN. For example, Microsoft Research Asia developed EZSetup system in order to let a mobile user find services provided by his/her neighbors [27]. The Nokia Instant Community offered an instant social networking platform to allow people in vicinity to communicate, get to know, and share information with each other [28]. Intel Berkeley Lab ran a project named Familiar Stranger based on mobile devices to extend social experiences with strangers that we regularly observe but do not interact with in public places [29]. However, the issue on unwanted traffic control was not considered in these systems.

Most existing research focused on the security issues in the network layer of PSN, but seldom paid attention to the unwanted content control issue in its application layer. Many solutions have been proposed to secure MANET, which mainly aim to ensure the security in MANET link layer (e.g., secure MAC protocols such as IEEE 802.11 series) and network layer, such as secure ad hoc routing protocols and packet forwarding solutions [30, 31]. Trust evaluation and management were also applied to ensure MANET security as a supplement to cryptographic measures for malicious node detection, information quality assessment, and node authentication [32].

4.3 TruCon: Unwanted Traffic Detection and Control Based on Trust Management

4.3.1 System Model, Assumptions, and Design Goals

4.3.1.1 System Model and Assumptions

We consider a trust management system for UTC in the networks, as illustrated in Fig. 4.1: The host tracks its own behaviors on unwanted traffic handling, monitors inbound traffic to detect potential intrusions, analyzes the collected data, detects different kinds of unwanted traffic, and reports the detection results to its NSP; the NSP monitors the traffic sourced from a local host or another NSP, if triggered by GTO or when the aggregated detection result is positive, collects detection reports, and cooperates with GTO to conduct UTC; the GTO, which is an authorized trusted party, collects trust evidence from NSPs to do trust evaluation, judge unwanted traffic sources (UTSs), and instruct personalized UTC by cooperating with NSPs.

Our research holds a number of assumptions based on our previous work (Routing Edge-to-Edge and Through Ethernets [33]).

Identity assumption. A source of unwanted traffic and its receiver in most cases can be identified with the accuracy of an IP address of the host or a network address translation (NAT) outbound IP address when a NAT hides the source host. The identity of the sender's NSP can be extracted from any of these addresses. Meanwhile, each traffic flow (i.e., a sequence of packets from a source to a destination) can be identified based on its hash code.

Tracking assumption. Normally, the UTS can be tracked by analyzing traffic logs and applying a traffic identification solution.

Fig. 4.1 A system model

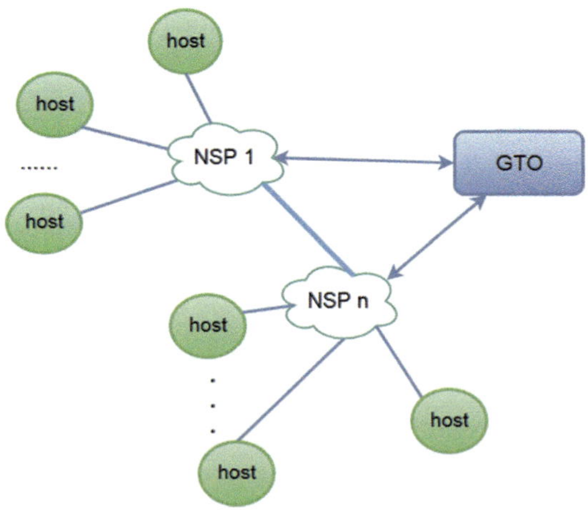

GTO assumption. A GTO behaves as an authorized trusted party to collect trust evidence and conduct global trust evaluation on different system entities. We assume that a secure and dependable communication channel is applied in the system for unwanted traffic reporting and controlling. Multiple GTOs could exist in the system, each supporting their own alliance of NSPs. The GTOs can collaborate together to exchange trust information and instruct UTC by applying a trustworthy collaboration protocol [34]. In this chapter, we treat all GTOs as one authorized trusted party and simplify it as one GTO in our presentation.

Traffic assumption. We assume that the unwanted traffic is sourced from a host and targets other hosts via other entities (e.g., NSPs) in the network.

4.3.1.2 Challenges in Identification in the Internet

When traffic over the Internet is carried using TCP, source identification is based on source address. Unfortunately, that address can be dynamically allocated, and the same address in the next session can actually belong to another host. The same is true for NAT outbound addresses. Prior to TCP session establishment or when UDP is used, it is often possible to spoof the source address.

In a line of work [33, 45], we propose cooperative firewalls as the edge nodes or the gateways through which all hosts would be connected to the Internet. These edge nodes would impose some rules of communication over the Internet, such as the following: (a) each host would have a stable host name that would be used for trust management as well as negotiation of the conditions for flow admission; (b) all flows would be admitted based on receiver and sender policies defined by the user; and (c) hosts have either just a private address or a globally unique address, and the interoperation is provided either by the cooperative firewalls or a realm gateway that makes servers in private addresses reachable to legacy Internet hosts.

The cooperative firewalls of a network share evidence of misbehavior and can make use of trust information (grey- and blacklists of entities) created by the GTO. Due to suitable policies, they can block spoofing and either block or slow down denial of service flows. Normally, we assume that the cooperative firewall is a network-based device, so it will apply its policy prior to the traffic that reaches the air interface of a wireless host saving the host's battery as well as the air interface capacity.

4.3.1.3 Threat Models

Three external unwanted traffic intrusion models commonly appear in the network:

- *Botnet infection*: a number of hosts are infected by unwanted traffic in the Internet; thus, they further send unwanted traffic to other hosts.

- *DDoS* via *reflectors* [35]: unwanted traffic could intrude a victim host from a number of attacked innocent hosts (reflectors). The unwanted traffic from different reflectors could be the same or different.
- *DDoS attack using spoofed source IP addresses*: for this type of attack, tracking the UTS is hard in today's network.

Moreover, internal misleading hosts and NSPs could also attack the trust management system as described below:

- *Internal misleading attacks by a host*: a misleading or indifferent host may report the unwanted traffic by applying a pattern, e.g., (1) hide evidence attack (indifferent hosts hide detection evidence and do not report to their NSP), (2) - bad-mouthing attack (misleading hosts intentionally frame a good traffic as unwanted), (3) on-off bad-mouthing attack (misleading hosts behave well or frame a good traffic alternatively, hoping that they can remain undetected while causing damage), and (4) on-off hide evidence and bad-mouthing attack (misleading hosts behave well or hide evidence/frame good traffic alternatively).
- *Internal misleading attacks by an NSP*: a NSP could maliciously perform an attack on the designed system. It behaves well to get a high trust value and then turns its resources against the system. The misleading NSP could perform a hide evidence attack by blocking all detection reports of its hosts or a bad-mouthing attack by framing a good traffic source.

4.3.1.4 Design Goals

We recognize the key desirable properties of the trust management system for UTC as below:

- Timely/efficient and accurate defense against unwanted traffic intrusion at hosts.
- Efficient recognition of UTSs during traffic intrusions, such as botnet intrusion and DDoS intrusion.
- Automatic maintenance of trust for each system entity.
- Robustness against attacks raised by misleading hosts and NSPs.
- Personalized control of unwanted traffic.

4.3.2 Notations

We propose a number of algorithms to implement UTC. For ease of reference, Table 4.1 summarizes the notations used in this section.

Table 4.1 Notions

Symbol	Description
$f(x)$	The Sigmoid function $f(x) = \frac{1}{1+e^{-x}}$; used to normalize a value into (0, 1)
U_k	The system entity, it can be either an NSP or a host
$\mathrm{tr}_k^{in}(t)$	The inbound traffic flow of host U_k at time t
φ^k	The ith content received by host U_k
e_i^k	The ith content received by host c
r_t^i	The receiving time of e_i^k at U_k
d_t^i	The discarding time of e_i^k at U_k
τ_i	The unwanted traffic indicator contributed by the process behaviors of unwanted traffic of a host regarding the ith content
T	The time window used to normalize the unwanted traffic process time
$v_k^i(t)$	The probability of e_i^k being an unwanted content indicated by U_k at time t, the unwanted traffic intrusion indicator
$s_i^k(t)$	The unwanted traffic detection result at time t by U_k about e_i^k
$S_i(t)$	The unwanted traffic detection result at time t about e_i^k
$\mathrm{sim\_in}_i^k$	The similarity of inbound traffic correlated to e_i^k
$\mathrm{sim\_in}^k$	The similarity of U_k inbound traffic by considering all similar traffic received by U_k
$\theta(I)$	The Rayleigh cumulative distribution function $\theta(I) = \left\{1 - \exp\left(\frac{-I^2}{2\sigma^2}\right)\right\}$ to model the impact of an integer number I, $\sigma = 100$ in our simulation
$\mathrm{tr}_k^o(t)$	The outbound traffic flow of U_k at time t
ut_k^t	The global trust of U_k at time t
thr	The threshold of the host to report to NSP
thr_0	The threshold to trigger traffic monitoring at local NSP
thr_1	The threshold of NSP to report to GTO
φ_{sp}^k	The unwanted traffic indicator contributed by the NSP traffic monitoring on U_k
$\mathrm{sim\_out}_i^k$	The similarity of outbound traffic of U_k correlated to e_i^k
$\mathrm{sim\_out}^k$	The traffic similarity of U_k by considering all similar traffic sent by U_k
$\mathrm{sp}_k^n(t)$	The unwanted traffic detection value about host U_k provided by the nth NSP SP_n at time t
$rt_{k'}^t$	The contribution of reports from the hosts to the evaluation of $U_{k'}$'s global trust at time t
$mt_{k'}^t$	The contribution of reports from the NSPs to the evaluation of $U_{k'}$'s global trust at time t
$\mathrm{d}t_k^t$	The detection trust value of U_k at time t
y	The detection performance indicator
δ	The parameter to control the adjustment of $\mathrm{d}t_k^t$
γ	The warning flag to record the number of bad detections
μ	The parameter to control bad detection punishment
thr_2	The threshold to put an entity into the greylist at GTO
thr_3	The threshold to determine on-off or conflict behavior attack
thr_4	The threshold to determine dishonest NSP
$fi_{k'}^{i'}$	The filtering indicator for $U_{k'}$ regarding traffic i' from a host in the greylist

4.3.3 The Proposed Scheme

UTC is implemented in the proposed scheme on the basis of the unwanted traffic detection at hosts, which triggers traffic monitoring at NSPs. According to trust evaluation at GTO based on collected detection reports from hosts and monitoring reports from NSPs, GTO authorizes NSPs to perform traffic filtering and control for specific hosts. The proposed scheme is a user-driven method. In European Union and based on its privacy law, NSP cannot act concrete monitoring on some host's traffic without sufficient evidence. NSP can also perform traffic monitoring and later on traffic control if needed, but the cost could be high and this approach wastes a lot of resources. Due to the above reasons, we propose initiating unwanted traffic detection from hosts.

4.3.3.1 Unwanted Traffic Detection at Host

Local traffic monitoring. The purpose of monitoring the inbound traffic of a host U_k $(k = 1, \ldots, K)$ is to detect if the host has been intruded. The increase of inbound traffic of a node indicates the possibility of being intruded. An unwanted traffic indicator φ^k contributed by the local traffic monitoring can be described as below by using the sigmoid function to normalize the traffic deviation into the interval of (0, 1):

$$\varphi^k = \left|1 - 2f\left\{d_t\left[tr_k^{in}(t)\right]\right\}\right| \tag{4.1}$$

The bigger φ^k is, the more probably U_k is intruded.

The subjective implication of receiving traffic. The behaviors of a node in processing the received traffic imply its wants or dislikes. This information can be applied to indicate whether the traffic is wanted subjective to personal needs. If the receiving time of a content e_i^k is r_t^i and its discarding time (e.g., the time to move it to a spam folder or specify it as unwanted) is d_t^i, the interval between r_t^i and d_t^i implies user need. The unwanted traffic indicator τ_i contributed by the node content processing behavior can be described as

$$\tau_i = 1 - {d_t^i - r_t^i}\big/{T}, \text{ when } d_t^i - r_t^i < T, \tag{4.2}$$

where T is the time window used to normalize the content process time. The bigger τ_i is the more possible e_i^k is unwanted by U_k. Note that if $d_t^i - r_t^i \geq T, \tau_i$ will not be counted.

Similarity check. In most network intrusions, similar contents could be sent many times to the same hosts. Therefore, we further check the similarity of contents received by U_k if $\varphi^k \geq \text{thr}_1$. For similar sized contents,

$E_k = \{e_i^k\}\ i = \{1, \ \ldots, \ I\}$ by U_k within a time window ($w = \left[t - \frac{T}{2}, \ t + \frac{T}{2}\right]$), we calculate their similarity as

$$\text{sim\_in}_i^k = {}^{\theta(I)}\!/_{I} - I\sum_{i' \neq i}^{I}\left(1 - \left|e_i^k - e_{i'}^k\right|\right), \tag{4.3}$$

where $\left|e_i^k - e_{i'}^k\right|$ is the difference between e_i^k and $e_{i'}^k$. It can be calculated based on a semantic relevance measure [36]. Obviously, U_k could receive multiple sets of similar traffic intrusion. The similarity of U_k inbound traffic by considering all similar contents is

$$\text{sim\_in}^k = \frac{1}{M'}\sum_{M'}\left(\left[{}^{\theta(I)}\!/_{I} - I\sum_{i' \neq i}^{I}\left(1 - \left|e_i^k - e_{i'}^k\right|\right)\right]\right), \tag{4.4}$$

where M' is the number of the sets of similar contents. We note that the bigger the number of similar contents I in a set I, the more possible it is that the similar content is unwanted. Thus, in Formulae (4.3) and (4.4), we consider the influence of integer I using the Rayleigh cumulative distribution function $\theta(I)$.

Unwanted traffic reporting. A host could complain about unwanted traffic to its local NSP. The complaint is based on the host behavior in the content process and local traffic auto-monitoring, as well as traffic similarity check. Thereby, we describe the unwanted traffic detection value $v_k^i(t)$ at time t by U_k about traffic e_i^k as:

$$v_k^i(t) = \text{sim\_in}^k \times \varphi^k \times \tau_i. \tag{4.5}$$

The detection reports are aggregated at NSP in order to decide whether traffic monitoring and check at NSP is needed for a local traffic source. Aggregation is also conducted at GTO for a remote traffic source in order to decide whether traffic monitoring and check at its NSP is needed. The aggregation is based on Formula (4.6) by applying the global trust and detection trust of the host as the credibility of its complaint:

$$s_i(t) = {}^{\Sigma_k v_k^i(t) \times ut_k^t \times dt_k^t}\!\Big/_{\Sigma_k ut_k^t \times dt_k^t}. \tag{4.6}$$

An unwanted traffic detection report containing $v_k^i(t)$ is automatically sent to the local NSP of the host if $v_k^i(t) \geq \text{thr}$.

4.3.3.2 Traffic Monitoring and Controlling at NSP

The purpose of monitoring a host U_ks' traffic at its local NSP is to find the sources of unwanted traffic with such credibility that the NSP can either take administrative action or impose contractual penalties on the sources. This traffic monitoring is triggered by a condition $s_i(t) \geq \text{thr}_0$ in order to save the running cost of the NSP. Particularly, it can detect an infected host that has become a source of unwanted traffic due to infection. U_k can be any entity that links to the NSP; thus, its traffic can be monitored by the NSP. It is efficient for the NSP to monitor its own subscribers because it sees all traffic sourced from them, while other NSP subscribers are numerous and the NSP can only see a fraction of their traffic. Therefore, for scalability, monitoring of other NSP subscribers should be very selective. Similar to Formula (4.1), an unwanted traffic indicator contributed by the NSP traffic monitoring on the outbound traffic of U_k is

$$\varphi_{\text{sp}}^{k}(t) = \left|1 - 2f\left\{d_t\left[\text{tr}_k^o(t)\right]\right\}\right|. \tag{4.7}$$

Similar to Formula (4.4), the similarity of multiple M different unwanted contents sent from U_k can be designed as

$$\text{sim\_out}^k = \frac{1}{M}\sum_{M}\left[{}^{\theta(I)}/_{I} - I\sum_{i' \neq i}^{I}\left(1 - \left|e_i^k - e_{i'}^k\right|\right)\right]. \tag{4.8}$$

We calculate the unwanted traffic detection value about U_k provided by the nth NSP at time t according to Formula (4.9) by monitoring the increase of outbound traffic and checking content similarity:

$$\text{sp}_k^n(t) = \varphi_{\text{sp}}^{k}(t) \times \text{sim\_out}^k. \tag{4.9}$$

NSP reports $\text{sp}_k^n(t)$ to GTO if $\text{sp}_k^n(t) \geq \text{thr}_l$.

4.3.3.3 Detection Trust: The Credibility of Detection

The credibility of detection reporting should be analyzed against misleading behaviors of reporters because of many reasons. For example, the complainer may be intruded; the host or NSP intentionally frames other hosts; the detection tools installed in the host are broken; the detection tools are poor and the detection is not qualified. Therefore, we apply the detection trust to indicate the quality of the reports since we can't ensure that the unwanted traffic detection is trustworthy. The detection trust is generated at GTO. If the detection reported by U_k doesn't match the final evaluation result, $y = -1$, and $\gamma++$; if the detection matches the fact, $y = 1$, and γ; if not changed. If no traffic detection report is provided, $y = 0$ and γ is not changed. Good detection performance will cause an increase of dt_k^t; otherwise, dt_k^t will be decreased. The detection trust dt_k^t of U_k at time t is

$$dt_k^t = \begin{cases} dt_k^t + \delta y \ (\gamma < \text{thr}_3) \\ dt_k^t + \delta y - \mu\gamma \ (\gamma \geq \text{thr}_3) \end{cases} = \begin{cases} 1 \ (dt_k^t > 1) \\ 0 \ (dt_k^t < 0) \end{cases} \tag{4.10}$$

where $\delta > 0$ is a parameter to control the change of dt_k^t. In order to detect on-off and conflict behavior of attackers [37], we further introduce a warning flag g to record the number of bad detections. The initial value of g is 0. It is increased by 1 each time when a bad detection happens. Parameter thr_3 is a threshold to indicate the on-off and conflict behavior attacks, and $\mu > 0$ is a parameter to control bad detection punishment. In [37], we have proved that this design is effective in trust evaluation against a number of internal misleading behaviors [38].

4.3.3.4 Trust Evaluation at GTO

The GTO evaluates the trust of each entity based on the collected reports from the hosts and NSPs in order to find the sources of unwanted traffic. Obviously, U_k $(k = 1, \ldots, K1)$ could report many times at different time t. Considering the time influence and potential on-off and ballot stuffing attacks, we pay more attention to the recent reports [37].

We use $e^{-|t - t_p|^2/\tau}$ to decay $v_k^i(t)$, while t_p is the trust evaluation time, and τ is a parameter to control the time decaying. We aggregate the reports $v_k^i(t)$ from $K1$ hosts who blamed U_k by considering both the global trust and detection trust, as well as time decaying as below:

$$rt_{k'}^{t_p} = \sum_{K=1}^{K1} dt_k^{t_p} \times ut_k^{t_p} \times v_k^i(t) \times e^{-|t - t_p|^2/\tau} \Big/ \sum_{K=1}^{K1} dt_k^{t_p} \times ut_k^{t_p} \times e^{-|t - t_p|^2/\tau}. \tag{4.11}$$

We further aggregate the reports from NSPs to calculate their contributions on the global trust evaluation of U_k. Since NSP reporting will trigger the trust evaluation at the GTO, we do not apply time decaying in Formula (4.12):

$$mt_{k'}^{t_p} = \sum_{n=1}^{N} dt_n^{t_p} \times ut_n^{t_p} \times sp_{k'}^n(t_p) \Big/ \sum_{n=1}^{N} dt_n^{t_p} \times ut_n^{t_p}. \tag{4.12}$$

The global trust value of the blamed entity k' can be calculated by deducting the original $ut_{k'}^{t_p}$ with $mt_{k'}^{t_p}$ and $rt_{k'}^{t_p}$. Meanwhile, we also consider the number of reporters by modeling its influence with the Rayleigh cumulative distribution function. Thus, the formula to update $ut_{k'}^{t_p}$ is designed as Formula (4.13):

$$ut_{k'}^{t_p} = ut_{k'}^{t_p} - \theta(K1) \times rt_{k'}^{t_p} - \theta(N) \times mt_{k'}^{t_p}. \tag{4.13}$$

4.3.3.5 Personalized Unwanted Traffic Filtering

The UTC can be personalized based on the correlation of detection reports. The previous reporting behavior of a host implies its preference on received contents. In order to decide whether to control a traffic flow i' sourced from an entity in greylist for U_k, we consider the previous reporting correlation between $U_{k'}$ and U_k $(k = 1, \ldots, K1)$ who reported i' as unwanted. We set a personalized filtering indicator $fi_{k'}^{i'}$ for U_k as below for i' according to the previous detection reports provided by both $U_{k'}$ and U_k:

$$fi_{k'}^{i'} = \sum_{k=1}^{K1} v_k^{i'}(t) \Big/ K1 \times K1 \sum_{k=1}^{K1} \left(1 - 1/I' \sum_{i=1}^{I'} \left| v_k^i(t) - v_{k'}^i(t) \right| \right) \tag{4.14}$$

where I' is the total number of traffic flows reported by both $U_{k'}$ and U_k. $v_k^i(t) - v_{k'}^i(t)$ denotes the deviation of the detection reports between U_k and $U_{k'}$. Algorithm 4 is used to provide personalized UTC.

This personalized filtering mechanism is a countermeasure against the system vulnerability caused by misleading reporting behaviors that could make extra intrusions and unexpected filtering. For example, a misleading host u could report differently from other hosts; thus, the unwanted traffic reported by these hosts may not be filtered for u due to the diverse behaviors. On the other hand, a good traffic flow reported as unwanted by other misleading hosts (collaborative bad-mouthing users) could not be filtered for u. Thus, applying this mechanism can encourage good reporting behaviors and fight against collaborative bad-mouthing attack. The proposed scheme can require a user's awareness of the nature of traffic to its host especially if they are required to delete it/report it as described in the traffic process step. Thus, it is possible that users can distinguish between "unknown and harmful traffic," "unknown but not harmful traffic," "unwanted and irrelevant traffic," "unwanted but relevant traffic," etc.

4.3.4 Game Theoretical Analysis on TruCon's Social Acceptance

The acceptance of TruCon by NSPs and hosts is crucial to its practical deployment and final success. In this section, we investigate the acceptance conditions of the TruCon system using game theory [43].

As mentioned earlier, TruCon system conducts unwanted traffic control based on the cooperation of NSPs and hosts. But there exists a social dilemma that a cost suffered by cooperative entities that adopt TruCon would generate a benefit shared by all, so entities involved would prefer to take a free ride (i.e., not adopt TruCon) as long as their utilities could be maximized. But this selfish behavior will make everyone worse off and finally degrade the performance of the TruCon system. To mitigate the dilemma, we propose a public goods-based TruCon game in network N, with the quality of the network environment treated as public goods. Since there exist two types of cooperation relationships in the TruCon system, we discuss respectively the game among the hosts and among the NSPs by considering their cooperation with GTO.

Games among hosts. A cooperative host sends effective reports, while uncooperative host sends no report or false reports or hides detection reports, degrading the performance of our TruCon system. We design a detection trust-based punishment mechanism to encourage hosts to behave in a cooperative manner. Concretely, we introduce a compromised device cleaning fee which is charged according to its detection trust value and should be included into the network access expense. When the detection trust value drops, the cleaning fee rises, providing the hosts incentive to cooperate.

Games among NSPs. NSPs are operating for profits and their strategies are directly correlated to the utilities they could obtain. Hosts, no matter cooperative or uncooperative, are all subscribers that can bring income to NSPs. If a NSP can still benefit when its subscribers are not cooperative, selfish/uncooperative NSPs would prefer not to punish them in order to prevent the loss of them and win the selection of potential customers who could switch from other NSPs. In contrast, a cooperative NSP suffers a potential loss of subscribed hosts (uncooperative ones). Thus, under such a condition, uncooperative NSPs can take a free ride of the contribution of cooperative NSPs. However, if all NSPs refuse to contribute, the TruCon system will be invalid, and the whole network will suffer from unwanted traffic sent from malicious, selfish, and compromised entities. According to our TruCon scheme, if a NSP cooperates, its detection trust value increases. Thus, we employ a detection trust-based punishment mechanism to provide motivation for NSPs to act cooperatively. Concretely, the traffic transit fee of NSP i decreases when its detection trust increases. With this punishment mechanism, NSP's motivation of being uncooperative will be greatly restrained.

We also strongly suggest NSPs developing cooperation with anti-spam vendors. Namely, we recommend that NSPs offer free antivirus toolkits to their subscribers to encourage their hosts to be cooperative. In this way, NSPs can earn good reputation due to qualified services and thus gain customers (hosts) and profits. Moreover, since transit charge should follow the rules of a country or a region, it could be difficult to apply the above punishment mechanisms in practice in a worldwide setting. Thus, in this case, the cooperation of NSPs with anti-spam vendors becomes crucial for applying TruCon in practice.

Naturally, it is also possible that the regulator imposes incident reporting obligations to certain types of customer networks as well as NSPs.

4.4 TruSMS: SMS Spam Detection and Control with TruCon

4.4.1 System Implementation

We implemented TruCon for the purpose of controlling SMS spam and evaluating its performance in the context of mobile networks [42]. We adopted Android phones as mobile devices that can send or receive SMS spam. Samsung S5690 and ZTE U985 were selected in our prototype. NSPs and GTO each were implemented as a server in desktop computers that communicate with mobile phones via mobile Internet through a WiFi connection. NSP and GTO servers were implemented in Windows 7 Professional operating systems. The prototype system was developed using Java language. The system applies SQLite as an example database, which is lightweight and both energy and memory efficient, thus suitable for applying to embedded devices, such as mobile phones.

4.4.2 Performance Evaluation

4.4.2.1 Evaluation Settings

In our prototype system, we have a total of $K = 1800$ devices subscribed to the same NSP. This is achieved by using a virtual machine technology. We set the initial global trust value of each system entity as 1 and the initial detection trust value of each entity as 0.5. We selected SMS spam distribution speed as 1 piece/100 ms, 1 piece/200 ms, 1 piece/300 ms, or 1 piece/400 ms. Normal SMS and SMS spam are extracted from the device database. For each SMS spam from a device, we apply the same SMS spam content for distribution.

We adopt commonly used metrics in information retrieval, recall (R) precision (P) and F measure (F) to describe the performance of TruCon. The bigger the F value, the better the detection result. F. Measure equals to 1 indicates all spam sources are detected. We denote the number of entities that are the sources of unwanted SMS messages (SUM) and are indeed detected as SUM as x; the number of entities that are not SUM but are thought as SUM as y; the number of entities that are SUM but are not detected as SUM as z. With these data we do a precision-recall evaluation. We define

$$R = {}^{x}/_{(x+z)} \tag{4.15}$$

$$P = {}^{x}/_{(x+y)} \tag{4.16}$$

$$F = {}^{2PR}/_{(P+R)}, \tag{4.17}$$

where $R, P, F \in [0, 1]$. R indicates the performance of false-negative detection (i.e., SMS spam goes unnoticed). P indicates the performance of false-positive detection (i.e., the blaming of innocent devices). Good system performance requests both high recall R and high precision P. Thus, we make use of the F measure to indicate system performance. Obviously, a high P value is desirable for good performance of the system.

4.4.2.2 Accuracy and Efficiency of SMS Spam Detection

We tested the accuracy of our system under the situation that there are 60, 70, 80, or 90 SMS spam sources (i.e., up to 5 % spam sources in the system) and all the rest of the devices are good ones that report detection results in an honest and timely way. We generate spam in the spam source device and send it to all the rest of the hosts. In this test, we set SMS spam dissemination with different speeds (i.e., 1 piece per 100, 200, 300, or 400 ms). Applying 1 piece/100 ms speed option means that the SMS spam source sends ten SMS spam messages in 1 s. The evaluation result in Fig. 4.2 shows that our implemented system can detect all

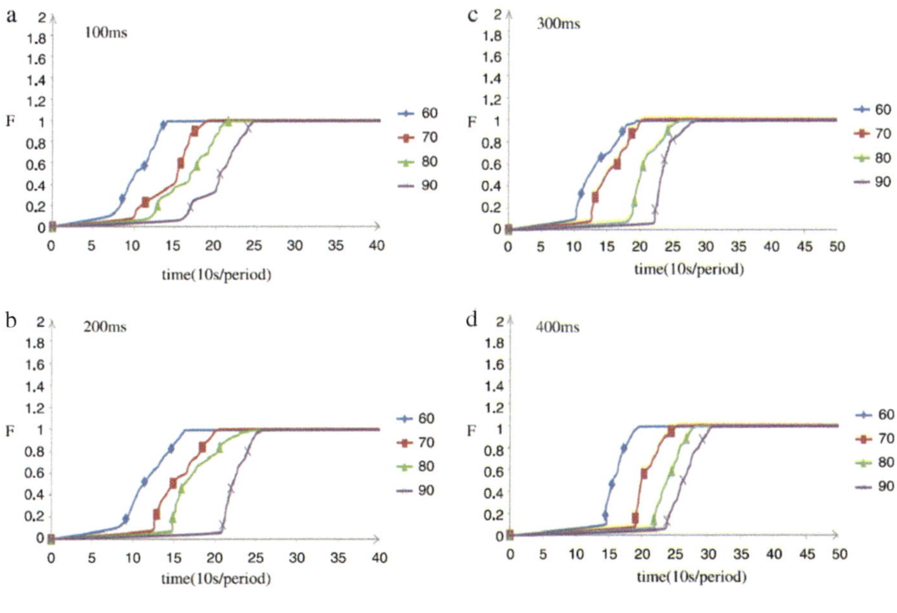

Fig. 4.2 Accuracy of SMS spam source detection with different SMS spam dissemination speeds: (**a**) 1 piece/100 ms; (**b**) 1 piece/200 ms, (**c**)1 piece/300 ms, and (**d**) 1 piece/400 ms

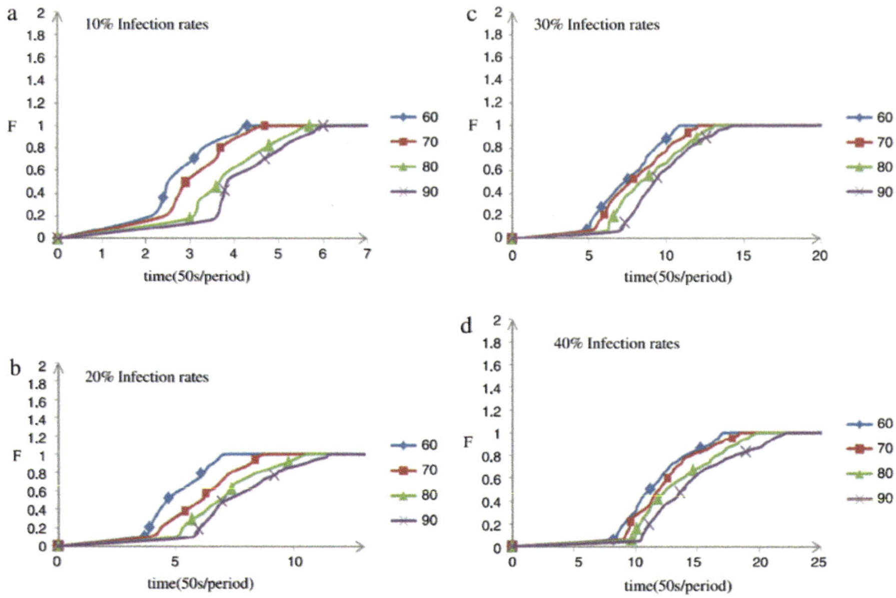

Fig. 4.3 Accuracy and efficiency of SMS spam source detection with different infection rates: (**a**) 10 %, (**b**) 20 %, (**c**) 30 %, and (**d**) 40 %

SMS spam sources accurately within 300 s in different spam distribution speeds. We observe that the lesser the number of spam sources, the faster the detection. In addition, the faster the spam dissemination speed, the quicker the detection. This result is easy to understand since the system can collect sufficient evidence for spam detection within a shorter period of time if the spam dissemination is faster.

Efficiency is an important system quality attribute. It can be measured by the detection speed, i.e., how fast the system can detect the sources of SMS spam. The F measure can also be used to test the efficiency in the following attack model: every time 10 % (20, 30, and 40 %) SMS spam destination devices are infected and super-distribute the same SMS spam to all the good devices, and this procedure continues until the system detects all SMS spam sources. We set spam dissemination speed option as "1 piece/100 ms" in this experiment. The result in Fig. 4.3 shows that our system can efficiently detect the SMS spam sources (F reaches 1 within 23 time periods, i.e., 1150 s), even when 40 % destination hosts are infected when the original SMS spam sources occupy no more than 5 % of the system devices.

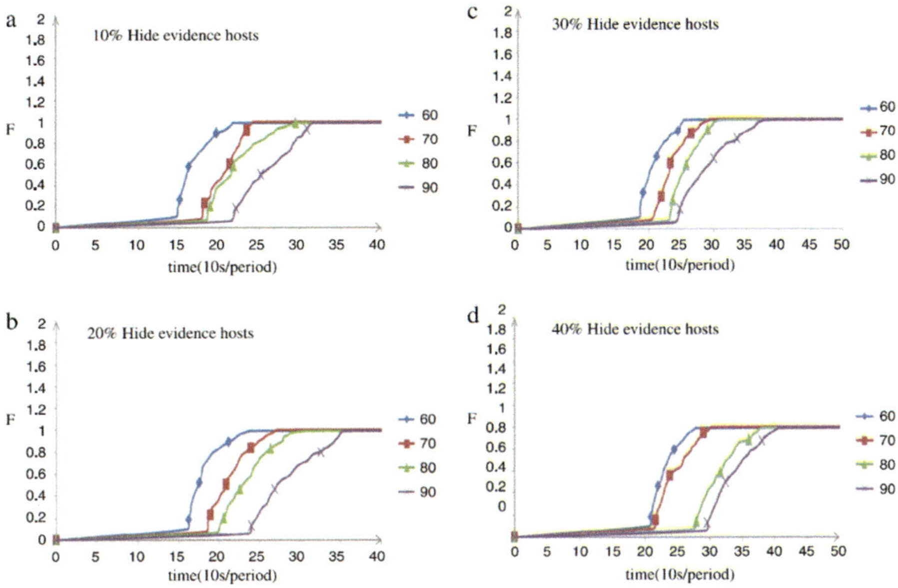

Fig. 4.4 Robustness of SMS spam source detection with hide evidence attack at a dissemination speed of 1 piece/400 ms: (**a**) 10 % hide evidence devices, (**b**) 20 % hide evidence devices, (**c**) 30 % hide evidence devices, and (**d**) 40 % hide evidence devices

4.4.2.3 Robustness of SMS Spam Control

We tested the robustness of our system under the following four internal misleading system attacks: hide evidence attack, bad-mouthing attack, on-off bad-mouthing attack, and on-off hide evidence or bad-mouthing attack. In this experiment, there are $L = 60$, 70, 80, or 90 (i.e., up to 5 % of the total) original independent SMS spam sources and a number of misleading devices (up to 40 % of the total) that do not report detection results in a good way:

(a) *Hide evidence attack*: misleading devices hide detection evidence by not reporting to their NSP. We respectively set 10, 20, 30, and 40 % of the total devices as misleading ones and tested system performance with spam dissemination speed of "1 piece /400 ms." The results in Fig. 4.4 show that our system performs well against the hide evidence attack. The F measure can generally reach one within 410 s in the situation that the SMS spam sources occupy no more than 5 % of the system devices.

(b) **Bad-mouthing attack**: misleading devices intentionally frame a good SMS as unwanted or unwanted SMS as a good one. We tested the situation that 10, 20, 30, and 40 % of the total devices are misleading ones. In this test, there are ten good SMS sources and L (= 60, 70, 80, 90) SMS spam sources. Figure 4.5 shows the experimental results with spam dissemination speed of

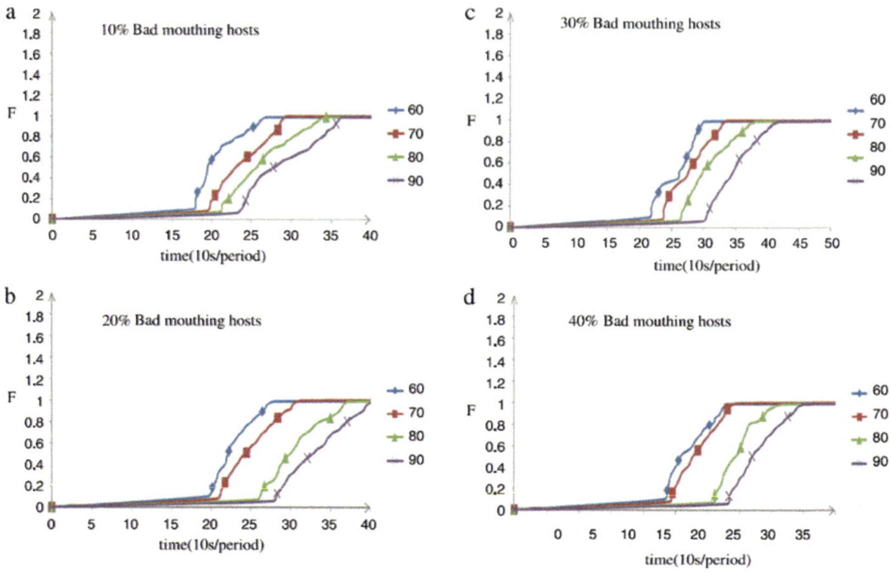

Fig. 4.5 Robustness of SMS spam source detection with bad-mouthing attack at a speed of 1 piece/400 ms: (**a**) 10 % hide evidence devices, (**b**) 20 % hide evidence devices, (**c**) 30 % hide evidence devices, and (**d**) 40 % hide evidence devices

1 piece/400 ms. The result shows that our system performs well against the bad-mouthing attack. The F measure can generally reach one within 440 s in the situation that the SMS spam sources occupy no more than 5 % of the system devices.

(c) *On-off bad-mouthing attack*: misleading devices behave well or frame a good traffic alternatively, hoping that they can remain undetected while causing damage. We simulate some good traffic in the system and test the situations that the proportion of on-off bad-mouthing devices is 10, 20, 30, and 40 % of the total, respectively. In this test, there are ten good SMS sources and L ($=$ 60, 70, 80, 90) SMS spam sources. The experimental results in Fig. 4.6 show that our system performs well against the on-off bad-mouthing attack. The F measure can generally reach 1 within 420 s with spam dissemination speed of “1 piece/400 ms” in the situation that the SMS spam sources occupy no more than 5 % of the system hosts. Comparing these results to those in Fig. 4.8, we find that TruCon performs better under on-off bad-mouthing attack than pure bad-mouthing attack. This is because it can collect sufficient good evidence for spam detection in a shorter period in the first situation than in the second situation.

(d) *On-off hide evidence or bad-mouthing attack (conflict behavior attack)*: misleading devices behave well or hide evidence/frame a good traffic alternatively. We simulate some good traffic in the system and test the situations that the proportion of on-off hide evidence or bad-mouthing devices is 10, 20,

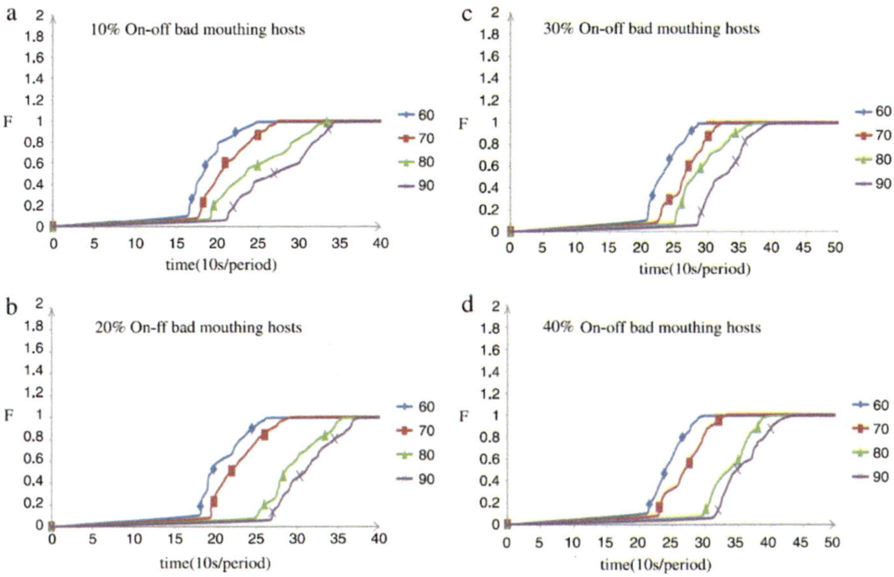

Fig. 4.6 Robustness of SMS spam source detection with on-off bad-mouthing attack at a speed of 1 piece/400 ms: (**a**) 10 % on-off bad-mouthing devices, (**b**) 20 % on-off bad-mouthing devices, (**c**) 30 % on-off bad-mouthing devices, and (**d**) 40 % on-off bad-mouthing devices

30, and 40 % of the total, respectively. In this test, there are ten good SMS sources and L (= 60, 70, 80, 90) SMS spam sources. We applied a flag for each misleading device to indicate its behavior model. If the flag is 0, the device adopts normal behavior model to report spam detection in a good way. If the flag is 1, the device performs hide evidence attack, and if the flag is 2, the device conducts bad-mouthing attack. The flag is updated based on the counter of received spam messages, which is set as 30 in our test. Concretely, the flag was initialized as 0. After the device receives 30 SMS spam messages, it turns to 1 and the system resets the counter to 0. After further receiving 30 new SMS spam messages, the flag was updated to 2. This flag updates from 0 to 1 and then 2 and back to 0 in a circulating way based on the counterchange. The test results in Fig. 4.7 show that our system performs well against this kind of conflict behavior attack. The F measure can generally reach 1 within 410 s with spam dissemination speed of "1 piece/300 ms" and 420 s in the situation that the SMS spam sources occupy no more than 5 % of the system hosts in the evaluation environment. The results show that TruCon can afford hybrid and integrated attacks well.

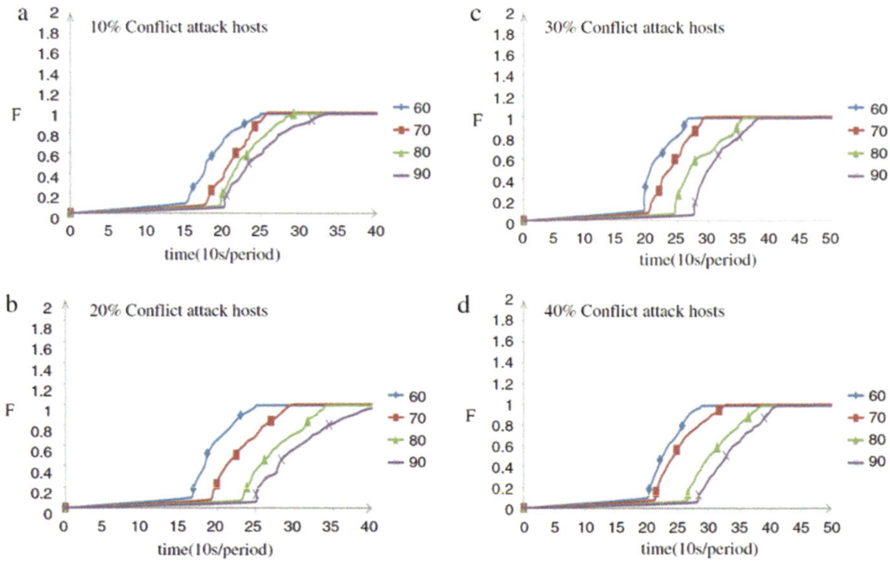

Fig. 4.7 Robustness of SMS spam source detection with on-off hide evidence or bad-mouthing attack at a speed of 1 piece/400 ms: (**a**) 10 % on-off hide evidence or bad-mouthing devices, (**b**) 20 % on-off hide evidence or bad-mouthing devices, (**c**) 30 % on-off hide evidence or bad-mouthing devices, and (**d**) 40 % on-off hide evidence or bad-mouthing devices

4.5 PSN Controller: Unwanted Content Control in Pervasive Social Networking with TruCon

4.5.1 System Implementation

We implemented TruCon in a pervasive social chatting scenario based on MANET which represents a typical application of PSN [41]. It is developed in Android phones using Java language. The system applies SQLite as an example database.

4.5.2 Performance Evaluation

4.5.2.1 Evaluation Settings

Since PSN supports instant social communications among a limited number of nodes in vicinity, we set a total of $K = 100$ node devices in PSN for performance evaluation. In each test, there were L sources of unwanted contents. Each unwanted content source randomly selects a number of good devices to intrude. The initial device trust value is set to 1; the initial detection trust value of each device is 0.5. Based on our previous evaluation through MATLAB simulations [39, 40], we

Table 4.2 Settings of system parameters

Symbol	Settings	Symbol	Settings
thr	0.8	σ	5
thr_3	5	τ	2
thr_2	0.1	δ	0.05
T	10 time slots	μ	0.1
$d_t^i - r_t^i$	1 time slot	w	0.1

found the settings shown in Table 4.2 are the most proper for achieving good system performance. w is the threshold to decide a bad detection and is newly introduced in the context of PSN (refer to [41] for more detail). Thus, we used the same settings of system parameters in the following tests. We assume the process of content deletion takes one time slot if the content is unwanted in our tests. Similarly, we adopt commonly used metrics in information retrieval, R, P, and F to indicate the performance of unwanted content detection.

4.5.2.2 Accuracy of Unwanted Content Intrusion Detection at a Device

We tested the accuracy of unwanted content intrusion detection in a PSN device. In this test, the devices did not provide subjective rates on contents. We evaluated the system performance under the situation that there are $L = 2, 6, 10$, or 20 unwanted content sources and all the rest of the devices are good ones. We conducted the evaluation under different distribution speeds of unwanted contents. The evaluation result is shown in Fig. 4.8. We observed that TruCon can detect all unwanted content sources accurately within 100 s when the unwanted content distribution speed is faster than 1 piece/400 ms. This implies that TruCon can quickly detect all the unwanted content sources in the context of PSN. We also found that the faster the distribution speed of unwanted contents, the more efficient the detection is. It took longer time for TruCon to detect all unwanted content sources if L is bigger no matter the distribution speed of unwanted contents.

4.5.2.3 Efficiency Against DDoS Intrusion at a Device

In this test, we evaluated whether a device can defend against DDoS intrusion when $L = 40$ sources of unwanted contents intrude together on one good device under different distribution speeds of unwanted contents. In this evaluation, PSN users did not handle the unwanted contents and provide subjective rating. Figure 4.9 shows the detection result. We observed that the TruCon system can quickly detect the intrusion under the experimental condition. It can defend against DDoS intrusion very well. The faster the speed of DDoS intrusion, the quicker the detection is.

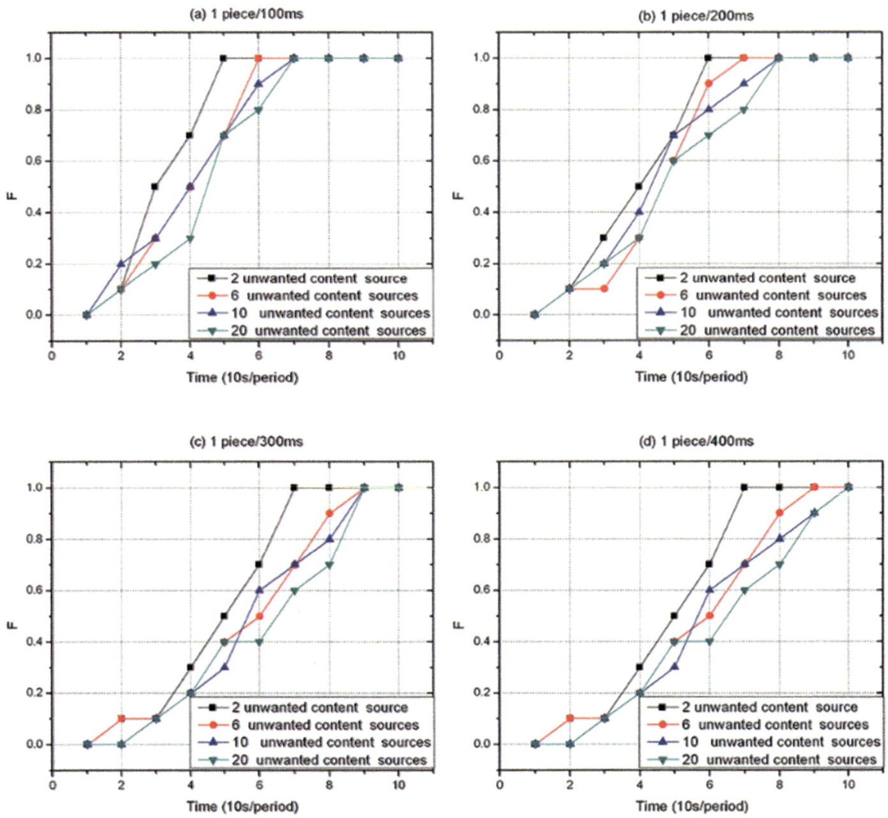

Fig. 4.8 Accuracy of unwanted content detection at a device with different transmission speeds: (**a**) 1 piece/100 ms, (**b**) 1 piece/200 ms, (**c**) 1 piece/300 ms, and (**d**) 1 piece/400 ms

Robustness of Unwanted Content Control

We tested the robustness of TruCon in PSN under the following two internal misleading system attacks: hide evidence attack and bad-mouthing attack. The F measure is used to indicate the robustness of unwanted content detection and control:

(a) *Hide evidence attack*. We tested whether the devices can detect unwanted content intrusion when there were $L = 5, (10, 20, \text{or } 40)$ sources of unwanted contents. In order to test the robustness of TruCon in an accurate way, we selected the slowest speed (i.e., 1 piece/400 ms) in the offered list because the faster the distribution of unwanted contents, the more efficient for TruCon to detect the unwanted content sources. Each unwanted content source randomly selected a number of users in PSN to intrude. A number of M selfish or indifferent users ($M = 10, 20, 30, \text{or } 40$) didn't broadcast complaints even though they detect intrusion. The good users complained the unwanted

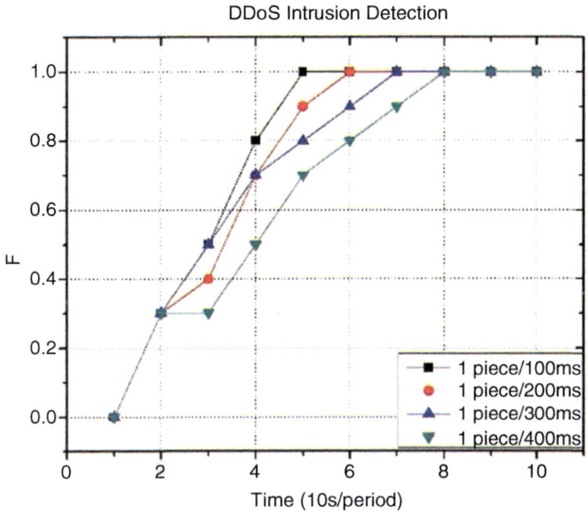

Fig. 4.9 Efficiency against DDoS intrusion at a device with different transmission speeds

traffic sources honestly and timely without subjective rating. Figure 4.10 shows the experimental result. We observed that our system performs well against the hide evidence attack. The F measure generally reaches 1 within 110 s in this test. This implies that TruCon can quickly detect all the unwanted content sources under the hide evidence attack. We also found that the bigger the value of M, the slower the detection is. It took longer time for TruCon to detect all unwanted content sources if the number of L is bigger no matter the value of M is.

(b) *Bad-mouthing attack*. In this evaluation, we tested whether the nodes can detect unwanted content intrusion in the following conditions: there were $L = 6, 10, 16, 20$, or 30 sources of unwanted contents and the same number of good content sources. Each content source randomly selected a number of nodes to send contents. Some misleading devices ($M = 10, 20, 30$, or 40) framed good content sources by broadcasting complaints on them in PSN. The speed of sending unwanted contents is 1 piece/400 ms, and the speed of sending normal contents is 1 piece/600 ms. In addition, we made a good node to rate an unwanted content as "very ugly" or "spam" if its local detection is positive. We investigated the performance of TruCon in this situation. The result is shown in Fig. 4.11.

We observed that when $L \leq 20$, the F measure reaches 1 quickly within 170 s. In this situation, TruCon performs well against the bad-mouthing attack. The difficulty of detection relates to the values of M and L. The smaller the value of M or L, the faster the detection is. But when $L = 20$ and $M > 10$, TruCon cannot detect all unwanted content sources. This is because there are too many misleading devices in PSN. When the number of content receivers in PSN is too small, the normal content sources mostly selected misleading devices to send contents and were framed as unwanted ones by them. In this case, the detection trust of the misleading devices

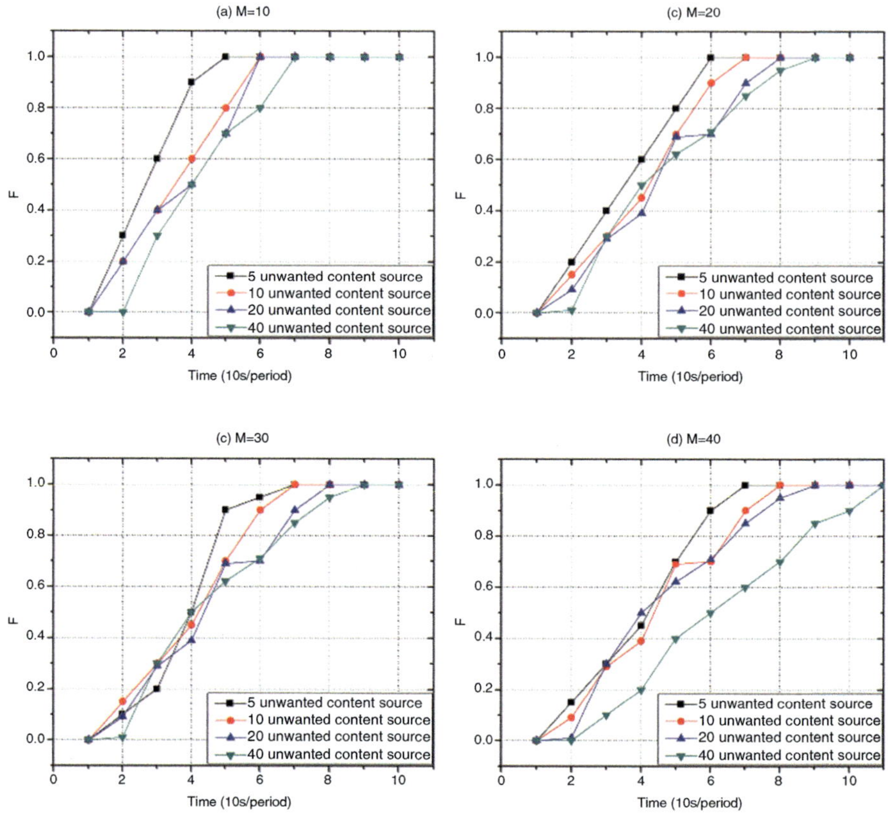

Fig. 4.10 Robustness of unwanted content source detection under hide evidence attack: (**a**) $M = 10$, (**b**) $M = 20$, (**c**) $M = 30$, and (**d**) $M = 40$

could increase, making the normal source devices being treated as bad ones, which causes that the unwanted content sources could not be detected correctly. An extreme case is a total of 40 devices receive contents from 30 good sources and 30 unwanted content sources. When the number of bad-mouthing users is 40, there are no any good users receiving contents in PSN, F becomes 0 always. Based on our additional tests, we found that TruCon can work effectively when the number of bad-mouthing misleading devices is less than 60 % of the total number of content receivers.

4.6 TruCon Implementation in SDN Infrastructure

We also implemented TruCon in SDN infrastructure, which is a current trend in communication networks due to its several advantages such as flexibility, high performance and efficiency, and ease of implementation and administration. We set

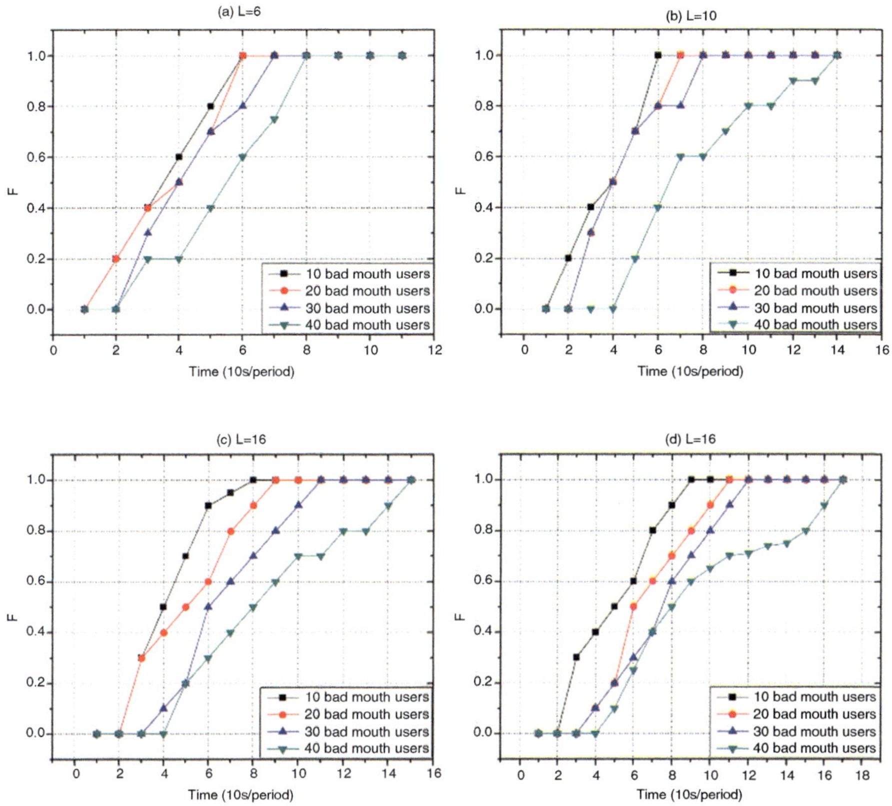

Fig. 4.11 Robustness against bad-mouthing attack: (**a**) $L = 6$, (**b**) $L = 10$, (**c**) $L = 16$, and (**d**) $L = 20$

up a SDN virtual platform in Ubuntu 14.04 and choose Mininet 1.0 as our experimental platform and Floodlight as external controller. Figure 4.12 shows the network topology in our implementation. As shown in Fig. 4.12, NSP and GTO are embedded in the controller as a virtualized network function. We have done some preliminary tests based on our SDN implementation. In these tests, we set the number of HostSe that is used to distribute traffic as 5 and the number of HostRe used to receive traffic as 1.

Accuracy. We assume HostRe reports unwanted traffic sources honestly and set the unwanted traffic sending speed as 1 piece/100 ms. Figure 4.13a shows the result of five HostSe sending unwanted traffic to HostRe. As shown in Fig. 4.13a, TruCon can detect all the unwanted traffic sources timely and accurately.

Efficiency. We tested the efficiency of TruCon under SDN with four different sending speeds, namely, 1p/100 ms, 1p/200 ms, 1p/300 ms, and 1p/400 ms, when all of the five HostSe send unwanted content to HostRe. As shown in Fig. 4.13,

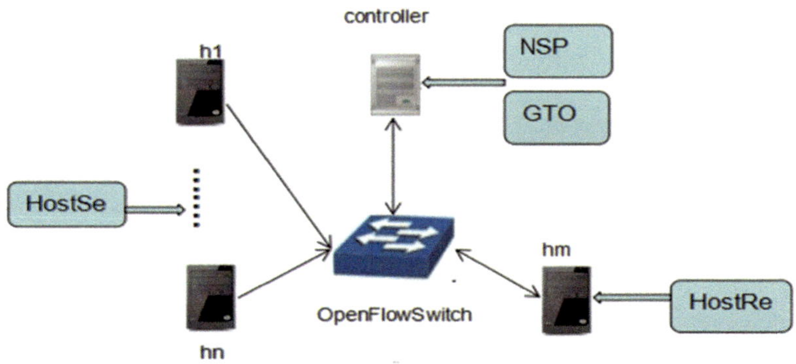

Fig. 4.12 Network topology of TruCon in SDN

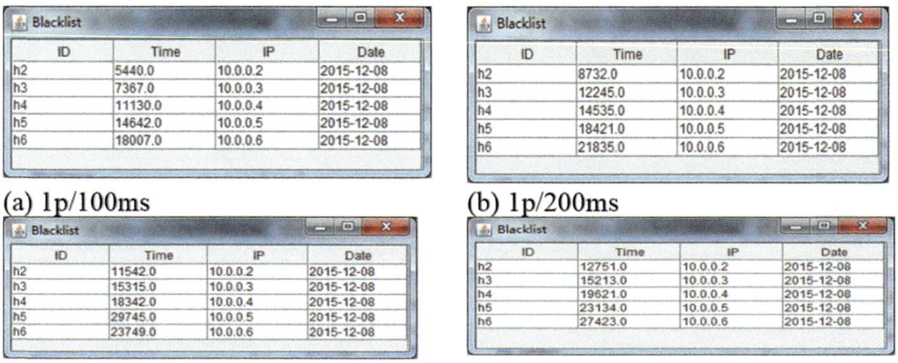

Blacklist

ID	Time	IP	Date
h2	5440.0	10.0.0.2	2015-12-08
h3	7367.0	10.0.0.3	2015-12-08
h4	11130.0	10.0.0.4	2015-12-08
h5	14642.0	10.0.0.5	2015-12-08
h6	18007.0	10.0.0.6	2015-12-08

(a) 1p/100ms

Blacklist

ID	Time	IP	Date
h2	8732.0	10.0.0.2	2015-12-08
h3	12245.0	10.0.0.3	2015-12-08
h4	14535.0	10.0.0.4	2015-12-08
h5	18421.0	10.0.0.5	2015-12-08
h6	21835.0	10.0.0.6	2015-12-08

(b) 1p/200ms

Blacklist

ID	Time	IP	Date
h2	11542.0	10.0.0.2	2015-12-08
h3	15315.0	10.0.0.3	2015-12-08
h4	18342.0	10.0.0.4	2015-12-08
h5	29745.0	10.0.0.5	2015-12-08
h6	23749.0	10.0.0.6	2015-12-08

(c) 1p/300ms

Blacklist

ID	Time	IP	Date
h2	12751.0	10.0.0.2	2015-12-08
h3	15213.0	10.0.0.3	2015-12-08
h4	19621.0	10.0.0.4	2015-12-08
h5	23134.0	10.0.0.5	2015-12-08
h6	27423.0	10.0.0.6	2015-12-08

(d) 1p/400ms

Fig. 4.13 The result of five HostSe sending unwanted traffic to HostRe

TruCon can detect unwanted sources efficiently in the context of SDN, and it takes shorter time to detect unwanted traffic sources when the sending speed is faster.

In addition, we implemented the cooperative firewall and the realm gateway in the SDN manner: the control plane is a virtual function, while the policy enforcement takes place using an OpenFlow switch [45]. In an SDN-controlled network, it is easy to direct all traffic from an infected host to a proxy that lets the user upgrade the host software but prevents it from doing any harm to other hosts.

4.7 Conclusions

This chapter proposed a generic and comprehensive UTC solution based on trust management by evaluating trust of each system entity at GTO and analyzing traffic and behaviors at hosts and NSPs. We successfully implemented the solution under mobile networks, PSN, and SDN and evaluated its performance under a variety of intrusions and attacks based on real datasets. The evaluation results verify that our scheme is accurate, robust, and effective on unwanted traffic control in a generic way.

Acknowledgment This work is supported in part by the National Key Foundational Research and Development on Network and Space Security, China, under Grant 2016YFB0800704, in part by NSFC under Grant 61672410 and U1536202, in part by the 111 Project under Grant B08038, in part by the Ph.D. Grant of Chinese Educational Ministry under Grant JY0300130104, in part by the Natural Science Basic Research Plan in Shaanxi Province of China under Program 2016ZDJC-06, and in part by Aalto University. The corresponding author is Zheng Yan.

References

1. J. McGibney and D. Botvich, "A trust overlay architecture and protocol for enhanced protection against spam", in The Second International Conference on Availability, Reliability and Security (ARES 2007), 10–13 April 2007, Vienna, pp. 749–756, 2007
2. Janecek, A.G.K., Gansterer, W.N., Kumar, K.A.: Multi-level reputation-based greylisting. In: Conference on availability, reliability and security ARES08, 4–7 Mar 2008, IEEE, pp. 10–17 (2008)
3. Tang, Y., Krasser, S., He, Y., Yang, W., Alperovitch, D.: Support vector machines and random forests modeling for spam senders behavior analysis. IEEE GLOBECOM 2008, 30 Nov 4 Dec 2008, New Orleans, LA, pp. 1–5, IEEE (2008)
4. Zheleva, E., Kolcz, A., Getoor, L.: Trusting spam reporters: a reporter-based reputation system for email filtering. ACM Trans. Inf. Syst. **27**(1), Article 3 (2008)
5. Liu, W., Aggarwal, S., Duan, Z.: Incorporating accountability into internet email SAC'09, pp. 875–882. ACM Press, New York, NY (2009)
6. Zhang, H., Duan, H., Liu, W., Wu, J.: IPGroupRep: a novel reputation based system for anti-spam. In: Symposia and workshops on ubiquitous, autonomic and trusted computing, 7–9 Jul 2009, Brisbane, QLD, pp. 513–518, IEEE (2009)
7. X. Zhang, B. Han and W. Liang, "Automatic seed set expansion for trust propagation based anti-spamming algorithms", Proceedings of WIDM'09, ACM, New York, pp. 31–38, 2009.
8. Zhang, J., Xu, W., Peng, Y., Xu, J.: MailTrust: A mail reputation mechanism based on improved TrustGuard", CMC10, 12–14 April 2010, Shenzhen, China, pp. 218–222. IEEE Computer Society, Washington DC (2010)
9. Bi, J., Wu, J., Zhang, W.: A trust and reputation based anti-spam method. In: INFOCOM 2008: the 27th conference on computer communications, 13–18 Apr 2008, Phoenix, AZ, IEEE conference publications, pp. 2485–2493 (2008)
10. Kolan, P., Dantu, R.: Socio-technical defense against voice spamming. ACM Trans. Autonom. Adapt. Syst. **2**(1), Article 2(44) (2007)
11. Page, L., Brin, S., Motwani, R., Winograd, T.: The Pagerank citation ranking: bringing order to the web. Technical report. Stanford University, Stanford, CA (1998)
12. Gyongyi, Z., Garcia-Molina, H., Pedersen, J.: Combating web spam with TrustRank. In: Proceedings of very large data bases VLDB, VLDB endowment, pp. 576–587 (2004)
13. Wu, B., Goel, V., Davison, B.D.: Topical trustRank: using topicality to combat web spam. WWW'06, pp. 63–72 (2006)
14. Liu, Y.T., Gao, B., Liu, T.Y., Zhang, Y., Ma, Z.M., He, S.Y., Li, H.: 'BrowseRank: letting web users vote for page importance. In: Proceedings of the 31st annual international ACM SIGIR

conference on research and development in information retrieval, Singapore, ACM, pp. 451–458 (2008)
15. Schwartz, A.: SpamAssassin, O'Reilly Media, Inc, Sepastopol, CA (2004)
16. Meyer, T.A., Whateley, B.: Spambayes: effective open-source, bayesian based, email classification system, CEAS (2004)
17. Wong, M, Schlitt, W.: Sender policy framework (SPF) for authorizing use of domains in e-mail, version 1 (online). RFC 4408 (Experimental). http://www.ietf.org/rfc/rfc4408.txt. Accessed 14 Jan 2013 (2006)
18. Allman, E., Callas, J., Delanym, M., Libbey, M., Fenton, J., Thomas, M.: DomainKeys identified mail (DKIM) signatures (online). RFC 4871 (proposed standard). http://www.ietf.org/rfc/rfc4871.txt. Accessed 14 Jan 2013 (2007)
19. Haskins, R., Nielsen, D.: Slamming spam: a guide for system administrators. Addison-Wesley Professional, Indianapolis, IN (2004)
20. Grandison, T., Sloman, M.: A survey of trust in internet applications'. IEEE Commun. Surv. **3** (4), 2–16 (2000)
21. Yan, Z., Prehofer, C.: Autonomic trust management for a component based software system. IEEE Trans. Depend Sec Comput **8**(6), 810–823 (2011)
22. Tufiq, M., Abdullah, M.F.A., Kang, K., Choi, D.: A survey of preventing, blocking and filtering Short Message Services (SMS) spam. In: Proceedings of international conference on computer and electrical engineering. IACSIT, November 2010, vol 1, pp. 462–466 (2010)
23. Deng, W., Peng, H.: Research on a naive Bayesian based short message filtering system. In: Proceedings of 5th IEEE international conference on machine learning and cybernetics, Aug 2006, pp. 1233–1237 (2006)
24. MobileSocial. http://mobisocial.stanford.edu. Accessed 30 Dec 2014
25. Sarigöl, E., Riva, O., Stuedi, R., Alonso, G.: Enabling social networking in ad hoc networks of mobile phones. Proc. VLDB Endow. **2**, 1634–1637 (2009)
26. Ott, J., Hyytiä, E., Lassila, P., Kangasharju, J., Santra, S.: Floating content for probabilistic information sharing. Pervas. Mobile Comput. **7**(6), 671–689 (2011)
27. EZSetup. from http://research.microsoft.com/en-us/groups/wn/mssn.aspx. Accessed 30 Dec 2014
28. Nokia-instant-community. http://conversations.nokia.com/2010/05/25/nokia-instant-community-gets-you-social/. Accessed 30 Dec 2014
29. Familiar Stranger. http://www.paulos.net/research/intel/familiarstranger/index.htm. Accessed 30 Dec 2014
30. Yang, H., Luo, H., Ye, F., Lu, S., Zhang, L.: Security in mobile ad hoc networks: challenges and solutions. IEEE Wireless Commun. **11**(1), 38–47 (2004). doi:10.1109/MWC.2004.1269716
31. Luo, H., Kong, J., Zerfos, P., Lu, S., Zhang, L.: URSA: ubiquitous and robust access control for mobile ad hoc networks. IEEE/ACM Trans. Netw. **12**(6), 1049–1063 (2004). doi:10.1109/TNET.2004.838598
32. Govindan, K., Mohapatra, P.: Trust computations and trust dynamics in mobile ad hoc networks: a survey. IEEE Commun. Surv. Tutor. **14**(2), 279–298 (2012). doi:10.1109/SURV.2011.042711.00083
33. Routing Edge-to-Edge and Through Ethernets. http://www.re2ee.org/. Accessed 14 Jan 2013
34. Yan, Z.: Security via trusted communications. In: Stavroulakis, P., Stamp, M. (eds.) Handbook on communications and information security, pp. 719–746. Springer, Rueil-Malmaison (2010)
35. Bossardt, M., Dubendorfer, T., Plattner, B.: Enhanced internet security by a distributed traffic control service based on traffic ownership'. J. Netw. Comput. Appl. **30**(3), 841–857 (2007)
36. Wang, J., Wang, F., Yan, Z., Huang, B.: Message receiver determination in multiple simultaneous IM conversations'. IEEE Intell. Syst. **26**(3), 24–31 (2011)
37. Yan, Z., Zhang, P., Deng, R.H.: TruBeRepec: a trust-behavior-based reputation and recommender system for mobile applications'. J. Pers. Ubiquit. Comput. **16**(5), 485–506 (2012)

38. Sun, Y., Han, Z., Liu, K.J.R.: Defense of trust management vulnerabilities in distributed networks'. IEEE Commun. Mag. **46**(2), 112–119 (2008)
39. Yan, Z., Kantola, R., Shi, G, Zhang, P.: Unwanted content control via trust management in pervasive social networking. In: Proceedings of the IEEE conference on TrustCom/ISPA/IUCC, pp. 202–209 (2013). doi:10.1109/TrustCom.2013.29
40. Shi, G.: Performance evaluation of PSN unwanted content control system via trust management. Master's thesis, Xidian University, Xi'an, China (2014)
41. Ma, S., Yan, Z.: An unwanted content control system in pervasive social networking based on trust management. ACM Trans. Multimedia Comput. Commun. Appl. **12**(1S), Article 17 (2015). 23 pages
42. Chen, L., Yan, Z., Zhang, W., Kantola, R.: TruSMS: a trustworthy SMS spam control system based on trust management. Fut. Gen. Comput. Syst. **49**, 77–93 (2015)
43. Shen, Y., Yan, Z., Kantola, R.: Analysis on the acceptance of global trust management for unwanted traffic control based on game theory. Comput. Secur. **47**(2014), 3–25 (2014)
44. Yan, Z., Kantola, R., Shen, Y.: A generic solution for unwanted traffic control through trust management. New Rev. Hypermedia Multimedia **20**(1), 25–51 (2014)
45. Kantola, R., Llorente Santos, L., Beijar, N.: Policy based communications for 5G mobile with customer edge switching. Wiley Security and Communication Networks (2015). doi:10.1002/sec.1253

Chapter 5
Characterization of Evolving Networks for Cybersecurity

Josephine M. Namayanja and Vandana P. Janeja

5.1 Introduction

Computer networks are vulnerable to varying cyber attacks that alter the structure and activity of the network. Hence, in order to define and understand the vulnerabilities associated to the network, one must have an understanding of the overall structure and nature of communication patterns within the network as well as the potential points of vulnerability. Network analytics provides the basis for how network structures are modeled, measured, and compared such that a network is modeled as a graph, which describes a collection of nodes or vertices and the communications between them, indicated by edges.

This chapter discusses approaches to change detection where the objective is studying how the network evolves over time and how these changes can be attributed to potential cyber attacks. Techniques such as change detection play a role in network characterization mainly because they detect shifts in network behavior over time. Changes in network behavior can be defined as sudden downtime of key points, for example, servers on the network during peak hours, existence of new or unidentified connections to the network, and specific time periods associated with shifts in network behavior. Such shifts in network behavior may come as a result of a cyber threat. This chapter discusses graph theory concepts to model network behavior and then utilizing analytics to understand the dynamics

J.M. Namayanja (✉)
University of Massachusetts Boston, Boston, MA, USA
e-mail: josephine.namayanja@umb.edu

V.P. Janeja
University of Maryland, Baltimore County, Baltimore, MD, USA
e-mail: vjaneja@umbc.edu

I.M. Alsmadi et al. (eds.), *Information Fusion for Cyber-Security Analytics*, Studies in Computational Intelligence 691, DOI 10.1007/978-3-319-44257-0_5

of the network. Cyber attacks are becoming increasingly sophisticated. One of the key challenges is knowing whether there is even an attack on the network in the first place. Let us consider the following scenario:

5.1.1 Cyber Attacks Are Unrelenting

Large computer networks comprised of tens of thousands of machines generate terabytes of network traffic each day. This traffic typically consists of hundreds of millions of connection records and poses a big data problem. Such significant volume and diversity traffic presents a daunting challenge in the detection of cyber attacks, particularly when it comes to small amounts of malicious activity. Additionally, attacks are increasingly becoming sophisticated and are designed to be undetectable. The behavior of such cyber attacks is extremely dynamic and thus changes over time. Furthermore, the continuous evolution of network structures such as the Web creates complexity in the efficient analysis of computing environments.

In an effort to establish a state of continuous awareness of network behavior, the Supercomputing Enabled Transformational Analytics Capability (SETAC) project at Lawrence Livermore National Laboratory aims to increase the ability to detect, characterize, and combat malicious attacks on large computer networks [1].

Several major incidents of cyber attacks have reported delayed detection of attacks. This delay can take from months to even years before the threat on the network is discovered. In 2014, it took organizations a median of 205 days to detect attackers in their network environments [2]. Such delays in attack detection can be due to the complexity of networks both in scale and dynamism which makes it difficult to keep track of what is taking place. From a graph perspective, networks are comprised of multiple dimensions, which include, nodes, edges, and time, where such dimensionality poses a challenge in identifying a vulnerability, detecting an attack, and potentially preventing an attack. Certain attacks are usually targeted to specific points in the network and are used in conjunction with advanced persistent threats. Such targeted attacks are designed to exploit and cause harm on the network.

The process of characterizing networks through change detection can be potentially useful to understand and control the dynamics of the network [3]. This chapter discusses state-of-the-art techniques in change detection that may be geared toward modeling network behavior and detecting patterns, which can indicate potential cyber threats such as the onset of a massive cyber attack which changes the way a network appears.

The rest of the chapter is organized as follows: Sect. 5.2 presents a detailed background on concepts in graph theory. Section 5.3 discusses fundamental concepts in network evolution. Section 5.4 presents an extensive overview on the fundamentals of change detection in temporally evolving networks. Section 5.5 discusses key applications for change detection. Lastly, Sect. 5.6 presents conclusions and future work.

5.2 Graph Theory Concepts

Each network presents specific topological features that characterize a network and its connectivity. Several different network measures can be calculated from a given graph. Network measures can be calculated for the entire graph or for each individual node. Node-level measures in the form of node centrality enable modeling the network to determining the role of a node in a network which can be useful in threat detection [4]. According to [5], an assessment of network vulnerabilities indicates that an attacker is likely to exploit the weak points such as critical nodes whose corruption greatly affects network performance. Additionally, graph-level measures such as density and diameter provide an overall picture of the impact on threats on individual nodes to the entire network. Let us consider the fundamental concepts in graph theory as they are utilized in network analytics for cybersecurity.

5.2.1 *Graph*

A graph is made up of nodes or vertices and edges that connect them. It is defined as:

A graph $G=(n, e)$, where $n=\{n_1...n_v\}$ is a set of nodes and $e=\{e_1...e_w\}$ is a set of edges, such that (n_i, n_j) is an edge between nodes n_i and n_j.

A graph can be directed or undirected. A directed graph G identifies the direction of the edge between the source and destination nodes, respectively. For example, $n_i \rightarrow n_j$ indicates n_i as a source node and n_j as the destination node as shown in Fig. 5.1a. On the other hand, an undirected graph G does not identify the direction of the edge between the nodes as shown in Fig. 5.1b.

This chapter discusses concepts that are applicable to both directed and undirected networks.

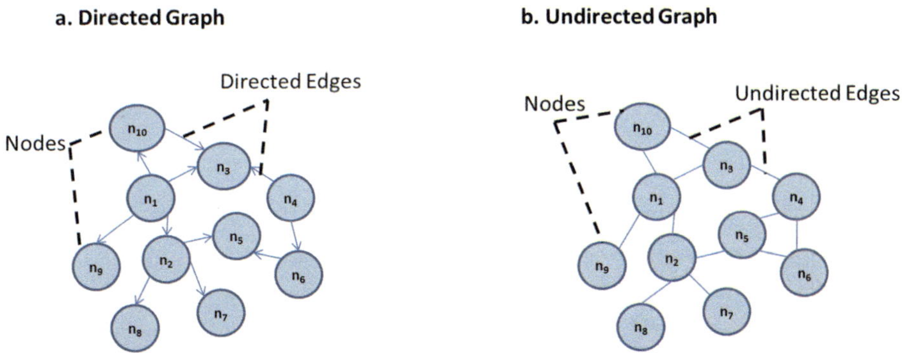

Fig. 5.1 Directed versus undirected graph

5.2.2 Node Centrality

The centrality of a node in a network determines a node's individual connectivity on the network. Here, we discuss selected centrality measures, namely, degree centrality [6] which is relative to the node, betweenness centrality [6], PageRank centrality [7], and eigenvector centrality [8, 9], which are individual node based but still relative to the rest of the network. Other measures include closeness centrality and Katz centrality to mention a few. These measures are applicable to both directed and undirected networks.

5.2.2.1 Degree Centrality

The degree of a node n_i is the number of edges incident on it. The degree centrality [6] is the most basic of all measures, and it counts how many times a node is involved in an interaction. It is defined, for a node n_i, as the number of edges that are incident on it.

Given x number of nodes in the network, the connectivity $a_{ij} = 1$ *if nodes i and j are connected by an edge and* $a_{ij} = 0$ *otherwise. Hence, the degree* d_i *of node* n_i *is the sum of all* a_{ij}*. The connectivity between nodes is represented through a v*v adjacency matrix A, where v is the number of nodes.*

If a node n_i is connected to a node n_j, then there exists an edge (n_i, n_j) between nodes n_i and n_j. We provide an example of an adjacency matrix in Fig. 5.2.

In Fig. 5.2, we see that if two nodes are adjacent or connected, then the row and column intersection is 1, else 0. For example, nodes n_1 and n_2 are connected and

	n_1	n_2	n_3	n_4	n_5	n_6	n_7	n_8	n_9	n_{10}
n_1	0	1	1	0	0	0	0	0	1	1
n_2	1	0	0	0	1	0	1	1	0	0
n_3	1	0	0	1	0	0	0	0	0	1
n_4	0	0	1	0	1	1	0	0	0	0
n_5	0	1	0	1	0	1	0	0	0	0
n_6	0	0	0	1	1	0	0	0	0	0
n_7	0	1	0	0	0	0	0	0	0	0
n_8	0	1	0	0	0	0	0	0	0	0
n_9	1	0	0	0	0	0	0	0	0	0
n_{10}	1	0	1	0	0	0	0	0	0	0

Fig. 5.2 An adjacency matrix representing an undirected computer network

nodes n_2 and n_3. Using this adjacency matrix, one can determine the degree centrality of nodes in a network. For example, the degree centrality of node n_1 is 4 based on the sum of connectivity a_{ij}, for n_2 the degree centrality is 4, for n_3 is 3, and so on.

5.2.2.2 Betweenness Centrality

Betweenness centrality [6] is a measure of how often a node lies along the shortest path or geodesic path between the two other nodes for all nodes in a graph.

Given x nodes, g_{jk} is the number of geodesic paths between nodes n_j and n_k; the betweenness of node n_i is defined as $g_{jk(i)}$ which is the number of geodesic paths that pass through n_i among g_{jk}.

5.2.2.3 Eigenvector Centrality

The eigenvector centrality [8, 9] can be understood as a refined version of the degree centrality in the sense that it recursively takes into account how neighboring nodes are connected.

Given λ as the largest eigenvalue, the eigenvector centrality e_i for a node n_i is the ith component of the eigenvector associated with the largest eigenvalue λ of the network and is proportional to the sum of the eigenvector centrality of the nodes it is connected to. λ assures the centrality is nonnegative.

While the eigenvector centrality of a network can be calculated via the standard method using the adjacency matrix representation of the network, it can be also computed by an iterative degree calculation [10].

5.2.2.4 PageRank Centrality

PageRank [7] is used to measure the relative importance of nodes on the network by computing a ranking for every node based on the connectivity on the network. Let A be a square matrix with the rows and column corresponding to nodes. Let $A_{u,v} = 1/N_u$ if there is an edge from u to v and $A_{u,v} = 0$ if not. If we treat R as a vector over nodes, then we have $R = cAR$. So R is an eigenvector of A with eigenvalue c. In fact, we want the dominant eigenvector of A. It may be computed by repeatedly applying A to any nondegenerate start vector.

Overall, metrics for node centrality are considered individual or local network measures. However, these can also be translated into graph-level measures by averaging them out over the count of nodes in the graph [11–13].

5.2.3 Graph-Level Measures

Graph-level measures account for connections in the entire network and not just individual nodes in a network.

5.2.3.1 Density

A network is called dense if its number of edges is roughly quadratic to its number of nodes.

Density of the network is defined as the proportion of the actual number of edges to the potential number of edges.

Network structures with high density are well connected internally. This may work well for information sharing; however, as the size of the network increases, a high-density measure may be undesirable because the corresponding high number of links for each node could lead to information overload. According to [14, 15], networks densify over time. This means that the number of edges is increasing superlinearly with the number of nodes. This superlinear increase in the number of edges can be measured through an increase in the average degree of nodes in a network over time. Therefore, as the average degree increases over time, then a network is said to obey the densification power law. Densification power law is defined as a relation $e(t) \propto n(t)^{a}$ where $e(t)$ is number of edges at time t and $n(t)$ is the number of nodes at time t, while $\boldsymbol{a}$ is the densification exponent [14, 15]. When $\boldsymbol{a} = 1$, then the average degree of nodes is constant over time, whereas if $\boldsymbol{a} = 2$, then average degree is increasing over time; hence, the network is becoming denser with time [14, 15].

5.2.3.2 Diameter

The diameter of a graph G is the shortest maximum distance between any two nodes in G. In order to find the diameter of a computer network, we first determine all possible paths p in G where $p = \{p_1 \ldots p_n\}$. A path $p_i = (n^{pi}, e^{pi})$, where $n^{pi} = \{n_0, n_1, \ldots, n_k\}$ and $e^{pi} = \{n_0n_1, n_1n_2, \ldots, n_{k-1}n_k\}$ such that nodes n_o to n_k are linked by p_i, and the number of edges in p_i or $|p_i|$ is the length of p_i. Thus, p_i is a simple graph whose nodes can be arranged in a linear sequence in such a way that two nodes are adjacent if they are consecutive in the sequence and nonadjacent if otherwise. We show an example of a path between nodes in Fig. 5.3.

In Fig. 5.3, we show a path p_i from nodes n_6 to n_9 in a computer network represented as p_i (n_6, n_9). The length of p_i (n_6, n_9) represented as $|p_i\ (n_6, n_9)|$equals to 4 based on the total number of edges between n_6 and n_9. Paths are used to determine the distance between nodes on the network defined as:

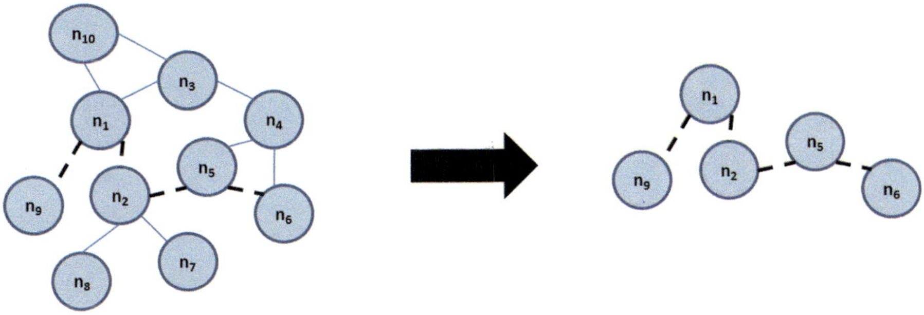

Fig. 5.3 Path between nodes

For any path p_i *where* $|p_i| = min|(n^{pi}, e^{pi})|$, *then* p_i *is shortest path between each pair of nodes* n_i *and* n_j, *and* p_i *is also referred to as distance d where d is the distance* $dist_G(n_i,n_j)$ *between* n_i *and* n_j.

This distance d is measured in terms of the number of edges between the nodes in question. In Fig. 5.3, the number of edges from nodes n_6 to n_9 is 4 such that $d = 4$. Hence, this is the shortest path between these two nodes and is thus the distance between these nodes. It should be noted that a computer network can have multiple distances since it is based on the shortest path between each pair of nodes on the network. However, the network can only have one diameter defined as:

For any path p_i *where* $|p_i| = min(max|(n^{pi}, e^{pi})|)$, *then* p_i *is the diameter h of G represented as diam(G).*

In order to determine the diameter of the network, we need to first determine all the shortest paths or distances *d* between each pair of nodes. The shortest maximum distance value between any pair of nodes is the diameter *h* of the overall network. According to Fig. 5.2, the distance between n_6 and n_9 is the shortest maximum distance between any pair of nodes in the network which makes it the diameter *h* of the network. The diameter of a network can be used to determine how dense or sparse a network is. Thus, if a network has a small diameter, then it is said to be well connected. On the other hand, if a network has a large diameter, then it is said to be sparse.

Both node centrality and graph-level metrics can be utilized to characterize how a network evolves over time.

5.3 Network Evolution

5.3.1 Node Evolution

The study of node evolution involves observing connections in a graph. From this, top central or influential nodes such as high-degree nodes as well as less popular nodes such as low-degree nodes can be identified [17–22], observed, and compared

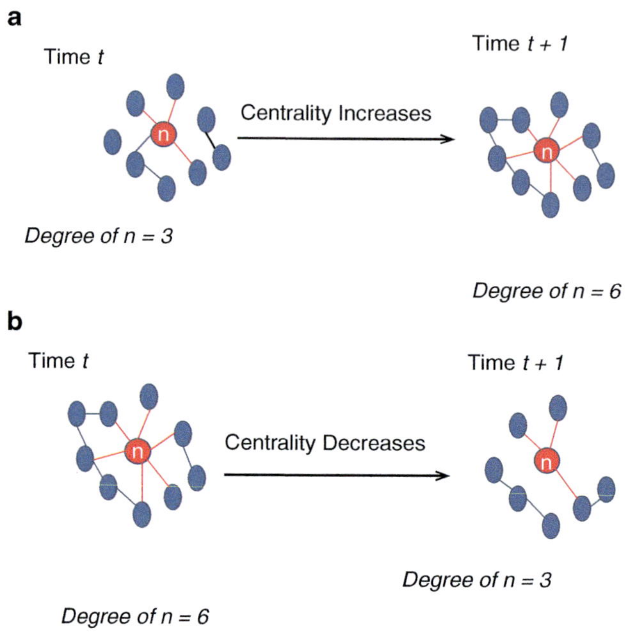

Fig. 5.4 (**a**) Centrality of a node increases over time. (**b**) Centrality of a node decreases over time

over time. Node evolution can also be observed in relation to neighborhoods as discussed in [23]. A certain set of numerical features of the neighborhood can be established for each node such as the number of neighbors (degree of a node) and the edges of the neighborhood, among others. Here it is possible that during network evolution, node centrality changes over time and that some nodes may disappear after sometime, or their centrality levels go higher and drop after a while for some, and that some nodes appear after a while and remain constantly present and maintain a high centrality level [19–22]. An example of changing node centrality is shown in Fig. 5.4.

5.3.2 Community Evolution

In order to detect community changes, [24–26] identify communities of nodes or communication patterns in the network and study how they evolve over time. For example, [25] study time-evolving networks where they analyze the evolution of network clusters through time to identify splits, merges, appearances, and disappearances of communities. On the other hand, [26] model the evolution of communities in heterogeneous networks where they study the size of communities to determine how they increase or decrease with time.

5.3.3 Graph Evolution

In graph evolution, [16–18, 27] observe key fundamental network properties to determine how networks grow and evolve over time. Particularly, such fundamental properties include densification power law, power-law degree, power-law eigenvector and eigenvalue distribution, edge-by-edge evolution, shrinking diameter, diameter, and radius. These properties are observed in relation to the degree of nodes.

For instance, [14, 15] clearly demonstrate that networks obey the densification power law where edges grow faster than nodes. First, the graph over time maintains a power-law degree distribution with a constant power law degree distribution exponent γ. If $\gamma < 2$ and is constant over time, then the graph is said to densify. An illustration is provided in Figs. 5.5, 5.6, and 5.7 for undirected and directed networks based on key subgraphs selected from network traffic data by the Center for Applied Internet Data Analysis (CAIDA) for the duration of December 2008 to January 2010 [37–39].

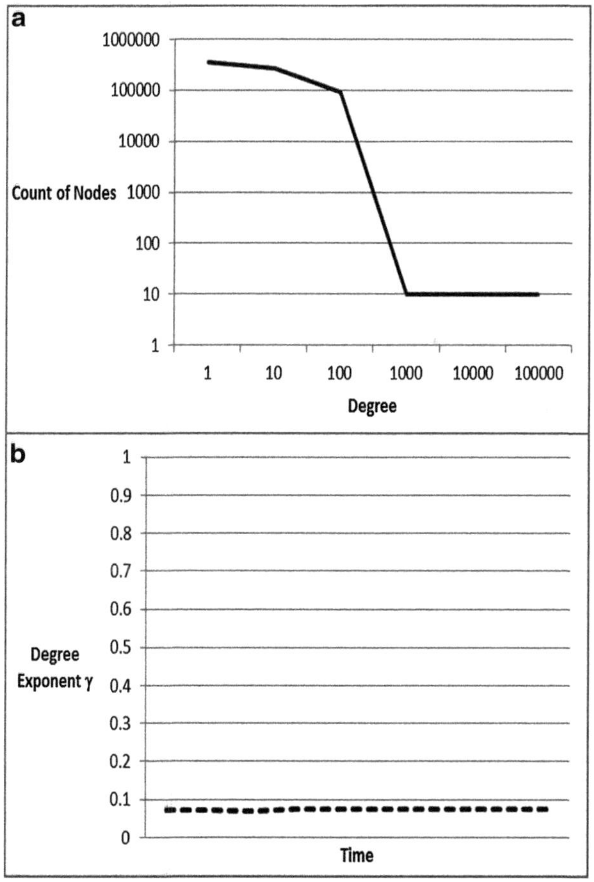

Fig. 5.5 (**a**) Example of a degree distribution in an undirected network. (**b**) Example of a degree exponent over time in an undirected network

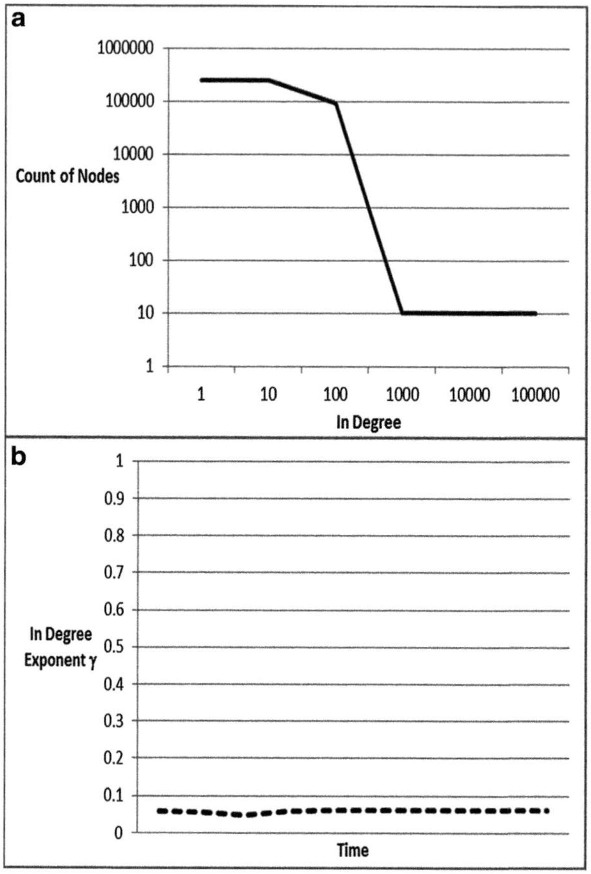

Fig. 5.6 (**a**) Example of an in-degree distribution in a directed network. (**b**) Example of an in-degree exponent over time in a directed network

Overall, Figs. 5.5, 5.6, and 5.7 show that the degree distribution has a long tailed distribution and thus follows a power-law distribution. Additionally, the power law degree distribution exponent $\gamma < 1$ in all cases and is constant over time. These [14, 15] also show that the diameter of the network shrinks over time such that as the network grows, the distances between nodes slowly decrease.

According to [15], in a temporally evolving graph, if the power law degree distribution exponent γ is constant over time, the densification exponent α is a function of γ such that $\alpha = 1$ if > 2, $\alpha = 2/\gamma$ if $1 < \gamma < 2$, and then $\alpha = 2$ if $\gamma < 1$. These properties can be used to clearly demonstrate how graphs densify over time.

5.4 Scientific Fundamentals for Change Detection

The study of network structures calls for an understanding of network structural features and fundamental network properties as described in graph concepts and network evolution, respectively. Such features and properties provide a basis for

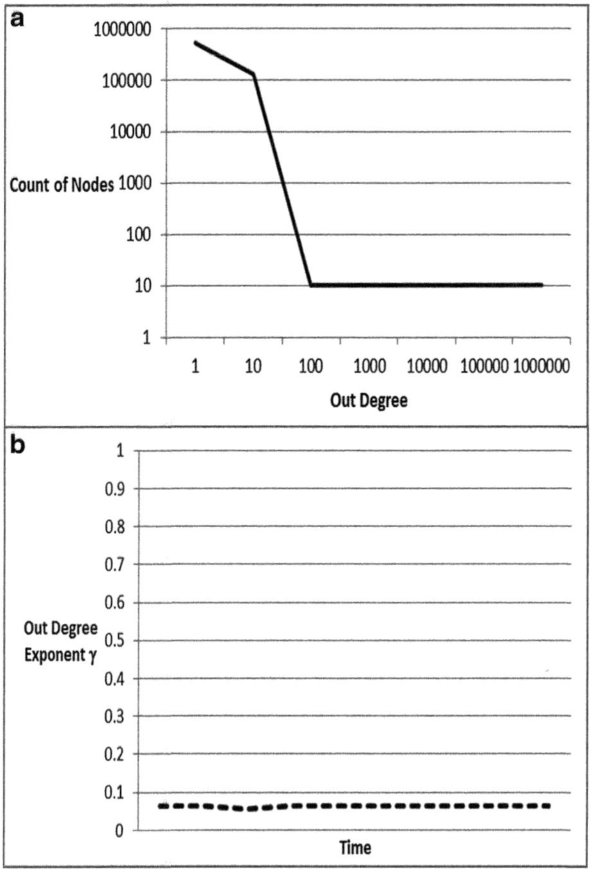

Fig. 5.7 (**a**) Example of an out-degree distribution in a directed network. (**b**) Example of an out-degree exponent over time in a directed network

analysis of network behavior associated to identifying patterns such as changes in the network over time. This section presents an overview on the concept of change detection in evolving networks. Here the focus of change lies on evaluating network behavior based on node features, network-level properties, or both. This chapter therefore discusses change detection in network structures from two perspectives: (1) uni-level change detection which focuses on detecting changes in either node-level behavior or network-level behavior, respectively, and (2) multilevel change detection which combines aspects of the network by observing both node-level and network-level behavior. A detailed description on each approach follows.

5.4.1 Uni-Level Change Detection

Uni-level change detection refers to the detection of change in a single network dimension where a single dimension is considered to be network level or node level, respectively. The analysis of macroscopic behavior in network structures has been widely applied in detecting changes at the network level based on structural differences in network-level properties such as density, diameter, average degree, as well as other node centrality measures by translating them into network-level metrics [3, 11–13, 22, 28–30] study techniques to detect a change or disorder in the state of a time process, usually from normal to abnormal [24] propose GraphScope, an approach to discover communities of graphs and identify any changes in the community structure over time. Their approach identifies new graph segments which mark an abrupt change in the community structure and are thus considered to be discontinuities in time.

The concept of change detection has been explored in network analysis in relation to the application of Statistical Process Control (SPC) using techniques such as sequential probability ratio test (SPRT), the cumulative sum (CUSUM) chart, the exponentially weighted moving average (EWMA), and the Shiryaev–Roberts (SR) procedure [11, 29, 30]. However, SPC operates on the assumption that the data is sequential or time sequenced [31]. Additionally, such techniques may not be suitable to identify changes in non-sequential data such as variations between graph elements such as nodes within the same time period. Furthermore, there are differences between change-point analysis and control charting where the latter is generally better at detecting isolated abnormal points and at detecting a major change quickly, while change-point analysis can detect subtle changes frequently missed by control charts. Interestingly, the two methods can be used in a complementary fashion [32] given that changes usually cause shifts, minor or major, that can be viewed as abnormal. On the other hand, pattern recognition techniques, spectral theory, and mean/median of graphs have been discussed in graph change detection for macroscopic analysis [3]. Also, distance measures such as Hamming distance and Euclidean distance have been applied in change detection, although they do not provide the statistical distribution of the data and are suitable for static networks [11, 12].

5.4.2 Multilevel Change Detection

Multilevel change detection identifies multiple dimensions of change defined as micro- and macro-level changes in evolving networks. Here micro-level changes refer to changes with respect to structural characteristics in the behavior of nodes [20, 21] such as the centrality of nodes in the network, and macro-level changes refer to changes with respect to structural characteristics in the behavior of network-level properties such as density, average degree, and diameter [22]. Detection of multidimensional changes presents unique benefits to challenges associated to big data and dynamism of large complex network structures. As such, it can be used to detect phenomena that may not be evident from a single perspective, such as only micro level or macro level, respectively. More so, multilevel change detection can be used to identify correlated network behavior that may prove useful in detecting cyber threats [22]. For example, changes at the macro level such as the diameter of the network may be associated to micro-level shifts in the behavior of key components within the network such as changes in the centrality level of nodes. Alternatively, changes in the centrality level of nodes such as a decrease in degree centrality indicates decrease in network connectivity which may thus lead to an increase in network diameter. In both micro- and macro-level changes, identifying time when such changes occurred indicates time points of change especially if they exist in a novel pattern [20–22].

Therefore, the studies described in [20–22] present a novel approach to characterizing large evolving networks and detecting changes in such evolving networks, which includes the following steps:

(a) **Selection of central nodes and subgraphs**: This utilizes a hybrid methodology that combines sampling, clustering, and stratified binning to select central nodes and key subgraphs associated to the central nodes from a network over time. This provides a selective analysis of large networks to reduce on the size and dimensionality. Most importantly, graph properties of selected subgraphs should emulate the established graph properties in the full graph. These properties as outlined by [15] specify that the networks are becoming denser over time and the average degree is increasing; hence densification follows a power-law distribution, and the diameter decreases as the network decreases in many cases.
(b) **Micro-level change detection:** For micro-level shifts in the network, the presence and centrality levels of the central nodes is observed to identify **Co**nsistent and **In**consistent (**CoIn**) central nodes where inconsistency marks changes in the presence and centrality of central nodes, respectively. Additionally, times associated to the changes in behavior of these central nodes are detected which are also referred to as CoIn **T**ime **P**eriods of **C**hange (**CoIn-TPC**). A node-level analysis drills down into the network and provides specifics on network activity that may be invisible on a larger scale.
(c) **Macro-level change detection:** Given that micro-level characteristics of the network do not relay information about the bigger picture in the overall network, the key subgraphs associated with the central nodes are used to identify times when the fundamental structural or network-level properties, particularly when significant changes in density, diameter, and average degree occur as a result of changes in the behavior of central nodes. These macro-level changes are referred to as Network Level CoIn (**NL-CoIn**) Change Points. Additionally, a correlation between CoIn central nodes and NL-CoIn is used to determine the impact of node-level changes on the network level as well as similarities between change points in CoIn-TPC and NL-CoIn. Here a network-level analysis describes a generic picture of underlying events in the network.
(d) **Validation based on real-world cyber events:** In order to ascertain that the changes identified are associated to real-world cyber events, CoIn-TPC and NL-CoIn are evaluated using ground truth in order to determine if the changes

are associated to existing cyber attacks [22]. The ground truth evaluation is based on real-world events from Internet-security reports by Akamai Technologies [35, 36]. Specifically, findings in [22] show high accuracy, precision, and recall levels in both node- and network-level changes associated to big cyber attacks such as the Conficker worm particularly during December 2008, January 2009, and February 2009.

5.5 Key Applications

The process of change detection to characterize network behavior can be potentially useful in the cybersecurity domain as discussed in the following sections.

5.5.1 Network-Intrusion Detection

Network-based intrusion detection attempts to identify unauthorized, illicit, and anomalous behavior based solely on network traffic. A network intrusion detection system (NIDS) is used to monitor traffic on a network looking for suspicious activity, which could be an attack or unauthorized activity. Change detection can be used to maintain a map of network activity by identifying and creating critical points on the network. For example, a large NIDS server can be set up on a backbone network, to monitor and audit all traffic; or smaller systems can be set up to monitor traffic and define a threshold on the behavior of central network elements, which can be a particular server, switch, gateway, or router. Specifically, a NIDS server can also detect changes in the connectivity levels of such central nodes on a network based on the number of connections at a particular time by looking for suspicious traffic or usage patterns that match a typical network compromise or threat. Such a server can also play a proactive role to identify potential exploits or for scanning live traffic to see what is actually going on within the network. The process of change detection can be used to develop a comprehensive list of network activity and structural organization in order to establish normal versus abnormal network activities.

5.5.2 Threat Mitigation

Security and technology teams must be ready for cyber attacks against critical infrastructure. With destructive cyber attacks on the rise, there is a need to practice troubleshooting processes for critical system restorations before outages occur [34]. Hence, it is important to know a system so well in order to quickly determine what process caused the outage by identifying what went wrong and why. The motivation behind computer crime can be anything: financial gain, curiosity, revenge, boredom, "street cred," delusions of grandeur, and more. But what if it is a cyber attack? Change detection can be used to reduce on the complexity surrounding network analysis by identifying vulnerable points on the network. Here, an attack profile can be developed to control and minimize the impact of an attack on the network. For example, taking down a highly connected node such as a server could put network communications on a halt. This essentially affects the connectivity on the network which is determined by density on the network, as well as the distance from one network point to another which is determined by the diameter. Hence, change detection can be used to identify the potential source of the problem and use it to trace any changes in network behavior.

5.5.3 *Network Design*

Computer attacks have been graphically modeled since the late 1980s by the US DoD [33]. With the support of advanced tools, network risks can be modeled based on an attack graph where each node in the graph represents an attack state and the edges represented a transition or a change of state caused by an action of the attacker. Such models can be used for network security evaluation. Preventing cyber attacks poses several challenges considering the complexities surrounding large evolving network structures. In order to alleviate such challenges, a wide range of strategies may require testing to identify network vulnerabilities and determine resource allocation on the network. Particularly, change detection can be used to ensure risk management on the network during network design. Similar to threat detection, it can be utilized in identifying vulnerable points such as central nodes that can be targeted to cripple the network. Based on this, network redundancy can be created where such central nodes are duplicated to maintain consistent network activity by redirecting communications in case of an attack.

5.6 Future Directions

This chapter has reviewed state-of-the-art techniques in change detection and network characterization utilizing essential graph-based knowledge. The future directions for this work include addressing challenges associated with sampling big data contained in large graphs by predicting the samples from a given range data in large evolving graphs while at the same time preserving the fundamental network properties. On the other hand, the process of change detection can be extended into predictive network modeling where change points detected as well as non-change points can be used as feature vectors for prediction of network behavior in order to determine if a persistent pattern exists in the micro- and macro-level changes.

Furthermore, given that change detection has been mainly explored in the context of time, it creates an interesting opportunity to adapt such techniques into the spatio-temporal paradigm particularly by identifying spatial regions associated with network changes as well as potential cyber threats.

References

1. Matarazzo, C.: Defending computer networks against attack. Lawrence Livermore National Laboratory. Research Highlights. https://str.llnl.gov/JanFeb10/pdfs/1.10.3.pdf. (2010)
2. Aleksandrova, D.: Detecting cyber attacks. How long does it take? IT Governance. http://www.itgovernance.co.uk/blog/detecting-cyber-attackers-how-long-does-it-take/. (2015)
3. Gaston, M., Kraetzl, M., Wallis, W.: Using graph diameter for change detection in dynamic networks. Australas. J. Combin. **35**, 299–311 (2006)
4. Scripps, J., Tan, P.-N., Esfahanian, A.-H.: Node roles and community structure in networks. Proceedings of the 9th WebKDD and 1st SNA-KDD 2007 workshop on Web mining and social network analysis, pp. 26–35. New York, NY, USA, ACM, (2007).
5. Shen Y., Nguyen, N.P., Xuan, Y., Thai, M.T.: On the discovery of critical links and nodes for assessing network vulnerability. IEEE Trans. Network. (2012)
6. Freeman, L.C.: Centrality in social networks: conceptual clarification. Soc. Network. **1**(3), 215–239 (1979)
7. Page, L., Brin, S., Motwani, R., Winograd, T.: The PageRank citation ranking: bringing order to the Web. Technical Report. Stanford InfoLab. (1999)
8. Bonacich, P.: Technique for analyzing overlapping memberships. Sociol. Methodol. **4**, 176–185 (1972)
9. Bonacich, P.: Power and centrality: a family of measures. Am. J. Soc. **92**, 1170–1182 (1987)
10. Lee, C.-Y. Correlations among centrality measures in complex networks. arXiv:physics/0605220 [physics.soc-ph]. (2006)
11. McCulloh, I., Carley, K.M., Horn, D.B.: Change detection in social networks. United States Army Research Institute for the Behavioral and Social Sciences. (2008)
12. McCulloh, I.: Detecting changes in a dynamic social network. Institute for Software Research School of Computer Science Carnegie Mellon University. Thesis (2009)
13. McCulloh, I., Carley, K.M.: Detecting change in longitudinal social networks. J. Soc. Struct. **12**(2011) (2011)
14. Leskovec, J., Kleinberg, J., and Faloutsos, C. Graphs over time: densification laws, shrinking diameters and possible explanations. In: ACM SIGKDD International Conference on Knowledge Discovery and Data Mining (KDD) (2005)
15. Leskovec, J., Kleinberg, J., Faloutsos, C.: Graph evolution: densification and shrinking diameters. In: ACM Transactions on Knowledge Discovery from Data (TKDD), vol 1 (2007).
16. Leskovec, J., Faloutsos, C.: Scalable modeling of real graphs using kronecker multiplication. In International Conference on Machine Learning (ICML) (2007)
17. Leskovec, J.: Dynamics of large networks (2008)
18. Kang, U., Tsourakakis, C., Appel, A., Faloutsos, C., Leskovec, J.: Radius plots for mining terabyte scale graphs: algorithms, patterns, and observations. In: SIAM International Conference on Data Mining (SDM) (2010)
19. Tong, H., Papadimitriou, S., Yu, P.S., Faloutsos, C.: Proximity tracking on time-evolving bipartite graphs. In: SDM (2008)
20. Namayanja, J.M., Janeja, V.P.: Discovery of persistent threat structures through temporal and geo-spatial characterization in evolving networks. ISI 191–196 (2013)

21. Namayanja, J.M., Janeja, V.P.: Change detection in temporally evolving computer networks: a big data framework. First International Workshop on High Performance Big Graph Data Management, Analysis, and Mining, Co-located with the IEEE BigData 2014 21. J. M. (2013)
22. Namayanja, J.M., Janeja, V.P.: Change detection in temporally evolving computer networks: changes in densification and diameter over time. ISI (2015)
23. Akoglu, L., McGlohon, M., Faloutsos, C.: Oddball: Spotting anomalies in weighted graphs. In: PAKDD, pp. 21–24 (2010)
24. Sun, J., Faloutsos, C., Papadimitriou, S., Yu, P.S.: GraphScope: parameter-free mining of large time-evolving graphs. In: Proceedings of the 13th ACM SIGKDD International Conference on Knowledge Discovery and Data Mining, pp. 687–696 (2007a)
25. Ferlez, J., Faloutsos, C., Leskovec, J. J., Mladenic, D., Grobelnik, M.: Monitoring network evolution using mdl. In IEEE International Conference on Data Engineering (ICDE) (2008)
26. Han, J., Sun, Y., Yan, X., Yu, P.S.: Mining heterogeneous information networks. In: KDD (2010)
27. Fabrikant, A., Koutsoupias, E., Papadimitriou, C.H.: Heuristically optimized trade-offs: a new paradigm for power laws in the Internet, volume 2380 of Automata, Languages and Programming, p. 781. Springer (2002)
28. Opsahl, T., Agneessens, F., Skvoretz, J.: Node centrality in weighted networks: generalizing degree and shortest paths. Soc. Network. **32**(3), 245–251 (2010)
29. Akoglu, L., Faloutsos, C. Anomaly, event, and fraud detection in large network datasets. In: Proceedings of the Sixth ACM International Conference on Web Search and Data Mining, pp. 773–774. ACM. (2013)
30. Tartakovsky, A.G., Polunchenko, A.S., Sokolov, G.: Efficient computer network anomaly detection by changepoint detection methods. IEEE J. Selected Topics Signal Process. **7**(1), 7–11 (2013)
31. Slavin, V.: Improper use of control charts: traps to avoid. (2006)
32. Taylor, W.A.: Change-point analysis: a powerful new tool for detecting changes, WEB: http://www.variation.com/cpa/tech/changepoint.html. (2000)
33. Abraham, S., Nair, S.: Cyber security analytics: a stochastic model for security quantification using absorbing markov chains. J. Commun. (2014)
34. Lohrmann, D.: Hacking critical infrastructure is accelerating and more destructive. http://www.govtech.com/blogs/lohrmann-on-cybersecurity/Hacking-Critical-Infrastructure-is-Accelerating-and-More-Destructive.html. (2015)
35. Akamai Technologies.: 4th Quarter 2008: The State of the Internet. **1**(4). https://www.stateoftheinternet.com/resources-connectivity-2008-q4-state-of-the-internet-report.html. (2009)
36. Akamai Technologies.: 1st Quarter 2009: The State of the Internet. **2**(1). https://www.stateoftheinternet.com/resources-connectivity-2009-q1-state-of-the-internet-report.html. (2009)
37. The CAIDA UCSD [Anonymized Internet Traces 2008]–[April 2011–December 2013], http://www.caida.org/data/[/passive-2008/
38. The CAIDA UCSD [Anonymized Internet Traces 2009]–[April 2011–December 2013], http://www.caida.org/data/[/passive-2009/
39. The CAIDA UCSD [Anonymized Internet Traces 2010]–[April 2011–December 2013], http://www.caida.org/data/[/passive-2010/

Chapter 6
Cybercrime: Concerns, Challenges and Opportunities

George S. Oreku and Fredrick J. Mtenzi

Abstract It is widely accepted that technology is an agent of change in the society. If used properly, it can increase productivity and improve our quality of life. However, the current rate of change in technology leaves room for it to be exploited and be used for things it was not meant to do. This includes criminal activities which are carried using technology on the cyberspace that can be classified as cybercrime. Cybercrime or computer crime can be defined as a criminal activity in which computers or computer networks are a tool, a target or a place of criminal activity. Cybercrime can also be defined as the leveraging of information systems and technology to commit larceny, extortion, identity theft, fraud and, in some cases, corporate espionage.

In this chapter, we explore the challenges of combating cybercrime given its dynamic, pervasive and international nature. We examine in detail the technical, social and legal factors that are continually shaping the landscape of cybercrime. Further, the chapter consolidates research and practical work in the area of cybercrime, emerging perspectives, paradigms and trends. A deliberate effort is made in making sure that the role of underground economy is clearly spelled out and its contribution to cybercrime is analysed. The impact of new laws which are being quickly enacted without much thought, discussion on the regulatory framework for combating cybercrime and ethical dimension of cybercrime given its global nature are examined. This chapter is expected to stimulate constructive discussions on novel ways of mitigating and promoting research in cybercrime.

G.S. Oreku (✉)
Tanzania Industrial Research Development Organizational, P.O. Box 23235, Kimweri avenue, Dar Es Salaam, Tanzania

Faculty of Economic Sciences and Information Technology, North-West University, Vaal Triangle Campus, P.O. Box 1174, Vanderbijlpark, South Africa 1900

Faculty of Science and Forestry, School of Computing, University of Eastern Finland, P.O. Box 111, 80101 Joensuu, Finland
e-mail: gsoreku@tirdo.org; George.oreku@gmail.com

F.J. Mtenzi
School of Computing, Dublin Institute of Technology, Kevin Street, Dublin 8, Ireland
e-mail: Fredrick.mtenzi@dit.ie

I.M. Alsmadi et al. (eds.), *Information Fusion for Cyber-Security Analytics*, Studies in Computational Intelligence 691, DOI 10.1007/978-3-319-44257-0_6

6.1 Introduction to Cybercrime

In the last few years, we have witnessed major changes in terms of our usage of information and communications technology (ICT) facilities. ICT has now become an integral part of our lives and Internet usage is growing fast. Internet usage is becoming irrepressible in our daily life. We are using ICT and especially the Internet in nearly all walks of our lives such as controlling traffic lights and financial and emergency services. The benefits of using ICT in this information age are enormous. However, any attack to ICT infrastructure would potentially have disastrous consequences for individuals and for the society.

There have been justifiable mounting concerns about the vulnerability of the ICT infrastructure and especially the Internet; this has been heightened by the proliferation in crime committed using ICT. The nature of the Internet has made committing crime easy and opened other avenues for carrying out crime with impunity. This state of affair has led end users, policymakers and law enforcement agencies to become increasingly concerned that adversaries backed by reasonable resources will attempt to exploit the cyber-vulnerabilities in the critical infrastructure, thereby inflicting substantial damage.

The Internet is a world of people, not technology, and no human society has endured without law and order. Therefore, the Internet being a human society has very little laws governing its usage. The laws being enacted to protect and promote secure usage of the Internet are lagging behind technology development, fragmented and inadequate. The law is a powerful weapon but one that takes time and care to aim and can only be moved with considerable difficulty. We have limited ammunition and must ensure that it is well spent [1].

Cybercrime or computer crime is broadly defined as a criminal activity in which computers or computer networks are a tool, a target or a place of criminal activity. Cybercrime can also be defined as the leveraging of information systems and technology to commit larceny, extortion, identity theft, fraud and, in some cases, corporate espionage [2]. The National Institute of Justice defines cybercrime as 'any illegal act for which knowledge of computer technology is used to commit the offence' [3]. Forester and Morrison define computer crime as 'any criminal act that has been committed using a computer as a principal tool' [4]. For some other analysts, computer crime refers to illegal activities in which a computer is involved either as an object, subject or instrument of the criminal act. Herman Tavani maintains that a criminal act is one that can be carried out only through the use of computer technology [5]. Nevertheless, not all computer crimes involve a high degree of technical skill and expertise. As H. Cornwall states, 'most computer crime is ordinary crime which at one stage happens to involve a computer' [6]. And the pervasive nature of cybercrime makes it even more difficult to have a concise definition.

Cybercrime is not a new concept; it is as old as computers or computer networks, but the Internet has given it a new twist and steered current practice, research and debate in new directions. It is coming from the fact that as soon as computers were recognised to store something of value (information), criminals saw an opportunity. This is realised in most cases when data begin to move from one computer to another over the networks. Networks provide an entry point for the data stored in computers to be accessed. The digital nature of information and its ability to move seamless between and within networks open an infinite number of possibilities to commit cybercrime.

Consequently cybercrime must be fought at personal, social and political fronts as well as the electronics front. These are specific steps which must be taken to achieve the goal of mitigating the threat from cybercrime which is a global problem. Cybercrime is pervasive, non-discriminatory and dramatically on increase. With minimal risk, bad people are turning to cybercrime in ever-escalating numbers because of its low skill entry requirements and promise of extremely high rates of financial return.

For example, despite all the dangers and warning signs of cybercrimes, it is surprising to see that financial institutions continue to encourage customers to embrace electronic commerce and banking to reduce their brick-and-mortar expenditures. Most of these institutions are failing to make commensurate expenditure for computer security; they provide their customers with static logins and passwords for authentication, making keystroke-logging and packet-sniffing malware highly effective against these victims. This penchant for convenience and ease of use typically sacrifices security, thus providing cybercriminals with a bountiful harvest.

There are five features of the online world which create new security risks and shape the kinds of user and criminal behaviour we find online'. The first is digitisation; common standards for data transmission that enables manipulation and modification. The second is anonymity—the ability to act and spy on others without disclosing one's identity; it is one of the major attractions of users and hackers to use the Internet, and removal of this sometimes leads to allegation of loss of human rights. The third is interconnectivity—the fact that everyone on the network is connected to everyone else. The fourth is decentralisation—the lack of centralised control over digital networks. The fifth is interdependence—the shared vulnerabilities that inevitably exist between and among all the people who use digital networks. These features create an environment of dynamic changes of laws, technology and policies and a redesign of digital architecture [7].

The digital environment is different from the offline world. In cyberspace, events occur almost instantaneously across large distances, network boundaries do not align with physical and political boundaries, and everyone on the network is your neighbour. In the digital environment, copying is costless, the cost of data acquisition approaches zero, and it is easier to retain information than to delete it selectively. The new virtual environment will reshape legal concepts like jurisdiction and ownership, it will require us to rethink doctrines for tort liability and duties of care, and it will change the way we think about privacy [8].

When we examine the digital crime worldwide now, it seems to be littered with new sophisticated crimes even though, on closer examination, many of these supposedly novel forms of crime (such as phishing in order to perform identity fraud) in fact share much in common with conventional and long-standing crimes and criminal techniques. They are the same old tradition crimes but more elaborate scams conducted on the digital space. These crimes inherit nearly all characteristics of the conventional crimes and then have extra dimensions coming from the digitisation process. The extra dimension from the digital space adds considerable complexities in terms of legal process and its politics.

6.1.1 State of the Art in Cybercrime

It is important to realise that how and when we touch software and how and when it touches us are less of our choice everyday. Therefore, the quality of this software matters greatly, because we have less control of it [9]. The level of protection this software affords us from harm and exploitation matters even more. This level of software intrusion in our lives coupled with its insecure nature opens innumerable avenues for criminals to commit cybercrime.

Cyberattacks are successful because networked computers with exposed vulnerabilities may be disrupted or taken over by a hacker or by automated malicious code. And networked computers with vulnerabilities are everywhere because most of the systems are built on insecure software base. Therefore, botnets scan the Internet to find and infect computer systems that are poorly configured or lack current software security patches. Compromised computers are taken over to become slaves in a 'botnet' which is remotely controlled to collect sensitive information from the victim's PC or to collectively attack as a swarm against other targeted computers [10]. Even computers that have updated software and have the newest security patches may still be vulnerable to a type of cyberattack called 'zero-day exploit'. This may occur if a computer hacker discovers new software vulnerability and launches a malicious attack before a security patch can be created. Zero-day vulnerabilities in increasingly complex software are regularly discovered by computer hackers. Recently, we have started to witness the zero-day vulnerabilities available on online auctions; this allows newly discovered vulnerabilities to be sold quickly to the highest bidder [10].

Vulnerabilities persist largely as a result of poor security practices and procedures, inadequate training in computer security or technical errors in software products. Inadequate resources devoted to staffing the security function may also contribute to poor security practices. Home PC users have little or no training in best practices for effectively securing home networks and equipment. Vendors of commercial software are releasing products with errors that create the computer system vulnerabilities. Many security experts agree that, regardless of investment on developing secure software, vulnerabilities will continue to persist because software is becoming complex and because of the rate of integration of different software.

Cybercriminals of today have adapted to fit in the environment which they operate and illustrate the philosophy of 'think global, act local' business strategy. In their network, each individual is rewarded differently, depending on the country of origin of the infected computer. This structure is highly effective in avoiding the chances of detection since multiple players are operating as stand-alone, having no contact with their 'colleagues' [11]. The commodity that is being sold in the cybercrime underworld by cybercriminals is stolen data. The data includes credit card numbers, bank accounts, healthcare-related information, single sign-on login credentials for organisations and email exchanges. The emerging trend towards stealing healthcare-related information is a worrying one due to this information being rich/multimedia and sensitive [11].

6.1.2 Current Challenges in Cybercrime

General technological developments have continually created new opportunities for criminal activity which in turn have driven the development of new technologies. It should be noted that just as technology is utilised by many people for legitimate reasons, so it will be by those who intent on committing crime.

Available data on all forms of security threats, attacks and cybercrime is little and unreliable. There is no incentive so far on reporting security threats, and cybercrime is no different as the data available is obtained using unreliable methods. For example, at any given time, it is hard to know if there are hackers in your network or computer. The percentage of hackers caught may not show a true portion of those who are involved in cybercrime. It is easy to get away having committed a cybercrime, and the chances of getting caught are slim. Software weaknesses and profit motives of attackers act as incentives, enticing attackers to commit acts of crime and espionage. And the ease of hacking tools and explanation on how to hack which is widely available online for those thinking of hacking further lowers the entry point.

Information that web users generate when they visit the web is valuable for companies involved in marketing. What we are witnessing now is companies collecting this information without the knowledge of users [12]. The concerns here are more on privacy as the identities of the users may be exposed leading to identity theft. Many users are likely to object to the idea of web tracking which makes privacy advocates queasy. Some even argue that selling the information to advertisers is paving way for the police or intelligence agencies to view the information without a warrant. There are a lot of companies involved in web tracking such as Phorm, NebuAd and Front Porch [13]. Web tracking activities are fairly widespread in Asia and the USA, and the data obtained is used for targeting advertisements [12]. There is no reason to believe that such practice is not being used by the intelligence community. Other web tracking data such as when people are looking to buy a domain name and check to see if it is free by typing the new name into a browser can cause a moral dilemma to DNS registers. That is, if the URL does not exist, an error message comes back called NXD, for non-existent domain. The NXD data can be valuable, as anyone can register the domain themselves, knowing that they are likely to be able to sell them to the people who looked them up [12].

One of the major challenges in combating cybercrime is collaboration, where multiple international law enforcement agencies working together share information and techniques to gather evidence, identify the perpetrators and arrest them. This level of collaboration is the exception rather than a rule, however. In far too many instances, political or cultural impediments preclude the level of cooperation and interaction. Breaking down these barriers to collaboration is essential to waging a comprehensive campaign against cybercrime worldwide. The international collaboration should include laws, international relations, conventions, directives and recommendations culminating in a set of international guidelines to fight cybercrime. International collaboration is important because cybercriminals are scattered in the world, and digital evidence relating to a single crime can be dispersed across multiple regions.

6.1.3 Status Quo and Modern Crimes

With the research conducted recently, the definition of 'cyber crime' in Tanzanian context varies across the industry as it covers a wide scope of overlapping crimes committed with the aid of growing technology. These crimes are classified as how majority of the people are involved to make use of the Internet and technologies to exchange ideas, keep in touch with family and friends, buy and sell products (including online transactions, M-Pesa) and access online services.

Cybercrime in Tanzania is mainly committed by two groups of people, namely, those who perform the act without the knowledge that what they are doing is wrong and those who know what they are doing but are determined to perform the act in order to distract the country's equilibrium in different angles from destabilising peace in a country through misuse of social media and other communication media to stealing money through online transactions.

Tanzania has a high rate of cybercrime, and hate speech was highlighted in [14]: 'There is fear of high rate of individuals who made use of the Internet to threaten national security (Tanzania) due to misuse of the blogs and social media along with mobile phones to spread hate speech among communities in Tanzania and the number of cyber criminals worldwide has now increased'. In the past, there were a few cases where criminals made use of technology to tamper with ATM machines in Tanzania. Nowadays the act is growing faster, and the fear among the ATM users has increased due to the fact that each day cybercriminals are coming up with new techniques to steal money from ATMs.

Everyone seems to be surprised and shocked when the opposition leader spokesman for the Ministry of Communication, Science and Technology Mr. Joshua Nassari announced that the country has lost 892.18 billion Tanzanian shillings on cybercrimes as the official reports of the police said [15]. Some sophisticated criminals have been stealing money from the banks directly; 250 million were reported in [16] being stolen from Uchumi Commercial Bank through the ATMs in Moshi region in Tanzania. Cybercrime has become a global threat which needs an urgent attention at national, regional and international level.

Cyber criminals are always ahead of us: this fact was backed up by the deputy minister for home affairs, Mr. Pereira Silima, on an open conference which took place in Tanzania [17]. 'Many people are ignorant of cybercrimes while our police force has low capacity', he said in Arusha where IT specialists, lawyers, police officers and others from the East African region had a meeting to devise ways to tackle the problem. He added that currently there were more than 300 cybercrime cases which are being investigated by the police in Tanzania but admitted that the exercise was slow because the police and other agencies were ill equipped and not conversant with such crimes.

The wave of hacking underscores the financial industry's battle to thwart cybercrime and comes as consumers and banks are reeling from several high-profile data breaches at retailers that have exposed millions of credit cards and debit cards to potential fraud.

From January to 9 April 2015, the number of attacks on debit cards used at ATMs reached the highest level for that period in the last 20 years, according to FICO, a credit-scoring and analytics firm. The company tracks such incidents through its card-monitoring service for financial institutions that represent more than 65 % of all US debit cards [18].

Debit-card compromises at ATMs located on bank property jumped 174 % from January 1 to April 9, compared with the same period last year, while successful attacks at nonbank machines soared by 317 %, according to FICO [19]. The article went further by clarifying that the incidents come as banks are racing to issue new credit and debit cards with computer chips that make it more difficult for thieves to create counterfeits. However, most ATMs don't yet accept the new technology, though the JPMorgan Chase & Co. and Bank of America Corp. have recently begun to install the more advanced machines.

Criminals send phishing emails or text messages via mobile phones and trick the victims by convincing them to provide their bank details and ATM pins. They use different telephone numbers and pretend to be bank officials requesting some of the client's details for bank detail amendments.

Card skimming is also happening in the country whereby cybercriminals use and record card details by using a device called a 'card skimmer' which is placed right over the card slot or over the keypad of the ATM machine and captures card and PIN information.

Another attack which is becoming more popular in Tanzania as well is web attacks which occur when criminals hack websites which has turned out to be a technique now commonly used in Tanzania. Cybercriminals hack the websites to gain unauthorised access to the personal information of clients or web visitors. This in Tanzania imitates how global hackers are now using the technique to destabilise websites by performing attacks to deny and or destroy information, steal information, manipulate information, alter the context in which the information is viewed or change the perceptions of people towards the information.

The findings also recognised that there is a woman in the Kilimanjaro region in Tanzania who steals money from mobile money agents and is very good at it. She apparently does so by approaching agents pretending to want to draw some money, but after a while the agent will find all his money gone after she left the place. It is hard to tell what she does, but most of the agents have confirmed that the criminal is the same woman (Unknown, 2013). Such a report comes after a study that shows that transactions through mobiles are growing rapidly in Tanzania.

The use of M-money in Tanzania started in 2005 when Airtel introduced phone-to-phone airtime credit transfer. In 2008 Vodacom launched its M-money service called M-Pesa. Zantel Tanzania also introduced its M-money service in the same year, which was first called Z-Pesa (now called EasyPesa). In 2010, Tigo, the first mobile network operator (MNO) in the country, launched its M-money service called Tigo Pesa. Currently there are four M-money service brands in Tanzania: M-Pesa, Tigo Pesa, Airtel Money and EasyPesa. Up to April 2013, the registered customer base of mobile payment services was 28.8 million in Tanzania, 8.5 million being active users [20].

The weak link that can let a hacker clone the so-called 'chip-and-pin' credit and bank cards stems from the fact that, as the Cambridge researchers showed, the EMV scheme has, in too many cases, not been carried out as planned. The authentication process, as originally envisioned, was supposed to depend on the issuing bank to generate a random number for every unique transaction. In practice, where saving money often trumps security, it was left to point-of-sale terminals or cash machines to generate the number. In the study conducted in some of the banks, the researchers discovered to their horror that in half of the machines they looked at, the supposedly random numbers were generated by counters or timestamps and were, therefore, not random at all. This makes it all too easy for a hacker. 'If you can predict [the number], you can record everything you need from momentary access to a chip card and play it back and impersonate the card at a future date and location' [21].

The mentioned attacks might seem outdated in the developed world but are the most common and fast-growing techniques used by criminals in Tanzania as well, and it is unfortunate that the majority of the populations are not aware of this. There have been an increasing number of incidents where victims of credit card fraud had their requests for refunds refused by the issuing banks on the grounds that there is no way to explain the card having been authenticated without the cardholder's involvement.

News about hackers getting your information is everywhere. When breaches happen on a large scale, they can cost hundreds of millions of dollars with millions of people affected. In many cases, as with hacks at retail stores or healthcare providers, you have little to no control over protecting your data, you have to trust someone else in custody of your information who is taking precautions. But in other cases, you're able to take precautions to protect your information.

Hackers still aren't quite as interested in hacking your cell phones, but it's likely only a matter of time until your desktop, iPhone and cloud are all at equal risk. As a result, your purchases on Amazon, the photos of your kids and even your banking information are at risk. About 32 % of people prefer to bank online, and 12 % of people prefer to use mobile banking instead of visiting a branch—increasing the risk to have your personal financial data at a hacker's fingertips. You can use common sense to make sure you're protecting yourself as much as possible, but today there are easy steps to take to make sure you're as private as you'd like to be.

The News of the World phone-hacking scandal in the early 2000s was perhaps the first big wake-up call for the public that your cell phone activities can be hacked and broadcast to anyone with the right skills and technology. In many cases, that would mean hackers overhear your weekly calls to your mother and a conversation with your spouse about what to have for dinner. But if you're worried about people listening into more sensitive conversations or want to have privacy for privacy's sake there are tools you can use to encrypt your phone. 'If messages are intercepted, they'll be unreadable', Snowden said [22].

One app Snowden recommends is Signal, created by Open Whisper Systems. It's free for iOS and Android phones, and both the creator and Snowden call the app 'low friction', meaning it's easy to use and won't completely disrupt the way you text or call now [23].

No matter what page you go to online, companies are invisibly collecting information about your movements and your data. Being smart about what information you provide is an easy way to make sure you're not giving too much away.

All cybercrime scenes are unique and potentially present new challenges to the investigators. Investigators face challenges of both understanding complex ICT technologies and the principles and practices of criminal investigation, reconstruction of crime scenes, collection and preservation of admissible evidence, detection and investigation [24].

A key issue for policymakers is how the government can effectively monitor private networks for intrusions without infringing the privacy rights of the citizens whose data flow through these networks. For example, it is important to note that a lot of Internet traffic is passing through the USA; therefore can the US government monitor the traffic without violating the privacy of users worldwide? This fact alone gives the USA a competitive intelligence and espionage advantage [25]. It is hard at this day and age for one country to be entrusted with managing the Internet traffic of nearly the whole world.

6.2 Social and Economical Impact of Cybercrime

The economic and social impact of cybercrime is increasing every day. It is hard to get the exact figures on the cost of cybercrime; however, even going by the information, which is grossly underestimated, the cost is very high. For example, in Ireland, some organisations are experiencing single cybercrime incidents which cost over €250,000, and up to 14 % of organisations are spending over ten working weeks responding to individual issues [26].

It is worth pointing out that the incidents of some cybercrimes may increase due the current credit crunch. When considering business changes such as redundancies or other business changes, organisations may be faced with increased intellectual property theft, financial fraud and authorised access. These changes call for organisations to give equal priority to internal and external threats. We are witnessing the increase in the role of the media in reporting cybercrime incidents.

6.2.1 Social and Cultural Dimension of Cybercrime

The reporting of cybercrimes to law enforcement agencies or the media is very low. There are a number of reasons which hamper organisations, such as unfavourable publicity, public embarrassment or loss of public confidence. However, the culture of believing that the cyberspace is lawless and nothing good comes from pursuing criminals online is misguided and misinformed. Organisations should open the channels of communications for reporting incidents of cybercrime, and each member of the organisation must be aware of it. They should also be willing and ready to seek assistance in the case of cybercrime incidence.

It should be emphasised that our notions and perceptions of reasonable expectations of privacy are determined by social norms, which in turn are shaped by our interactions with the technology we see and use around us. Our privacy expectation in the cyberspace is influenced by the rate of technology change and change in the legal and regulatory environment and people. As the society becomes more conversant with the use of the Internet for everyday activities, the ethical security culture may become a norm, and this will help in mitigating cybercrime.

6.3 Cybercrime Ethics and Legislation

6.3.1 Legislations to Combat Cybercrime

Most security experts believe that tough legislation against cybercrime may go a long way in combating it. For example, there are very few federal laws enacted against cybercrime such as the Computer Fraud and Abuse Act of 1986 and 2001 (threats to computers), National Information Infrastructure Protection Act of 1996 (criminal intent), Computer Security Act of 1987 (federal agency information security), Federal Privacy Act of 1974 (privacy), Digital Millennium Copyright Act (protection of technology copyrights) and USA PATRIOT Act of 2001 (terrorism) that have been used to prosecute cybercriminals using spyware as a means of getting identity of unsuspecting users. Other laws of the same nature include the Internet Spyware Prevention Act (I-SPY) of 2005 and Securely Protect Yourself Against Cyber Trespass Act (Spy Act) of 2005. Many companies that write antispyware software want the government to protect them from frivolous lawsuit brought against them by the spyware companies (case of Lavasoft v New.net) [27].

The next major problem in the fight against cybercrime is the harmonisation of international cyberlaws and securing of electronic evidence until the criminals are brought to justice and other complex jurisdictional issues and procedures that arise at each step. In this respect, the EU Convention on Cybercrime is a step in the right direction, despite its weaknesses.

The role of the United Nations, the G-8 Subgroup on High-Tech Crime, the OECD and the Europol and Interpol in combating cybercrime is crucial. However, its effectiveness is hampered by countries not willing to collaborate. In order to convict cybercriminals, they must have broken laws; it is interesting to note that for the same offence, different countries may have different laws. Further, it is difficult to harmonise procedures, guidelines and laws especially cyberlaws to combat and prosecute cybercriminals. Countries with low penetration of ICT do not appreciate the importance of it. Historically countries in the world do not agree on everything, and disagreement in one area such as global warming may lead to disagreement in cyberlaws.

Finally, we seem to have all the most important elements for reducing the incidence of cybercrime. We have laws, tools and widespread awareness that is sometimes the most difficult component of a crime prevention effort. The question that we need to ask ourselves is 'why, then, is cybercrime not only going away, but steadily increasing'.

6.3.2 Ethical Considerations in Cybercrime

Users may imagine that the process of discovering vulnerabilities is easy. However, as the software and hardware have become complex over years, discovering bugs has turned out to be difficult. This has resulted in the emergence of rewarding schemes for those who discover bugs [28]. There is currently a race to discover bugs between white hat hackers and black hat hackers which may be won by black hat hackers. The question of those who will discover bugs and find it ethical to sell them is worth investigating. It is now known that the effort it takes to discover vulnerabilities in software involving therefore any incentives which are skewed may lead to unintended consequences in the fight towards cybercrime.

When it comes to cyberspace, the attackers and defenders use the same tools, network, hardware and software, and there is a fundamental tension between cyberattack and cyberdefence. The NSA defines the 'equities issue' as the moral dilemma when a military discovers vulnerability in a common product; they can either alert the manufacturer and fix the vulnerability or not tell anyone. It is not an easy decision. This has an ethical dimension even to users when they discover vulnerability.

6.4 Cybercrime: A Borderless Crime

Warfare has changed forever; never again will we see a major warfare without cyberwarfare being a crucial component. Some countries are advocating the use of offensive cyberwarfare due to their increased dependence on the Internet. However, before carrying out offensive action, a country or company must be able to determine where the attack came from, and this is difficult using current technology. And for the offensive to be effective/deterrent, it must be made public, and this may be difficult for intelligence agencies where some of this information may be classified.

We believe that militaries throughout the world have personnel developing tools to wage cyberwarfare against other nations. They have collections of vulnerabilities in common operating systems, generic and specialised applications or even custom military software that their potential enemies are using and code to exploit those vulnerabilities, and they are keeping these as secret to have a competitive edge [29]. The best thing to do is to infiltrate the enemy's computers and networks, spy on them and surreptitiously disrupt select pieces of their communications when appropriate. This can be achieved by carrying out passive eavesdropping and performing traffic analysis.

For example, in mid-2007, Estonia, believed by many as 'the most wired nation in Europe' because of its pervasive use of computer networks for a wide array of private and public activities, had a significant portion of its national infrastructure crippled for more than 2 weeks by cyberattacks launched from hundreds of thousands of individual computers that had previously been hijacked by Russian hackers. Estonia was so overwhelmed by the attacks that Estonia leaders literally severed the country's connection to the Internet, along with it the country's economic and communication lifeline to the rest of the world. The reason for the attack was the Russian government's objection to Estonia's removal of a Soviet-era war memorial from the centre of its capital, Tallinn, to a military cemetery. The success of the Estonia attack was made possible in large part because of insecure software. It should be noted that traditional defensive measures employed by users such as firewalls, antivirus and software patches did little to help Estonia [9, 29].

The popular media belief is that there is a coordinated attempt by the Chinese government to hack into US and other nation's computers, military, government, and corporate and steal secrets. The truth is a lot more complicated. Most of these attacks seem to originate from China and are carried out by individuals who are doing this for nationalistic reasons, against the big power and for monetary reasons. The fact that they are not centrally coordinated makes them take more risks and act stupidly. However, the Chinese government is aware of their existence and probably buys some military or trade secrets from them.

According to IDC (a global market intelligence firm), 75 % of computers having access to the Internet have been infected and are actively being used without the owner's knowledge to conduct cyberattacks, distribute spam and support criminal activities. This deplorable state of affair cannot be blamed only on innocent computer users or hackers but on insecure software to a large extent. Insecure software is a big problem than most people would imagine, and insecure software is everywhere, from G-phone to laptops and from game consoles to public utilities. Insecure software is interconnected and woven more tightly into the fabric of civilisation with each passing day and with it comes an unprecedented level of vulnerability. Insecure software is making us fragile, vulnerable and weak, and this state of affairs is a fertile breeding ground for cybercrimes.

What is happening right now is that the worldwide interconnection of insecure software gives social problems once limited by geography and a new destructive range. Cybercriminals, terrorist and even nation-states are currently preying on millions upon millions of computer systems (and their owners) and using the proceeds to underwrite further crime, economic espionage, warfare and terror. We are only now beginning to realise the enormity of the storm set upon us by the tiny fluttering of software manufacturing mistakes and the economic and the social costs such mistakes impose.

Anonymity and absence of frontiers make the Internet an efficient weapon in the hands of criminals. In the virtual space, criminals usually act from sites in other countries. They use this strategy to avoid detection and prosecution. The money accrued from cybercrime had surpassed even that made from the global drug trade, which is very difficult to conduct nowadays, and the penalties once caught are heavy [30]. This means nations around the world should brace themselves for a wave of cybercrime. The low barrier to entry in order to commit cybercrime is a major incentive to cybercriminals; for example, a laptop or Internet connection costs a lot less than a tank. The complex society dependency on computers and networks provides another major opportunity to criminals because computer disruption has economic, logistical and emotional effect. Cybercrime is moving at such a high speed that law enforcement cannot catch up with it.

6.4.1 International Collaborations

The EU Convention on Cybercrime is one treaty that will help foster and secure international cooperation and easier assistance in combating cyberterrorism. National borders cannot restrict cybercrime due to the nature of the Internet. This means that a country will have great difficulty in addressing crime committed in a country by an individual in a foreign country. The process of prosecuting criminals is long. However, it is expected that the treaty will deny safe havens to cybercriminals.

The type of the cooperation required for combating cybercrime must involve all or nearly all countries in the world. The EU Convention on Cybercrime has few countries that have ratified the treaty; this makes it hard to use the convention effectively in combating crime. It is not easy to have global consensus in issues that pertain to the whole world such as global warming or human rights. And cybercrime is no different; there are countries who think that this is not a problem to them at their level of development, and we believe that there are those who are hoping of making a fortune by assisting cybercriminals. Therefore, the issue of international collaboration, while plausible, may not be achieved.

Countries need to cooperate because cybercriminals are not confined by national boundaries, and digital evidence relating to a single crime can be dispersed across multiple regions. While it is important for developing countries to have cybercrime laws in place, it is equally necessary that countries have the legal authority to assist foreign countries in an investigation, even if that country has not suffered any damage itself and is merely the location of the intruder or a pass-through site [31]. International measures to help these countries enact cybercrime laws are crucial. Given that some of the viruses, such as the Love Bug, originated in developing countries, the industrial world cannot solve cybercrimes without their help.

6.5 Combating Cybercrime

Realising the importance of having a central point of coordinating efforts to combat cybercrime, the USA established the Department of Homeland Security (DHS) for assessing and evaluating the vulnerability of the nation's critical infrastructure as a whole and cyber security threats in particular and for synchronising its actions with other federal, state, local and private entities to ensure the most effective response. This level of awareness and organisation is supposed to be shown by other countries and the same be demonstrated at the international level. We do not have any choice if the society dependence on information and Internet will continue to increase.

6.5.1 Tools, Skills and Techniques

Malicious tools enable attackers to gain access to a variety of valuable resources such as identities, credentials, hacked hosts and other goods and services. Some malicious tools and services are designed to counter security measures such as antivirus software to increase the lifespan of a malicious code sample in the wild. The result is a cycle whereby malicious tools must be continuously developed and used to produce other goods and services. The profits from these goods and services may then be reinvested into the development of new malicious tools and services [21, 32].

Tools for carrying out cybercrime range from kits that automatically scan and exploit vulnerabilities to botnets. These tools may be used to provide services such as denial of service (DoS) attacks, spamming and phishing campaigns and finding exploitable websites and servers [32]. Botnets are becoming a major tool for cybercrime, partly because they can be designed to effectively disrupt targeted computer systems in different ways and because a malicious user, without possessing strong technical skills, can initiate these disruptive effects in cyberspace by simply renting botnet services from cybercriminals. Botnets have been described as the 'Swiss army knives of the underground economy' because they are so versatile.

Exploits are another effective malicious tool. Exploits constitute vulnerability information and exploit code. They differ from the other categories of attack tools in that they are not automated by nature. When exploits are incorporated into automated tools, they can then be classified as attack tools. The exploits available in the underground economy are typically tailored to specific market demands. The market for exploit code and vulnerability is geared towards attackers.

Spam and phishing tools and related goods and services are marketed on underground economy servers. Products include spam software, spam relays, compromised computers to host phishing scams and content such as phishing scam pages and phishing scam letters. Spam is often used to advertise black-market products, such as pharmaceutical drugs, to distribute malicious code and launch phishing attacks that steal credentials, personal information and credit card numbers [32].

Another set of tools which are becoming prevalent now is crimeware. Crimeware is a software that performs illegal actions unanticipated by a user running the software; these actions are intended to yield financial benefits to the distributor of the software [33]. Crimeware software include keyloggers and screenscrapers, redirectors, session hijackers, web Trojans, transaction generators, system reconfigurators and data stealers.

Cybercriminals use an arsenal of highly effective crime tools, deploying sophisticated criminal-to-criminal (c2c) business models for their operations and heavily borrowing and copying from the legitimate business world. The Russian Business Network is a prime example of the highly organised cybercrime gang, which at its peak provided web hosting to online criminals and created fake anti-spyware and virus removal tools. There are different categories of tools that are available for committing cybercrimes such as 'how-to' software packages that instruct users step by step on how to infect a system. 'Do-it-yourself' toolkits enable cybercriminals to easily gain access to a wide array of sensitive and valuable information; crimeware toolkit creators also deploy 'crimeware as a service (CaaS)'; a classic example is Neosploit toolkit [11].

6.6 Future of Cybercrime

The increase in collaboration among hackers in the design and implementation of new attack vectors is a worrying trend as it is going to make future malware to be more robust, reliable and resistant to computer security countermeasures [34]. This, in turn, is expected to help increase the demand for malware services in the future. Unless deliberate measures are taken, there is very little collaboration between countries and among law enforcement agencies in combating cybercrime [18]. For example, cybercriminals have reportedly made alliances with drug traffickers in countries such as Afghanistan, Columbia and elsewhere where profitable illegal activities are used to support terrorist groups. While existing evidence seems to suggest that cyberterrorism is not advanced enough to conduct well-coordinated cyberattacks, it is well documented that cyberterrorism is heavily involved in cybercrime [35].

It is estimated that only 5 % of cybercriminals are ever arrested or convicted because the anonymity associated with the web activity makes them hard to catch and the trail of evidence needed to link them to a cybercrime is hard to unravel. The challenge of identifying the source of attack is complicated by the unwillingness of organisations to report attacks, owing to potential liability concerns.

Cybercrime requires less personal contact, less need for formal organisation and no need for control over a geographical territory. Some security experts argue that the classical hierarchical structures of organised crime groups may not be suitable for organised crime on the Internet. Consequently, online criminal activity may emphasise lateral, loose and dynamic relationships and networks instead of hierarchies [10]. Instead of assuming stable personal configurations that can persist for years, online criminal organisations may incorporate the 'swarming' model, in which individuals coalesce for a limited period of time in order to conduct a specific task, or set of tasks, and afterwards go their separate ways [36]. It is widely accepted now that, as opposed to the fixed, hierarchical organisational models found in the 'real world', criminal organisation in the cyberspace will be transient, lateral and fluid, all of which can pose real challenges for law enforcement.

6.6.1 Laws, Standards and Ethics

The emerging system of online law enforcement is largely preventive and strongly decentralised, involves a hybrid of public and private enforcement and is highly automated. This new model is far more pervasive than the traditional offline law enforcement by state actors; it tries to achieve ubiquitous policing of online activities to monitor, control, deter, deflect, detect, prevent or pre-empt risky and potentially malicious activities [7]. Very few law enforcement members are trained to deal with digital evidence even at a basic level. Experience has identified the need for stronger sharing of training material and common certification across countries. Since this is a global problem, the training scheme should be adapted worldwide. The training should include expertise on recovering, examining, analysing and reporting digital evidence [24]. For example, cybercrime training in the EU is inconsistent and fragmented, with no mutual recognition of training standards and little or no academic recognition of qualifications and experience.

The EU Convention on Cybercrime was viewed as an attempt to 'harmonise laws against malicious hacking, virus writing, fraud and child pornography on the net'. It also aims to ensure that police forces in separate countries gather the same standard of evidence to help track and catch criminals across borders [37]. The treaty accomplishes three key goals. The first goal is the establishment of a specific list of domestic criminal offences and conduct that are prohibited by the treaty. The second goal is to adopt a set of procedural tools and powers to properly and effectively investigate crimes. The third goal is to establish strong mechanisms for fostering international cooperation [38]. The convention also allows for any nation when signing or ratifying to state declarations or reservations towards any of the obligations tied to the provisions of the convention. However, at the time of writing this chapter, only 23 countries had signed and ratified the convention.

One of the core criticisms of the convention is that it lacks a 'dual criminality' provision. A dual criminality provision would require that for an offence to be considered a crime under the convention, it would have to be a crime in all countries it was committed and in the countries whose assistance is being ask for. This makes cooperation among countries in combating cybercrime to be difficult. The main concern from the EU Convention on Cybercrime is that it is left to the interpretation of the member country. They can decide on which parts to ratify and which not to. This makes the harmonisation efforts to face major difficulties. The regulatory framework lags behind the state of the art in cybercrime. It is not strange to see crimes being prosecuted using laws that are not related at all to cybercrime. In a lot of cases, there are simply no laws to prosecute a specific crime. Cybercrime will present the law enforcement community with new investigative challenges, particularly of technical nature.

6.6.2 Proposed Solutions

There are a number of countries and companies that are researching and developing tools for a cyber offensive. The rationale for developing cyber offensive tools comes from the difficulty in apprehending those involved in cyberattacks because of the anonymity offered by the web and the ability offered by the tools they use in disguising their tracks. The question in using offensive tools is how hard the victims of these attacks should hit back. However, the ethical question is how sure the victims can be that they are hitting the right attacker. This is a problem because a machine of an unaware user may be compromised and used to attack another machine.

New tools are available to public and private actors to detect, investigate and prevent criminal behaviours. These tools can provide sophisticated means for surveillance and powerful ways to analyse a vast amount of information. These tools can deter illegal activities, investigate crime, collate personal information and track criminals with increasing efficiency (with the ease with which these tools are available, it is going to be hard to distinguish between criminals and normal users [39]). In addition, advanced analysis software and intelligent data mining tools can pre-empt crimes by identifying suspicious patterns of behaviour allowing law enforcement agencies to neutralise threats before they are realised. At the same time, these new tools can be a threat to our civil liberties, particularly personal privacy. When they fail to give correct analysis, the consequences are not worth imagining. Some of these tools and methods have gone as far as claiming to be able to predict crime or read criminal thoughts. While these are interesting research agendas, their impact to our civil liberties and human rights must be thought long and hard. For example, the fact that some of these tools are automated and can make decisions without human interventions is a worrying trend [7].

Wiretapping of telephone conversations has been a common practice for a long time; this practice is being extended to even VoIP, and security experts worry that this will stifle innovation or move innovation to other countries with less stringent measures [40]. Dictating the design of new technologies to facilitate law enforcement has its own cost and may even be self-defeating.

In order to help improve the fight against cybercrime, there will also be a criminalisation of hacking tools including possession, creation and distribution, where the conduct is (1) intentional, (2) 'without right' and (3) done with the intent to commit an offence of the type described in Articles 2–5 of the convention [39, 41]. It is further said that Article 6 of the treaty is a cause for concern, due to the fact that it makes production, distribution and use of 'hacking tools' and exploit code illegal [42].

Software industry initiative such as the Software Assurance Forum for Excellence in Code (SAFECode) [43] is a step in the right direction in improving the trust in information and communications technology products and services through the advancement of effective software assurance methods [44]. SAFECode leads the effort to identify and promote best practices for developing and delivering more secure and reliable software, hardware and services.

Digital evidence is an important part in nearly every modern cybercrime investigation, and proper handling of this evidence can mean the difference between a conviction and a cybercriminal walking free [45]. The use of computer forensics in cyberspace criminal investigation is crucial in solving cybercrime.

6.6.3 Paradigm Shift in Combating Cybercrime

The nature and characteristics of cyberattacks give attackers asymmetric advantage because of the following reasons [46]:

1. Attackers do not have sufficient financial support to create or introduce new technology. They use existing technology in innovative ways, finding flaws and using them against the defenders.
2. Attackers operate alone or in small groups and are mobile and flexible in their tactics.
3. Attackers mingle with the general population and have greater mobility and speed than law enforcement and other forces. The absence of consistent international regulations and cooperation and coordination between law enforcement agencies adds to the problem.

4. Attackers strive to take the enemy by surprise, maintain the initiative and leave before the defender is able to fight back.
5. Attackers seek support and cooperation from local population either by sympathy or intimidation. Although one might question the validity of this premise in cyber security, the low investment in security training and expertise and the reluctance to communicate security incidents to law enforcement actually support the attackers' actions.
6. Attackers are knowledgeable about the target environment prior to the encounter. This is especially true when elements within the defending organisation are collaborating with or conducting the attack.
7. The attacker's information service is better than the defender's. In the current environment of cyber security, the defending organisations typically respond in isolation and rarely communicate the attack to law enforcement or other groups. The only information services on the side of the defenders are public organisations and the media, which are shared with attackers.
8. Attackers maintain command of the situation during the whole engagement. The defending organisations can rarely regain control during the attack—first, the attack would need to be detected in real time, then near-perfect coordination of incident response procedure would need to be in place, and the response team would need to be fully alert, trained and supported in the decision-making process. The cost of maintaining such capabilities is overwhelming for most organisations.

The characteristics of the Internet give massive advantage to the attacker. The advantage comes from the original design of the Internet in which security was not a main concern. Then you have loopholes in the law, lack of collaboration and sharing information at the international level and the pace of technological change. All these strongly suggest the need for paradigm shift. Our current approach of patching the Internet, whenever there is a security breach, will not work in the long run.

6.6.4 Crime Convergence

The cost and convenience of carrying business online is bringing other illegal businesses such as drug smuggling and robbery. The best way to solve many of these crimes is to monitor network traffic or to examine the evidence stored within computers. Investigators may need a higher level of technical knowledge. In the end, all criminal activities will use the Internet in one shape or form; when that day arrives, combating cybercrime must become a multifaceted affair incorporating technical prowess, political will, economic muscle and social norms. It will take more than research curiosity of a few organisations or nations. A global concerted effort will be required then, given our inability to cooperate in solving cybercrime now. It is not worth even imagining the change and menace of the cybercrime in the future, should we fail to realise the danger and start acting.

6.7 Discussion and Conclusions

One of the most significant challenges in cybercrime is the sharing of information on cyberthreats or attacks among law enforcement agencies within and between different jurisdictions. This is made even harder by the sensitivity of information on the national security or firm competitive edge. Some of this information shared has resulted in either no conviction because of the different laws in different countries or has led to a country or company being a victim of espionage. The fight against cybercrime will benefit a lot if all stakeholders (intelligence community, law enforcements, government, regional bodies and victims) share information they have on crimes. In some cases, this information can be de-identified in case it is used for research or law/policy formulation

Cybercrime is highly organised nowadays as its motives have changed from individual hacker for fame, seeking limelight, to professional hackers, deploying sophisticated cybercrime models to maximise their profit while avoiding detection. It should be noted that cybercrime is highly sensitive to location, language and regional economic trends; these campaigns enable them to operate locally, focusing their attacks on specific geographic locations and target selected business. The organisational structure of cybercrime is loosely centralised with strong cells distributed around the world specialising in specific tasks that contribute to the overall network. These networks are amorphous making the effort to destroy them difficult as they are loosely coupled. Catching a few criminals creates more gangs which are reorganised differently. Therefore, as our effort to fighting cybercrime intensifies, we are unknowingly creating more cybercriminals with potent ability to fleece us and our financial institutions.

We have to be aware that security which is meant to mitigate the occurrence of cybercrime is a part of a system which is always complex than the individual components [29]. Therefore, when analysing the occurrence of cybercrime, we should do so in the context of the broader system. And the complexity of today's systems does little in mitigating cybercrime. Sean Price [47] proposes a new paradigm on how to win cyberwar which will be based on assurance and knowledge of our complex IT systems. There are some security experts who believe that complexity is one of the biggest threats in cyberspace, because complex systems break in complex ways [48]. Cybercrime causes national security concerns especially when it comes to critical national infrastructure. The growing threat to national security will occur due to more sophisticated web-based espionage. The growing attacks are threatening online services and eroding public trust in Internet services.

The rise of a sophisticated market of software flaws used to carry out attacks and espionage on networked systems will lead to a black market (cybermarket of vulnerabilities) [19]. For example, the Microsoft's stance on piracy together with the implementation of the Windows Genuine Advantage (WGA) further makes all the Windows systems vulnerable as those who have pirated copies of the Windows operating system cannot patch their computers [49, 50].

The proliferation of broadband and advanced wireless technologies has made more users always connected to the Internet and conduct their daily activities electronically; as a result, computer users have become the target of an underground economy that infects hosts with malware or adware for financial gain. Unfortunately, even a single visit to an infected website enables the attacker to detect vulnerabilities in the user's applications and force to download a multitude of malware binaries. Frequently, this malware allows the adversary to gain full control of the compromised systems leading to the exfiltration of sensitive information or installation of utilities that facilitate remote control of the host. This behaviour is similar to the traditional understanding of botnets. However, the main difference is that web-based malware infections are pull based and that the resulting command feedback loop is looser. To characterise the nature of this rising threat, four prevalent mechanisms used to inject malicious content on popular websites are identified: web server security, user-contributed content, advertising and third-party widgets [51].

Laws are written in a very general manner that they are no longer useful to users and investigators. Therefore, when using or obeying the law, users and investigators have to interpret what the law means in their particular situation. However, it is not easy to write laws that will be prescriptive covering all situations in case of violation.

The EU Convention on Cybercrime at its core was enacted to promote international collaboration in combating cybercrime. The fact that crime laws are not harmonised worldwide makes it impossible for the convention to achieve its initial objectives. To date not all EU countries have ratified the treaty because of reservations they have on certain articles of the treaty. We believe that with cybercrime becoming an almost integral part of our cyberspace lives, more countries will ratify the treaty. As the treaty does not provide a one-size-fits-all solution to all cybercrime, it will serve as a framework for discussion and future treaties.

The challenges of cybercrime far outweigh any other challenge that we as society have so far faced. Therefore, it is important that resources are mobilised to combat this scourge. We believe that cybercriminals will continue to shift away from relative visible web presence to private channels that are not easily monitored. The web and other forums online are being targeted by undercover sting operations [32]. Increasing awareness of the Internet users in the effort of combating cybercrime may go a long way in mitigating the problem. For example, the National Strategy to Secure Cyberspace details that user awareness is an important component in the fight against cybercrime [52]. However, an innovative global strategy of carrying out user awareness programmes is required, before Internet users' trust completely disappears. In the future, we would like to see metrics being developed to assess how awareness reduces the effect of cybercrime. However, the search for ways to improve the security of the cyberspace must be increased and fully funded by individuals, organisations and governments.

Countersurveillance in order to study and understand the motives of attackers has not been very productive. Attackers now are mapping the resources (such as sensor networks and honeypots) used by security vendors and research communities and poisoning them with false data. Consequently, the effectiveness and response time of the information gathering resources may be eroded [46]. This further complicates the fight against cybercrime.

The problem for many IT-based companies is that similar norms and understandings common in other industries simply have not emerged. While regular retail security has had over 150 years to develop, web-based e-commerce is, at most, 8 years old and continues to evolve rapidly. IT-dependent companies need to understand this: you cannot routinely expect the police to investigate crimes and recover assets where you yourself have failed to take reasonable precautions. Each player must know what is expected of them in terms of maintaining and achieving individual, organisational and national security [53].

The Internet is also becoming a tool of running the political process as politicians become technological savvy. Internet usage issues are getting more attention, and this brings more unwanted attention from criminals who may be working to sabotage the political process such as election [54]. Bringing the political process to the Internet gives the opportunity of having security problems highlighted to a wider audience. For example, when the email account of US Republican vice-presidential candidate Sarah Palin was hacked, nearly the whole world knew as news media spend hours covering the issue.

Alliances between the public sector and private enterprise are at an initial stage of development with limited success. Public-private partnerships (with government participation or sponsorship in some instances), such as the National Security Telecommunications Advisory Council (NSTAC), the National Security Information Exchange (NSIE) and the Cyber Security Industry Alliance (CSIA), have had some measure of success in executing overarching policies and practices for securing cyberspace in a critical sector. Other nations, such as the UK, have instituted the NSIE model for their own public-private cooperative relationships. Initial attempts at paralleling this collaborative structure in the international law enforcement community are under way but as of yet lack sufficient resources and skills to have substantial impact on the cybercrime juggernaut [2].

Flexibility, collaboration and responsiveness are paramount in combating cybercrime. The miscreants are predisposed to being mobile and elusive, given their freedom of movement within cyberspace. In such an environment, they can operate with impunity and arrogance. Countless conversations in open and public forums demonstrate their absolute lack of fear of law enforcement. Evidence indicates that miscreants are paid to compromise networks, or write tools to compromise networks, or just sell that which they have compromised [2]. In far too many instances, political or cultural impediments preclude the level of cooperation and interaction required to defeat cybercrime.

No matter how hard you try, attackers will, in the long run, have the upper hand. It's a simple fact of life, not an excuse. Security is humans pitted against humans. Therefore, it's critical that you improve your code's security posture at the design, implementation and testing levels.

Inconsistency and fragmentation in laws and policies leave most of the critical services vulnerable to cyberattack. For example, American industry and government are spending billions of dollars to develop new products and technology that are being stolen at little or no cost by adversaries [55]. Things that are being stolen include pharmaceuticals, biotech, IT, engine design and weapons design.

References

1. Hallam-Baker, P.: THE DOTCRIME MANIFESTO: How to Stop Internet Crime. Addison-Wesley, Upper Saddle River, NJ (2008)
2. Cymru, T.: Can we protect ourselves from the hazards of an online world? ACM Queue **4**(9), 24–28 (2006)
3. McEwen, T.J.: Dedicated Computer Crime Units. Diane Pub Co., Darby, PA (1993)
4. Forester, T., Morrison, P.: Computer Ethics: Cautionary Tales and Ethical Dilemmas in Computing. MIT Press, Cambridge, MA (1994)
5. Tavani, H.T.: Ethics and Technology: Ethical Issues in an Age of Information and Communication Technology. Wiley, Hoboken, NJ (2004)
6. Cornwall, H.: Hacker's Handbook. E.A. Brown Co., Alexandria, MN (1986)
7. Balkin, J.M., et al.: Cybercrime: digital cops in a networked environment. In: Balkin, J.M., Noveck, B.S. (eds.) The Information Society Project at Yale Law School. New York University Press, New York (2007)
8. Geer, D.E.: The physics of digital law: searching for counterintuitive analogies. In: Balkin, J. M., et al. (eds.) CYBERCRIME: Digital Cops in a Networked Environment, pp. 13–36. New York University Press, New York (2007)
9. Rice, D.: GEEKONOMICS: The Real Cost of Insecure Software. Addison-Wesley, Reading, MA (2008)
10. Wilson, C.: Botnets, cybercrime, and cyberterrorism: vulnerabilities and policy issues for congress. In: CRS Report for Congress, pp. 1–40 (2008)
11. Ben-Itzhak, Y.: Organized cybercrime. ISSA J. **6**(10), 37–39 (2008)
12. Giles, J., Biever, C.: Gathering privacy storm as ISPs sell web clicks to advertisers. New Sci. **198**(2653), 24–25 (2008)
13. Jesdanun, A.: Targeting ads by tracking Web surfing habits hits a wall. http://www.usatoday.com/tech/products/2008-09-02-web-tracking_N.htm. (2008) [cited 21 Dec 2008]
14. Mtanzania, a local newspaper, cited on July 26, 2013, page 6
15. http://www.tech360magaz.com/2012/07/tanzania-lost-89218-billions-on.html line. Cited on July 2012
16. Mwananchi local news paper on Monday, cited on 16 July 2012
17. http://www.thecitizen.co.tz/News/Hitches-in-fighting-cybercrime/-/1840392/2455418/-/x9w9oq/-/index.html. Source: Wednesday, September 17 2014 at 12:59
18. Martinelli, N.: Think FICO is a credit scoring company? Nope: it's about large-scale analytics. 27 May 2015 (2015)
19. Sidel, R.: Theft of debit-card data from ATMs soars, Thieves are stealing information to make counterfeit plastic. The Wall Street Journal, May 19, 2015 7:41 PM (2015)
20. Bank of Tanzania 2013, Monetary Policy Statement 2013/2014, ISSN 08556-6976

21. Bond, M., Omar Choudary, O., Murdoch, S.J., Skorobogatov, S., Anderso, R.: Chip and Skim: Cloning EMV Cards with the Pre-Play Attack. Computer Laboratory, University of Cambridge, UK, Cambridge, UK (2012)
22. Murphy, N.: Edward Snowden reveals tips for protecting your privacy. http://www.cheatsheet.com/money-career/edward-snowden-reveals-tips-for-protecting-your-privacy.html/?a=viewall on March 2016. 19 Jan 2016
23. Andy Greenberg, Security, Signal, the Snowden, Approved Crypto App Comes to Adroid Time of Publication: 8:06 pm. 8:06 pm, Source: Thinkstock.
24. Stephen, P., Induruwa, A.: Cybercrime investigation training and specialist education for the European Union. In Second International Workshop on Digital Forensics and Incident Analysis (WDFIA 2007), Samos-Greece (2007)
25. Markoff, J.: Internet traffic begins to bypass the U.S. http://www.nytimes.com/2008/08/30/business/30pipes.html. (2008) [cited 21 Dec 2008]
26. O'Connor, O.: The 2nd ISSA/UCD Irish Cybercrime Survey, pp. 1–25. Information Systems Security Association Ireland Chapter/UCD, Dublin (2008)
27. Loibil, T.R.: Identity theft, spyware and the law. In: Information Security Curriculum Development (InfoSecCD) Conference, Kennesaw, GA, USA (2005)
28. Biever, C.: Murky trade in bugs plays into the hands of hackers. New Sci. **194**(2608), 30–31 (2007)
29. Schneier, B.: Schneier on Security. Wiley Publishing, Inc., Indianapolis, IN (2008)
30. Robel, D.: International Cybercrime Treaty: Looking Beyond Ratification. SANS InfoSec Reading Rooms (2007)
31. Cerezo, A.I., Lopez, J., Patel, A.: International cooperation to fight transnational cybercrime. In Second International Workshop on Digital Forensics and Incident Analysis (WDFIA 2007). IEEE Computer Society, Samos-Greece (2007)
32. Fossi, M., et al.: Symantec Report on the Underground Economy. Symantec (2008)
33. Emigh, A., Ramzan, Z.: Overview of crimeware. In: Jakobsson, M., Ramzan, Z. (eds.) Crimeware Understanding New Attacks and Defenses, pp. 1–36. Symantec Press, Cupertino (2008)
34. Espiner, T., Sullivan, D.: Botnets, cybercrime, and cyberterrorism: vulnerabilities and policy issues for congress. In: Wilson C. (ed.) CRS Report for Congress, Washington (2007)
35. Denning, D.: A view of cyberterrorism 5 years later. In: Himma, K.E. (ed.) Internet Security: Hacking, Counterhacking, and Society, pp. 123–140. Jones and Bartlett, Sudbury, MA (2007)
36. Brenner, S.W.: Organized cybercrime? How cyberspace may affect the structure of criminal relationships. N. C. J. Law Technol. **4**(1), 29 (2002)
37. Ward, M.: Cybercrime treaty condemned. http://news.bbc.co.uk/1/hi/sci/tech/1072580.stm. (2000) [cited 21 Dec 2008]
38. Swartz, B.: Multilateral Law Enforcement Treaties. Senate Foreign Relations Committee (2004)
39. Sommer, P.: Criminalising hacking tools. Digit. Investig. **3**, 68–72 (2006)
40. RIPA. Regulation of Investigatory Powers Act 2000. http://www.opsi.gov.uk/acts/acts2000/ukpga_20000023_en_1. (2000) [cited 21 Dec 2008]
41. Council of Europe. Convention on cybercrime. http://conventions.coe.int/Treaty/Commun/QueVoulezVous.asp?NT=185&CL=ENG. (2001) [cited 21 Dec 2008]
42. Meinel, C.P.: Cybercrime treaty could chill research. IEEE Secur. Priv. **2**(4), 28–32 (2004)
43. SAFECODE. Software Assurance Forum for Excellence in Code (SAFECode). http://www.safecode.org/index.php. (2008) [cited 21 Dec 2008]
44. Howard, M.: Fundamental Practices for Secure Software Development: A Guide to the Most Effective Secure Development Practices in Use Today, pp. 1–20. Software Assurance Forum for Excellence in Code (SAFECode), Arlington, VA (2008)
45. Laliberte, S., Gupta, A.: Defend I.T.: Security by Example, p. 384. Addison-Wesley Professional, Boston (2004)
46. McNeil, K., Etges, R.: Cyber warfare and defense strategies. ISSA J. 6–10 (2008)

47. Price, S.: How to win the cyber war. ISSA J. **6**(12), 7 (2008)
48. Schwartz, J.: Who needs Hackers? http://www.nytimes.com/2007/09/12/technology/techspecial/12threat.html?_r = 1&adxnnl = 1&adxnnlx = 1229803518-cYQBQCp4uGjb5UpVgVGNHw. (2007) [cited 12 Sept 2007]
49. Day, O.: Microsoft's stance on piracy affects us all. http://www.securityfocus.com/columnists/484. (2008) [cited 30 Nov 2008]
50. Arbaugh, W.A., Fithen, W.L., McHugh, J.: Windows of vulnerability: a case study analysis. Computer **33**(12), 52–59 (2000)
51. Provos, N., et al.: The ghost in the browser analysis of web-based malware. In: Proceedings of the First Conference on First Workshop on Hot Topics in Understanding Botnets. USENIX Association, Cambridge, MA (2007)
52. Valentine, D.W.: Practical computer security: a new service course based upon the national strategy to secure cyberspace. In: SIGITE'05. ACM, Newark, NJ (2005)
53. Sommer, P.: The future for the policing of cybercrime. Comput. Fraud Secur. **2004**(1), 8–12 (2004)
54. Shannon, M.M.: Shaking hands, kissing babies, and blogging? Commun. ACM **50**(9), 21–24 (2007)
55. Nakashima, E.: Cyber attack data-sharing is lacking, Congress told. (2008) [cited 19 Sept 2008]

Chapter 7
Intrusion Prediction Systems

Mohamed Abdlhamed, Kashif Kifayat, Qi Shi, and William Hurst

Abstract In recent years, cyberattacks have increased rapidly in huge volumes and diversity. Despite the existence of advanced cyber-defence systems, attacks and intrusions still occur. Defence systems tried to block previously known attacks, stop ongoing attacks and detect occurred attacks. However, often the damage caused by an attack is catastrophic. Consequently, the need for improved intrusion detection systems and proposed robust prediction system is more urgent these days. In this chapter, we investigate the intrusion prediction systems to show the need for such system, the insufficiency of the current intrusion detection systems and how prediction will improve the security capabilities for defence systems. A survey of intrusion prediction systems in cybersecurity, the concepts of work and methods used in these systems is presented.

7.1 Introduction

Cyberattacks have the ability to disrupt or destroy computer systems and networks or the information stored on them [1]. Using various degrees of sophistication, an attacker can access a system remotely by the Internet to gain unauthorised privileges or misuse its privileges. This is known as an intrusion and is the attempt to manipulate the confidentiality, integrity or availability of security methods [2].

In recent years, cyberattacks have increased rapidly in huge volumes and diversity. In 2013, for example, over 552 million customers' identities and crucial information were revealed through data breaches worldwide [3]. This growing threat is further demonstrated in the 50,000 daily attacks on the London Stock Exchange [4]. It is predicted that the economic impact of cyberattacks will cost the global economy $3 trillion on aggregate by 2020 [5]. These immense effects and implications have urged the United States Department of Defense to categorise cyberattacks as an act of war, meriting physical military force in response [4]. Such

M. Abdlhamed (✉) • K. Kifayat • Q. Shi • W. Hurst
School of Computing and Mathematical Sciences, Liverpool John Moores University, Byrom Street, Liverpool L3 3AF, UK
e-mail: M.A.Abdlhamed@2013.ljmu.ac.uk

I.M. Alsmadi et al. (eds.), *Information Fusion for Cyber-Security Analytics*, Studies in Computational Intelligence 691, DOI 10.1007/978-3-319-44257-0_7

a categorisation depicts the severe view countries around the globe now have of cyberthreats.

The varieties of systems used to counter intrusions are operated in order to detect, prevent and mitigate attacks and their consequences. Among the existing traditional defence systems, intrusion detection systems (IDSs) are playing a major role. Typically, they are composed of software or hardware that automatically monitor the computer system or the network and perform an analysis function to detect intrusions [2]. There are many types of IDSs as we discuss in this report. However, IDSs are generally grouped into different classes: host-based intrusion detection systems, network-based IDS and distributed intrusion detection systems. All of these systems operate behind firewalls to increase the level of security.

Two of the main methodologies applied in IDS include signature-based and anomaly detection. Signature detection functions by identifying intrusions through correlating real-time data with a known malicious behaviour database. The anomaly-based approach detects intrusions by comparing the current behaviour to a profiled normal behaviour [6].

In cloud computing, for example, intrusion detection and prevention systems are used in different levels and technologies, where each of these technologies has its own limitations and issues [7]. Despite the advantages, like the possibility to use resources remotely and huge volumes of storage, the growth in cloud computing has further exacerbated the threat posed by cyberattacks, as critical infrastructures and digital service providers now have multiple access points to protect [8]. Hence intrusion detection systems in a cloud environment are in need of significant research efforts to enhance performance and overcome the challenges and issues.

Regardless of the existence of advanced cyber-defence systems, attacks and intrusions still occur. Defence systems try to block previously known attacks, stop ongoing attacks and detect occurred attacks; however, often the damage caused by an attack is catastrophic [1, 9]. Consequently, the need for improving intrusion detection systems is more urgent these days. We argue that proposing robust prediction system will help in improving the detection and prevention capabilities of defence systems.

7.2 The Role of Intrusion Prediction

Prediction is used in many areas such as stock market, weather forecasting, health sector and many others. Although plenty of research has been produced by academia, however, intrusion prediction systems are still not widely used in the cyber industry. In this section we will clarify the importance of intrusion prediction from related terms, namely, detection and prevention. Intrusion detection is 'the process of monitoring the events occurring in a computer system or network and analysing them for signs of *intrusions*, defined as attempts to compromise the confidentiality, integrity, availability, or to bypass the security mechanisms of a computer or network' [10]. Intrusion prevention is the process of detecting intrusions and trying

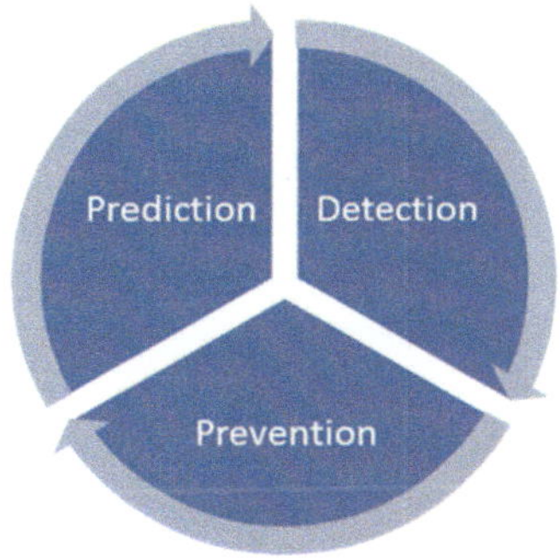

Fig. 7.1 The relationship among prediction, detection and prevention

to stop them [11]. Therefore, according to these definitions, prevention is depending on detection, and both are relying on monitored and identified security incidents, which imply that security incidents already happened or are in the course of happening. It is important to notice that detection and prevention systems need direct information to raise an alarm of security violation or prohibit known attacks. Furthermore, these systems failed to identify multiple step attacks, because detection concept is to capture a single incident at a time, not the whole series. In this context, the need for prediction systems appeared as a tool to support both detection and prevention systems. The basic idea behind prediction concept is the attempt of providing information of events that have not happened yet depending on historical information and gained knowledge of similar events or same events happened in the past. Figure 7.1 shows the relationship among the three concepts.

Prediction as a theory implies a lack of information about what might happen; therefore, security systems use the little amount of information that is available for processing and producing knowledge about potential futuristic incidents. The more data available, the more accurate prediction can be produced, and the certainty of futuristic incidents became more realistic. Prediction role stands where there is not enough information to consider incidents as an attack. This is because detection depends on the solid information to capture security violations; this information might be signatures or noticeable anomaly in the status of the system or network [11]. So the hierarchy of prediction, detection and prevention will be in the form in Fig. 7.2.

Cyberattacks are normally identified as single and multistage attacks, according to the number of steps that is required for achieving the attack successfully. In case of multistage intrusions, prediction is playing a significant role, as the probability of produced correct predictions is far higher than other attack vectors. This is because multistage intrusions involve many steps and attacks in order to achieve the intrusion goal, so there are more likely chances to predict the ultimate attack by observing the initial steps or prior attacks before achieving the attack goals and harms.

Fig. 7.2 Information in intrusion-related systems

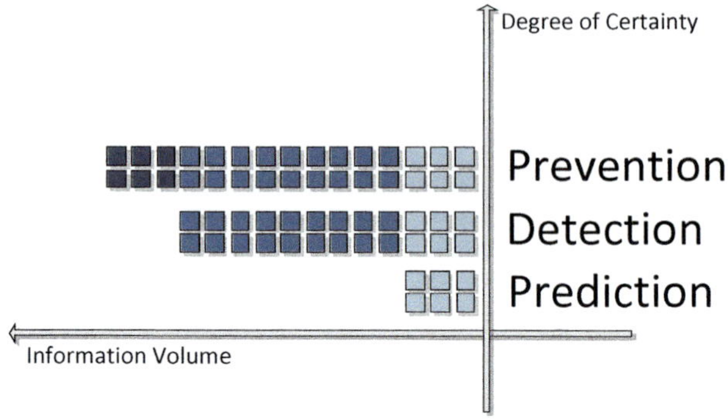

7.3 Prediction in Non-computer Science Areas

Prediction as a concept has been used in many aspects of life in order to get knowledge about future. In some disciplines prediction played a vital role as compared to others, especially when it concerns health or finance such as in medical care or stock markets. As this chapter focuses on the methods of prediction systems, we only present a glib of prediction methodologies used in other sectors. In healthcare sector, many methods have been used such as statistical analysis [12], probabilistic methods [13] and soft computing methods [14] to predict health issues. Prediction is also used in software engineering in order to predict software defects to estimate the number of faults in software before it is deployed in order to measure software quality and decrease maintenance efforts [15]. Based on the fact that states there is a logarithmic relationship between the number of faults and the size of the software's codes [16], most of the prediction models depend on arithmetic formulas and probabilistic models. Arithmetic formulas predict defects in software using size and complexity metrics, testing metrics and processing quality data. The probabilistic models mostly used are multivariate approaches and Bayesian belief networks [15]. In financial sector, the use of different prediction methodologies is not very different from health sector. Many methods has been used to predict stock markets such as time series analysis [17], machine learning algorithms [18] and hidden Markov model [19]. These approaches and others aim to forecast the prices of the stocks and the movement of the market. Failure prediction is another use of prediction concept in financial sector, where models are proposed to calculate the probability of bankruptcy of firms and companies by means of mathematical formulas. These formulas vary from one model to another, but in general, they apply functions and mathematical operations on the collected data and facts to produce the probability [20]. Approaches like market-based models and Z-score model are such examples of failure prediction models which transform over time [21]. The most popular usage of prediction that relates to people's day-to-day use is the weather forecasting and its related rainfall and flood volumes. To predict rainfall, many techniques have been used such as time series analysis, machine learning methods and statistical methods [22].

7.4 Intrusion Prediction Systems in Computer Security

Research on the prediction of cyberattacks and intrusion area has been enriched by many solutions and models over the last few years. In this section, we will review some of such solutions to draw a picture of the possible methods, which could be exploited.

7.4.1 Prediction Methodologies

Prediction systems use different methodologies to infer futuristic potential threats. Methodology means the main concept of work that the system built upon to achieve prediction. Methodologies are varying according to the type of data and source of data. Most of prediction methodologies can be classified as the following:

7.4.1.1 Alerts Correlation

This methodology can be defined as 'interpretation of multiple alarms so as to increase the semantic information content associated with a reduced set of messages' [23]. The goal of alert correlation is to identify the casual relationships between alerts [24]. Basically, alert correlation is achieved in the context of grouping alerts from intrusion detection systems, processing them to be put in a unified form and perform correlation to find a causal relationship between them. In literature, alert correlation is used to build different types of models such as alert optimisation [25], attack scenario [26, 27], attack strategies through finding connection between causes and consequences of attack [28], attack track using a suffix tree [29] and attack plan [27].

7.4.1.2 Sequence of Actions

In this methodology, security incidents are put into an ordered set to reflect the casual or temporal relationship among them. The serial sort of the set will make prediction relatively clear, as breaking the chain or emerging of malicious acts in the series will be considered as intrusion. This concept is used in models that are based on system call sequences [30, 31] and network packet sequences [32]. This methodology includes any set of things or actions that are sorted to accomplish a specific task in the system and any anomaly in this sort will give a prediction of a malicious activity. Alert correlation by concept can be classified under this category. However, alerts are only notifications that hold information about security incidents, which could produce a prediction upon the relationship nature among their information whether it is a temporal, casual or any type of relationships.

7.4.1.3 Statistical Methods

Literature showed that statistical methods are widely for the prediction and forecasting in computing and other areas. Statistical analysis tries to investigate and present the collected data to reveal the underlying patterns and trends [33]. Although statistical analysis might not be suitable for all cyber-related problems because of the linear nature of statistical analysis [34], however, these methods have been used effectively in many fields of cybersecurity such as in data mining [35]. The criteria for using statistical models are the data nature, attack mechanism and system architecture. Upon these crireria, literature showed using different statistical models to suit different solutions such as time series analysis, linear regression, moving average, weighted moving average, exponential smoothing, reliability analysis, etc. [36, 37].

7.4.1.4 Probabilistic Methods

This methodology is used when the amount or type of collected data is not sufficient to produce direct predictions, so these methods used to assign initial probabilities for system variables and then calculate initial outputs. This situation normally occur in the absence of historical data. Another use of this method is to compound with other models in the same system in order to improve the prediction result. Literature showed that the hidden Markov model (HMM) and Bayesian network are the most probabilistic method being used. HMM is ‘a tool for representing probability distributions over sequences of observations’ [38] and has been used in many solutions such as in [25, 39, 40], etc. Bayesian network is ‘a graphical model for representing conditional independencies between a set of random variables’ [38] and has been used in many solutions such as in [31, 41, 42].

7.4.1.5 Feature Extraction

This method is used when the system focuses on a specific piece of information out of the gathered data. Feature extraction is ‘a set of techniques that transform and simplify data so as to make data mining tasks easier’ [43]. Feature extraction can be decomposed into two parts: feature construction that standardises, normalises, filters and extracts local features from the data. The second part is the feature selection that selects relevant and informative features [44]. Although this methodology is embedded in all data mining-based solution, it has been used as a key methodology such as to build the Cyber Attacker Model Profile (CAMP) [45] and real-time intrusion prediction in [46].

7.4.2 *Prediction Systems in Computer Security*

In this section, we will demonstrate the existing solutions and approaches proposed to predict intrusions. We will arrange these solutions according to the methods being used.

7.4.2.1 Hidden Markov Model

HMM is a probabilistic method that is widely used in many applications such as voice recognition and medical diagnostics. HMM offers a solid mathematical structure that could effectively shape a robust theoretical foundation for problem modelling. In HMM, it is easy to translate theoretical ideas to mathematical model, which is an advantageous point over other modelling methods such as neural networks. This is reflected in practice through successful applications that uses HHM [39].

In their research, Haslum et al. [47] propose a distributed intrusion prevention system. This system is composed of many IPSs spread over networks, which communicate through a central controller. The system uses an HMM to identify intrusion and prepare for prediction process. In addition, a fuzzy inference system, for online risk assessment, is employed. HMM functions by passing a set of system states and probability indicators to the fuzzy inference system to assess the risk simultaneously, whereas a fuzzy inference system analyses the risk using threat level, vulnerability and asset values.

Continuing the investigation on prediction methodologies, specifically HMM, we identify that Sendi et al. [25] propose a framework for intrusion prediction, again using the HMM and alert correlation to predict distributed denial-of-service (DDoS) attacks. In their research, HMM is used to derive the interactions between the attacker and the network. The alert correlation is modified as an alert severity to generate prediction alarms of the effective steps of the attack and to enhance the accuracy. The results showed an accurate prediction of DDoSs and a capability to detect other multistep attacks.

Lai-Cheng propose a technology which can predict an intrusion based on the Markov chain [40]. Their research claims to function with high efficiency in real time by using a load-balancing algorithm and statistical prediction model. The heavy network traffic is divided into smaller segments using a balance-paralleling architecture to overcome the problem of handling heavy traffic loads. When considering the prediction, a series of states of the traffic is checked if it occurs in the order of the Markov chain model, and the probability of this occurrence is computed.

Zhengdao et al. investigate a host-based intrusion prediction system, in this case using the hidden semi-Markov model (HsMM) [30]. Firstly, in their research, they define the intrusion detection using the HsMM. Secondly, a case study using a prediction model using HsMM was put forward. This model depends on a differentiation between the normal system call sequences and the intruder.

7.4.2.2 Bayesian Networks

Bayesian networks are powerful graphical methods for modelling and reasoning under uncertainty and complexity of information. It is comprised of a directed acyclic graph (DAG) enabling smooth representation of the domain knowledge as an influence network and a conditional probability table (CPT) that allows to specify the uncertainty related to the relationships among domain variables [48].

In their research, Wu et al. propose a model to predict cyberattacks using a Bayesian network [41]. Specifically, an attack graph is used for vulnerability representation, as well as for detecting possible attack paths. The main aim of their research is to consider environmental factors that influence the probability of an attack taking place in order to calculate the vulnerabilities. The factors, which are considered in their research, include assets in the network, the attack history of the network and the usage condition of the network. Using the attack probability algorithms employed by a Bayesian network, Wu et al. claim to achieve accurate results, although no experiments or results are put forward in their paper.

Continuing the investigation into the Bayesian network for IDS, Feng et.al. propose a method to predict the abnormal events using the plan recognition approach [31]. This approach uses the dynamic Bayesian network theory to predict the intrusion by monitoring the system call sequences. The goal of the system call sequences is to classify data into both normal and intrusion with a high level of accuracy. By using this classification technique, system calls are represented by the states of Bayesian network. The main challenge of their work is scalability; in a big data environment, the performance and efficiency are questionable.

Ishida et.al. propose an algorithm which is able to forecast the increment or decrement of attack levels using the Bayesian inference [42]. The idea behind this research is to predict whether the attacks will increase or decrease based on patterns of daily or weekly activities. Their research claims to achieve good prediction results; however, their approach is fairly primitive and is unable to predict attack type or time.

7.4.2.3 Genetic Algorithms

Genetic algorithms are mathematical procedures used in artificial intelligence applications to learn the natural process of things. In this subsection, the focus of our investigation is on the use of genetic algorithms for enhancing intrusion detection. The aim, again, is to identify techniques that can influence our methodology.

Sindhu et al. propose a framework for the intrusion prediction in networks using an artificial neural network (ANN) and a genetic algorithm combination

[49]. A genetic algorithm is used to train the ANN, where the ANN is structured according to weight optimisation. The results showed that this technique can speed up the learning process and perform better than traditional backpropagation (BP) networks in prediction accuracy. However, the proposed system functions by checking alarm rates rather than vulnerabilities, which is still a statistical prediction and not an actual vulnerability testing. The framework itself is an intrusion detection system that uses the neural network to classify network traffic as a normal or abnormal behaviour. This approach could automatically improve the detection ability. The neural network is trained using the KDDCup'99 data set (which is a widely used data set for the experimental intrusion detection prepared by MIT Lincoln Labs [50]), and the features used in training were selected by the genetic algorithm. The genetic algorithm is used to select the most important features of the network traffic for detection and to adjust the neural network parameters. The results showed an effective outcome of detection.

7.4.2.4 Artificial Neural Networks

Artificial neural networks are defined as parallel and distributed processing systems constructed of a huge number of simple and enormously connected processors [51]. ANNs enhance the intrusion prediction system's ability to generalise data and classify it into normal or abnormal [7]. This means ANNs try to overcome the uncertainty of data by attempting to identify multiple shapes of data.

Zhang and Sun [32] propose a novel network-based model to predict intrusions using a fuzzy neural network and coefficient correlation. In their research, the focus is on a backpropagation algorithm in the fuzzy neural network, to train the network on KDD99 knowledge data set. The key focus of their work is on a solution to predict the intrusion by analysing the network traffic information and forecasting the following. In their research, they claim to produce acceptable levels of prediction to detect different attacks such as DoS, probing, U2R and R2L. However, the network traffic feature selection should be more specific, such that the selected features might not be an indicator of an intrusion, and the prediction probability could be accounted more specifically.

Continuing the case study, we focus on the work by Tang et al. [52], who propose a method for predicting a network security situation based on backpropagation neural with covariance. The aim of this research is depending on training the neural network with situation sequences as inputs, and self-learning parameters adjustments used in backpropagation training. The prediction process was done by the use of a historic and current value situation of the services and host fed to a BP neural network. This method succeeds in using the hyper-plan construction for prediction, because of the prediction decision being based on high-level information.

7.4.2.5 Data Mining

Data mining is another approach to identify the data and relationships among them to produce information for the prediction approach. Data mining is defined as a method for investigating data from multiple perspectives and summarising it in a useful information [53]. Li et al. [54] propose an approach for intrusion prediction using attack graphs generated by data mining techniques. The data mining association rule generates several scenarios for the attack graphs, depending on multistep attack patterns that are built using previous intrusion alerts. This algorithm computes the probability of attack incidence in each attack graph, which indicates all potential attacks. Ranking the attack scenarios according to predictability scores and correlating with real-time intrusion alerts can assess the future attack and predict intrusions.

Kim and Park [26] propose a model to predict the network-based advanced persistent threat (APT) attacks. This model depends on extracting intrusion detection events, analyses the correlation among events and then predicts the intrusion according to the context. The first intrusion data is gathered and treated to extract information about attacks threads and sessions. Subsequently, correlation analysis was done by creating sequential rules of detected intrusion events. The correlation is achieved by the use of a continuous association rule mining algorithm (CARMA). At this point, the researchers suggest two equations to predict the attempted time of intrusion and events occurring. By analysing and correlating intrusion detection events, their research uncovered an association between some specific attacks. Therefore, they could predict the ATPs according to the existence of some attacks. The results showed the possibility of predicting ATP attacks depending on the intrusion detection events.

Cheng-Bin [46] proposed a method for the intrusion prediction based on feature extraction. The researcher's method depends on a proposed algorithm that first filters the relevant data; then a support vector machine (SVM) is used to classify the data into normal or abnormal. The researcher claims that this method is able to predict the intrusion online. The proposed method exploits a machine learning method that is a point of power. However, this method is actually used to detect already existing attacks rather than forecast potential attack, in addition to its moderate accuracy.

Continuing the investigation, we identify that Onolaja et al. propose a conceptual framework for monitoring the trust dynamically and the prediction of the behaviour of network nodes [55]. Their framework employs the paradigm of Dynamic Data-Driven Application Systems (DDDAS) and a trust-based model. The framework consists of two parts: a physical system representing the actual network and its nodes and a controller representing a simulation of the entire network. The controller contains components, which collect data relating to the nodes' behaviour within the network. In addition, it collects a trust value (TV) calculation and performs a data mining and prediction service. A trust value (TV) is attached to each node. Initially this value is neutral and is changed according to the node behaviour. Nodes are distributed according to the trust value to conceptualise three levels of trust: high risk, medium risk and low risk in the network. The prediction was achieved in the controller by comparing the actual behaviour of the node with the previous behaviour stored in the controller, which changes the trust value and detects the misbehaviour.

Jayasinghe et al. [56] proposed an approach to predict a drive-by download attack in the web browser environment using a memory-friendly dynamic analysis. The approach depends on the monitoring of the bytecode stream produced by the web browser engine to render the web page. This is achieved by extracting the features from the stream and then predicts the attack by a data-mining algorithm. The process of capturing and analysing the data stream depends on extracting the intermediate call trace by using opcode function (a java built-in function) into n-gram keywords. These n-grams represent an individual feature and are fed into the data-mining algorithm, which is a support vector machine (SVM), to predict the attacks. The prediction is performed by the SVM to detect new traces that do not exist in the vector space (hyperplane). Results showed this approach's efficiency and effectiveness. However, this approach is not a universal solution since it is limited to a java library and is affected by the code complexity. In addition, the time required to train the classifier may affect the efficiency of the solution against new attacks.

7.4.2.6 Algorithmic Methods

In this subsection, the focus is on the use of algorithmic methodologies and various research areas, which employ these techniques to enhance intrusion prediction system methodologies.

For example, Kannadiga [57] propose an event-based system for predicting network intrusion. This system depends on the idea that some attacks could lead or be the start of a more severe intrusion attempt. The system they propose operates by distributing the attacks into categories, each representing a penetration level of the network. The proposed system collects information from databases about new and previous attack events, hardware and software events information and other attack reports based on the external network. An intrusion prediction engine is employed to store information and predict future attacks by mapping the attack events into relevant categories. A network penetration scenario from correlating attack categories is also built up. The prediction engine calculates the probability of future attacks that belong to the next category of attacks.

Another approach, which uses algorithmic methodologies, is proposed by Feng et al. [31], who detail an approach to predict the abnormal events using the plan recognition technique. This approach uses the dynamic Bayesian network theory to predict the intrusion by monitoring the system call sequences. System call sequences are classified into normal and intrusive, with a clear definition for both. According to this classification, system calls are represented by the states of the Bayesian network.

Pontes et al. [28] proposed a two-stage system to forecast a cyberattack in computer systems. The first stage is comprised of an event analysis system (EAS) that performs a multi-correlation process by analysing and correlating alerts and logs of operating system and intrusion detection and prevention system's logs. Researchers suggested a standard principle of causes and consequences within PC-correlation method, such principle based on the connections between the factors contributed in the attack (causes) and the effects of this attack. The second stage of the proposed system is the forecasting stage, where researchers employed the exponential weighted moving average (EWMA), which is a probabilistic technique to forecast the attack according to the output of stage one. Researchers present a definition for causes and consequences via the PC-correlation method.

Pontes and Guelfi proposed an architecture for forecasting intrusion using collaborative architecture [58]. Researchers did not mention any forecasting methods or techniques, but reference their other Portuguese written work. The basic idea of this research is to divide the process of analysis into four levels, each specialised in their specific part of the network: sensing, analysing and forecasting all over the network and sharing the forecast results. Theoretically, collaborating among different resources is an effective principle to improve the prediction and detection of intrusion. However, doing analysis and forecasting on a multilevel is computationally infeasible. In addition to the ambiguity of collaboration process especially researchers produced different formats of analysed data in different levels, and the variables of network should be sensed.

Grunske and Joyce [59] proposed a risk-based approach to forecast attacks on components of the system. The idea of this method is to construct attack trees for all system modules, so that each attack tree demonstrates all vulnerabilities in that specific module. These attack trees are used to measure the probability of a successful system security breach. Within each attack, three researchers defined the constraints and metrics. The predefined metrics are attacker motivation and ranks and attack risk and cost, and the information to calculate these metrics are collected from the proxy. This approach is applied to the concept of attack profile, where every possible attack is predefined along with the system status and environment factors that enable the attack. Prediction is done by comparing the calculated probability with the attack profile to decide whether the risk of an attack is in an accepted level or considered intrusive.

Park et al. [37] proposed a mechanism (FORE) for forecasting the cyber weather. FORE predicts worm attacks by analysing the randomness in the incoming network traffic. The basic idea behind this mechanism is the incremental nature of the worm propagation that produces more randomness in the network traffic, which is intrusive if exceeding a threshold. Researchers showed that this mechanism is faster than the previous method [60].

Fachkha et al. [36] proposed a model for forecasting the short-term future impact of the current distributed denial of service attack and its features. This model aimed to predict the intensity of the attack (number of packets) on the rate (period in seconds) from the number of compromised machines (size of the attack). The basic idea is collecting backscattered data and session flows from the darknet traffic [61], then applying DDoS detection parameters on collected data to anticipate the DDoS attacks. The anticipation result determined if the DDoS data is predictable and hence to apply the prediction methods. This research's power of point is the exploitation of the darknet traffic.

Abdlhamed et al. [62] proposed a system for intrusion prediction in cloud computing. Researchers employed the concept of using multiple sources of data to produce the prediction. They incorporate multiple powerful techniques that form an efficient system such as game theory concepts, behaviour profiling, risk assessment and statistical methods. These concepts are integrated in component-based framework system to perform prediction. Each concept is represented in one or more components in the framework; for example, components of data acquisition, game-based behaviour builder, dynamic risk builder and historian (Table 7.1).

7.5 Discussions

As stated earlier, the modern system environments need to be equipped in advance for attacks and intrusions, which have increased and become more sophisticated in techniques. Intrusion prediction systems should be the solid base that any proactive defence plan can rely on. This is because of the promising performance that prediction systems achieved on both directions: the number of attacks that have been predicted successfully and the false alarm rate that has been decreased [25, 48, 49].

The variety of methods used to implement the prediction process, as seen in the literature, comprises of many factors; the variables of the system that the solution exploits to make the prediction are crucial for selecting the appropriate method. Another factor is the degree of certainty, where a high degree of certainty means there is a lot of information available and a low degree of certainty means there is a lack of information. In case of a low degree of certainty, we notice prediction systems depend on hidden Markov models, whilst with a high degree of certainty, they tend to use Bayesian networks. Furthermore the type of prediction is an important factor to use a particular method rather than others. For example, solutions dedicated to predict attacks normally use HHMs, whilst devoted for forecasting, the intentions or the abnormal events mostly exploit the Bayesian networks. Other factors might be the simplicity of the theoretical system and HHM's ease in translating the well-planned system into a mathematical model.

Performance is one important factor, where any defence system (including prediction) should not decrease the performance of the protected system and the usage of its resources. There is an efficiency trade-off dilemma; in other words, the main prediction system should be as efficient as possible whilst preserving the high level of performance. The intrusion system should preserve the availability and performance of the protected system.

To enhance the predictability of prediction systems, these systems should be able to recognise attacks in its multiple shapes and variations. At the same time, these systems should have the ability to classify the data into normal and abnormal

Table 7.1 Intrusion prediction systems in cybersecurity

Approach	Solution	Methodology[a]					Tech type	Source of data	Performance	Scalability	Advantages	Drawbacks
		AC	SA	SM	PM	FE						
Hidden Markov model	Haslum et al. [47], Sendi et al. [25], Lai-Cheng [40], Zhengdao et al. [30]	✓	✓		✓		N-based H-based	Prevention systems, network traffic, system logs	High/medium	Static	Easy to translate the well-planned system into a mathematical model, simple computations	Not easy to extend, depends on predefined thresholds. Insufficient for new attacks or variation of old attack
Bayesian networks	Wu et al. [41], Feng et al. [31], Ishida et al. [42]		✓		✓		N-based H-based	Network logs, system logs, detection systems	High	Static	Ability to handle complex distributions in networks	Needs historical data
Genetic algorithms	Sindhu et al. [49]						N-based	Detection systems	Medium	Static	Improve neural network training	Not a standalone approach
Artificial neural networks	Zhang and Sun [32], Tang et al. [52]		✓			✓	N-based	System data, network traffic	Medium	Dynamic	Ability to identify attacks and its variants, ability to give more than single prediction	Time to train, insufficient for online prediction
Data mining	Kim and Park [26], Cheng-Bin [46], Onolaja et al. [55], Jayasinghe et al. [56]	✓				✓	N-based	Network traffic, network situation	High/medium	Dynamic	Ability to predict unknown attacks	High computational complexity

Algorithmic methods	Kannadiga [57], Feng et al. [31], Pontes et al. [28], Pontes and Guelfi [58], Grunske and Joyce [59], Park et al. [37], Fachkha et al. [36]	✓	✓	✓			N-based H-based	Network traffic, network reports, system logs, proxy logs.	High	Dynamic	Predict known and unknown attacks	Uses mixed approaches might combine their disadvantages in the same solution.

[a]Abbreviations: *AC* alert correlation, *SA* sequence of actions, *SM* statistical methods, *PM* probabilistic methods, *FE* feature extraction, *N-based* network-based, *H-based* host-based

(intrusive). This takes the research interest into different areas of computer and mathematical techniques and methods such as data mining and soft computing techniques.

Prediction philosophy is based on work in the situation of uncertainty of intrusion. This means there is a lack of information that the prediction system can use and rely on. The basic idea behind the prediction concept is the attempt of providing information of events that have not happened yet depending on historical information and gained knowledge of similar events or same events that happened in the past.

Artificial neural networks are used because of its various benefits for the prediction process. ANNs have the ability to generalise data and identify multiple shapes of data; the ability to produce more than a single output, which gives the prediction system a wider space of choices; and finally the ability of ANN to train and improve its performance over time. Data mining contributed to the prediction systems by its ability of summarising data and finding relations among data to produce new information that bridge the gap of information deficiency.

There are many contributions in the field of intrusion prediction and detection systems, whether it was on the methodology level or technique level. However, literature showed a very little solid solution for a holistic intrusion detection system and hence prediction system. This is because of the limitations and shortcomings of each model. Neural network suffers from the time required for training, data mining and the issue of computational complexity and so forth. We discuss the intrusion detection system because the vast majority proposed prediction systems are actually dependent on IDSs. This is why any improvements achieved on IDSs are reflected as a consequence on the prediction system.

Detection systems also pose problems such as the false alarm ratio, which researchers work on to decrease it to a very little amount. Anomaly intrusion detection systems adopted the threshold technique as the unique mechanism when it comes to making decisions, which means that these systems are static, since it relies on a fixed threshold. The process of defining threshold is ambiguous, and most of researchers depend on experts to define and measure the threshold. On the other hand, signature-based detection systems are not capable of detecting new or unknown attacks; since attacks should be discovered and known, then a signature of the attack details is then made to prevent future attacks of that signature.

7.6 Conclusions

The rising number of cyberattacks, the sophistication in techniques used to perform those attacks and the various types of attackers that initiate attacks for different purposes in a rapidly moving computer revolution are the major challenges not only for security specialists but also for all computer environment-related parties. The escalation of threats to cybersecurity also indicates the insufficiency of current defence systems. This implies to improve the working systems and develop our ways to defend. Defeating cyber intrusions involves enormous efforts and planning. The goal of prediction systems is to have sufficient knowledge of potential security threats (and hence possible countermeasures) prior to attacks launching against the target system. The challenge is to obtain this kind of information in advance to the intrusion. One way to do that is the use of prediction.

This chapter presents an introduction to intrusion prediction systems and clarifies the importance of such defence systems. As stated earlier, the need for gaining information about potential attacks or hazards is becoming crucial for computer systems. The limitations and shortcomings of intrusion detection systems (IDSs) magnify the need for prediction systems. Many solutions have been proposed to tackle this problem. Different methods have been used such as machine learning and data mining and probabilistic, statistical and algorithmic methods. Prediction systems use different types of data such as IDS alerts and notifications, system logs, memory dumps, network traffic, historical data, system/user behaviour, etc. Researchers claimed the ability to predict intrusions with a proportion of the false alarm ratio.

References

1. Waxman, M.C.: Cyber-attacks and the use of force: back to the future of article 2(4). Yale J. Int. Law **36**, 421–458 (2011)
2. Garrett, B.N.: Taming the Wild Wild Web: twenty-first century prize law and privateers as a solution to combating cyber-attacks. Univ. Cincinnati Law Rev. **81**(2), 684–706 (2013)
3. Wood, P., Nahorney, B., Chandrasekar, K., Wallace, S., Haley, K.: Internet Security Threat Report, vol. 19. Symantec Corp, Mountain View, CA (2014)
4. Tomaso, M.: BP fights off up to 50,000 cyber-attacks a day: CEO. http://www.cnbc.com/ [Online]. http://www.cnbc.com/id/100529483#. Accessed 19 Nov 2014
5. Chinn, D., Kaplan, J., Weinberg, A.: Risk and Responsibility in a Hyperconnected World: Implications for Enterprises. McKinsey Co., New York City, NY (2014)
6. Wu, S.X., Banzhaf, W.: The use of computational intelligence in intrusion detection systems: a review. Appl. Soft Comput. **10**(1), 1–35 (2010)
7. Modi, C., Patel, D., Borisaniya, B., Patel, H., Patel, A., Rajarajan, M.: A survey of intrusion detection techniques in Cloud. J. Netw. Comput. Appl. **36**(1), 42–57 (2013)
8. Wang, H., Zhou, H.: The research of intrusion detection system in cloud computing environment. Adv. Multimedia Softw. Eng. Comput. **1**, 45–49 (2012)
9. Ginsburg, A., Santos, L.J., Scoboria, E., Scoboria, K., Yeoh, J.: The Notorious Nine: Cloud Computing Top Threats in 2013, pp. 1–14. Cloud Security Alliance, San Jose, CA (2013)
10. Bace, R., Mell, P.: NIST special publication on intrusion detection systems NIST special publication on intrusion detection systems. Natl. Inst. Stand. Technol. **800-94**, 1–51 (2011)
11. Scarfone, K., Mell, P.: Guide to Intrusion Detection and Prevention Systems (IDPS) Recommendations of the National Institute of Standards and Technology. National Institute of Standards and Technology, Gaithersburg, MD (2007)
12. Shorr, A.F., Zilberberg, M.D., Micek, S.T., Kollef, M.H.: Prediction of infection due to antibiotic-resistant bacteria by select risk factors for health care-associated pneumonia. JAMA Intern. Med. **168**(20), 2205–2210 (2008)

13. Yang, Q., Khoury, M.J., Botto, L., Friedman, J.M., Flanders, W.D.: Improving the prediction of complex diseases by testing for multiple disease-susceptibility genes. Am. J. Hum. Genet. **72**(3), 636–649 (2003)
14. Sudha, A.S.A., Gayathri, P., Jaisankar, N.: Utilization of data mining approaches for prediction of life threatening diseases survivability. Int. J. Comput. Appl. **41**(17), 51–55 (2012)
15. Fenton, N.E., Centre for Software Reliability, London, UK, Neil, M.: A critique of software defect prediction models. IEEE Trans. Softw. Eng. **25**(5), 675–689 (1999)
16. Hatton, L.: Reexamining the fault density-component size connection. IEEE Softw. **14**(2), 89–97 (1997)
17. LeBaron, B., Arthur, W.B., Palmer, R.: Time series properties of an artificial stock market. J. Econ. Dyn. Control **23**(9–10), 1487–1516 (1999)
18. Shen, S., Jiang, H., Zhang, T.: Stock Market Forecasting Using Machine Learning Algorithms, pp. 1–5. Department of Electrical Engineering, Stanford University, Stanford, CA (2012)
19. Hassan, M.R., Nath, B.: Stock market forecasting using hidden Markov model: a new approach. 5th International conference on intelligent systems design and applications (ISDA'05), pp. 192–196 (2005)
20. Agarwal, V., Taffler, R.: Comparing the performance of market-based and accounting-based bankruptcy prediction models. J. Bank Finance **32**(8), 1541–1551 (2008)
21. Režňáková, M., Karas, M.: Bankruptcy prediction models: can the prediction power of the models be improved by using dynamic indicators? Proc. Econ. Finance **12**(14), 565–574 (2014)
22. Toth, E., Brath, A., Montanari, A.: Comparison of short-term rainfall prediction models for real-time flood forecasting. J. Hydrol. **239**(1–4), 132–147 (2000)
23. Gardner, R.D., Harle, D.A.: Methods and systems for alarm correlationProc. GLOBECOM'96. IEEE Glob. Telecommun. Conf. **1**, 136–140 (1996)
24. Sadoddin, R., Ghorbani, A.: Alert correlation survey : framework and techniques. Proceedings of the 2006 international conference on privacy, security and trust: bridge the gap between PST technologies and business services, pp. 1–10 (2006)
25. Shameli Sendi, A., Dagenais, M., Jabbarifar, M., Couture, M.: Real time intrusion prediction based on optimized alerts with Hidden Markov Model. J Netw. **7**(2), 311–321 (2012)
26. Kim, Y.-H., Park, W.H.: A study on cyber threat prediction based on intrusion detection event for APT attack detection. Multimedia Tools Appl. **71**(2), 685–698 (2014)
27. Farhadi, H., Amirhaeri, M., Khansari, M.: Alert correlation and prediction using data mining and HMM. ISC Int. J. Inf. Secur. **3**(2), 77–101 (2011)
28. Pontes, E., Guelfi, A.E., Kofuji, S.T., Silva, A.A.A., Guelfi, A.E.: Applying multi-correlation for improving forecasting in cyber security. In: The sixth international conference on digital information management (ICDIM), pp. 179–186 (2011)
29. Fava, D.S., Byers, S.R., Yang, S.J.: Projecting cyberattacks through variable-length Markov models. IEEE Trans. Inf. Forensic Secur. **3**(3), 359–369 (2008)
30. Zhengdao, Z., Zhumiao, P., Zhiping, Z.: The study of intrusion prediction based on HsMM. 2008 I.E. Asia-Pacific services computing conference, pp. 1358–1363 (2008)
31. Feng, L., Guan, X., Guo, S., Gao, Y., Liu, P.: Predicting the intrusion intentions by observing system call sequences. Comput. Secur. **23**(3), 241–252 (2004)
32. Zhang, G., Sun, J.: A novel network intrusion attempts prediction model based on fuzzy neural network. Lect. Notes Comput. Sci. **3991**(2002), 419–426 (2006)
33. Bienkowski, M., Feng, M., Means, B.: Enhancing Teaching and Learning Through Educational Data Mining and Learning Analytics: An Issue Brief, pp. 1–57. SRI International, Washington, DC (2012)
34. Ramasubramanian, P., Kannan, A.: Quickprop neural network short-term forecasting framework for a database intrusion prediction system. Artif. Intell. Soft. Comput. **3070**(1), 847–852 (2004)
35. Alampalayam, S.P., Kumar, A.: Predictive security model using data mining. Globecom **502**, 2208–2212 (2004)

36. Fachkha, C., Bou-Harb, E., Debbabi, M.: Towards a forecasting model for distributed denial of service activities. In: 2013 I.E. 12th international symposium networking and computer application, pp. 110–117, Aug 2013 (2013)
37. Park, H., Jung, S.-O.D., Lee, H., In, H.P.: Cyber weather forecasting forecasting unknown internet worms using randomness analysis. IFIP Adv. Inf. Commun. Technol. **376**, 376–387 (2012)
38. Ghahramani, Z.: An introduction to hidden Markov models and Bayesian networks. Int. J. Pattern Recognit. Artif. Intell. **15**(1), 9–42 (2001)
39. Baruah, P., Chinnam, R.B.: HMMs for diagnostics and prognostics in machining processes. Int. J. Prod. Res. **43**(6), 1275–1293 (2005)
40. Lai-cheng, C.: A high-efficiency intrusion prediction technology based on Markov chain. In: International conference on computational intelligence and security workshops, pp. 522–525 (2007)
41. Wu, J., Yin, L., Guo, Y.: Cyber attacks prediction model based on Bayesian network. In: 2012 I.E. 18th international conferences parallel and distributed systems, pp. 730–731, Dec 2012 (2012)
42. Ishida, C., Arakawa, Y., Sasase, I., Takemori, K.: Forecast techniques for predicting increase or decrease of attacks using Bayesian inference. In: 2005 I.E. Pacific Rim Conference on communications, computers and signal processing, 2005. PACRIM, pp. 450–453 (2005)
43. Liu, H., Motoda, H.: Feature Extraction, Construction and Selection: A Data Mining Perspective. Springer, New York, NY (1998)
44. Guyon, I., Elisseeff, A.: An introduction to feature extraction. In: Guyon, I.M. (ed.) Feature Extraction, Foundations and Applications, p. 24. Springer, Berlin (2006)
45. Watters, P.A., McCombie, S., Layton, R., Pieperzyk, J.: Characterising and predicting cyber-attacks using the Cyber Attacker Model Profile (CAMP). J. Money Laund. Control **15**(4), 430–441 (2012)
46. Cheng-Bin, L.: A new intrusion prediction method based on feature extraction. In: Second international workshop on computer science and engineering, pp. 7–10 (2009)
47. Haslum, K., Abraham, A., Knapskog, S.: DIPS: a framework for distributed intrusion prediction and prevention using hidden Markov models and online fuzzy risk assessment. Third Int. Symp. Inf. Assur. Secur. **2007**, 183–190 (2007)
48. Tabia, K., Leray, L.: Bayesian network-based approaches for severe attack prediction and handling IDSs' reliability. In: 13th international conference, IPMU 2010, Dortmund, Germany, 28 Jun to 2 Jul 2010. Proceedings, part II, pp. 632–642 (2010)
49. Sindhu, S.S.S., Geetha, S., Sivanath, S.S., Kannan, A.: A neuro-genetic ensemble short term forecasting framework for anomaly intrusion prediction. 2006 International conference advanced computing & communication, pp. 187–190 (2006)
50. KDD-CUP-99 Task Description. [Online]. https://kdd.ics.uci.edu/databases/kddcup99/task.html. Accessed 27 Apr 2015
51. Poojitha, G., Kumar, K., JayaramiReddy, P.: Intrusion detection using artificial neural network. Second international conference on computing, communication and networking technologies, pp. 1–7 (2010)
52. Tang, C., Xie, Y., Qiang, B., Wang, X., Zhang, R.: Security situation prediction based on dynamic BP neural with covariance. Adv. Control Eng. Inf. Sci. **15**, 3313–3317 (2011)
53. Jaiganesh, V., Mangayarkarasi, S., Sumathi, P.: Intrusion detection systems: a survey and analysis of classification techniques. Int. J. Adv. Res. Comput. Commun. Eng. **2**(4), 1629–1635 (2013)
54. Li, L., Lei, J., Wang, L., Li, D.: A data mining approach to generating network attack graph for intrusion prediction. In: Fourth international conference on fuzzy systems and knowledge discovery (FSKD 2007), no. Fskd, pp. 307–311 (2007)
55. Onolaja, O., Bahsoon, R., Theodoropoulos, G.: Conceptual framework for dynamic trust monitoring and prediction. Proc. Comput. Sci. **1**(1), 1241–1250 (2012)

56. Jayasinghe, G.K., Shane Culpepper, J., Bertok, P.: Efficient and effective realtime prediction of drive-by download attacks. J. Netw. Comput. Appl. **38**, 135–149 (2014)
57. Kannadiga, P., Zulkernine, M., Haque, A., Canada, B.: E-NIPS: an event-based network intrusion prediction. In: Proceedings of the 10th international conference, ISC 2007, Valparaíso, Chile, 9–12 Oct 2007, pp. 37–52 (2007)
58. Pontes, E., Lsi, P., Paulo, S.: IFS – intrusion forecasting system based on collaborative architecture. In: Fourth international conference on digital information management, 2009. ICDIM 2009, pp. 216–221 (2009)
59. Grunske, L., Joyce, D.: Quantitative risk-based security prediction for component-based systems with explicitly modeled attack profiles. J. Syst. Softw. **81**(8), 1327–1345 (2008)
60. Park, H., Lee, H.: Detecting unknown worms using randomness check. Inf. Netw. Adv. Data Commun. Wirel. Netw. **3961**, 775–784 (2006)
61. Bailey, M., Cooke, E., Jahanian, F., Myrick, A., Sinha, S., Arbor, A.: Practical darknet measurement. In: 2006 I.E. conference on information sciences and systems, 2007, pp. 1496–1501 (2007)
62. Abdlhamed, M., Kifayat, K., Shi, Q., Hurst, W.: A system for intrusion prediction in cloud computing. In: Boubiche, D.E., et al. (eds.) Proceedings of the International Conference on Internet of Things and Cloud Computing (ICC 2016), pp. 1–9. University of Cambridge, Cambridge (2016)

Chapter 8
Analytics for Network Security: A Survey and Taxonomy

Kaj Grahn, Magnus Westerlund, and Göran Pulkkis

Abstract IT operations produce data such as log files, events, packets, configuration data, etc. Security attacks, for example, an intrusion, can be detected and mitigated by analyzing and finding abnormal patterns from collected data. Intelligent and effective algorithms are needed for analyzing the massive amount of unstructured data created within computing networks. This has motivated research on and development of information analytics like tools, solutions, and services for network security.

Processing of the vast amount of monitoring data from agents and sensors in an intrusion detection system (IDS), updating a database with this data, combining this data with this rapidly growing database, and applying a decision support database for real-time responses require the use of technologies for processing unstructured big data in agents, sensors, and management servers. The best defense against intrusion is prevention. Forensic information about intrusion attack sources can be used for blacklisting corresponding network addresses and for reconfiguring network firewalls to prevent communication from these sources. An IDS should prevent malicious communication to a network. Anomaly-based intrusion detection identifies deviations from normal and typical user activity, communication patterns, application behavior, etc. Different users can have different profiles of normal activity. Normal behavior in a network must be learned from a training dataset or from monitoring communication and activity in a network. Anomaly identification can therefore detect also previously unknown intrusion attempt types. Forensic investigations after intrusion/intrusion attempts manage at least the same dataset that was processed before an intrusion response. By capturing, recording, and analyzing network events, the attack source can be found by using big data tools and real-time analysis techniques.

An advanced persistent threat (APT) is a targeted attack with a low profile during a long time. The purpose is to keep a target network unaware of the ongoing intrusion. APT attackers often use stolen user credentials and/or zero-day exploits to avoid triggering IDS responses. Big data analytics tools are particularly suitable for APT detection. To detect APT attacks, collection and correlation of large

K. Grahn • M. Westerlund • G. Pulkkis (✉)
Arcada University of Applied Sciences, Helsinki, Finland
e-mail: kajg@arcada.fi; westerma@arcada.fi; goran@arcada.fi

I.M. Alsmadi et al. (eds.), *Information Fusion for Cyber-Security Analytics*, Studies in Computational Intelligence 691, DOI 10.1007/978-3-319-44257-0_8

quantities of diverse data including internal data sources and external shared intelligence data is a necessity. Long-term historical correlation to incorporate a posteriori attack information in the history of a network must be performed.

Network security analytics uses big data software technologies like *Hadoop* and *Apache Mahout* which extend the *MapReduce* programming model. *MapReduce* decomposes data into smaller pieces, which are processed on their own network hosts instead of being moved to other network nodes for processing. *Hadoop* can process stored big data but cannot process data streams. The open-source, distributed, scalable, and fault-tolerant real-time processing system *Storm* and some commercial platforms can process big data streams. *Apache Mahout*, implemented on the top of *Hadoop*, is a machine learning software library.

Taxonomies have been proposed for information analytics and for intrusion detection/response systems. Our proposed security analytics taxonomy includes *descriptive analytics* for identification of network security threats, *diagnostic analytics* for forensics, *predictive analytics* for proactive intrusion prevention, *prescriptive analytics* for protection against malware, firewalls, protected network communication, recovery after security incidents, and *network security visualization*.

8.1 Introduction

Network and information security means the protection of networks and network hosts against threats like intrusion, eavesdropping, phishing, malicious programs, malicious communication, etc. The traditional network and information security tools, solutions, and services scale insufficiently, when protection against such threats requires management of very large unstructured datasets. As current networks are becoming rapidly more pervasive and ubiquitous, the amount of data to be managed by security solutions is also growing. Verma et al. [1] motivate current significance of data analytics for cyber security with:

- The lack of exact algorithmic solutions to security problems such as intrusion detection and resilience to malware, phishing, and other threats.
- Design and development of much current security software before the Internet age by developers unaware of Internet-related security threats.
- The noise and the necessity of good natural language understanding related to text data in emails and in social media.
- The huge amount of source data in automatic network activity logs, from network-connected sensors, and in processing traces of network-connected processing devices.

Therefore, new data analytics like security tools, solutions, and services are needed.

In the near future, we will have self-securing networks. The Cloud Security Alliance (CSA) describes the evolution of data analytics of security in three stages in the following way [2]:

- *First generation* is intrusion detection systems—Work toward layered security with reactive security and breach response is implemented because 100 % protective security is impossible.
- *Second generation* is security information and event management (SIEM) systems which aggregate and filter alarms from intrusion detection sensors and apply rule-based responses to them.
- *Third generation* is adding big data tools to SIEM—Big data tools shorten response time by combining in real-time various security data sources using big data analytics, including historical data for forensic purposes. Network forensics can provide evidence pointing to the source of a data breach and can be valuable in highlighting anomalies and pointing to a data breach in process or at least soon enough to trigger a response.

8.1.1 Role of Big Data in Network Security

Real-time processing of the vast amount of monitoring data from agents and sensors in an IDS, combining new monitoring data with stored monitoring data in a history database, updating this rapidly growing database with the new monitoring data, and applying a decision support database for real-time responses and alerts require the use of technologies for processing unstructured big data. The number of network events to monitor explodes in large dynamic networks. The size of logged information to be real-time analyzed can grow even exponentially.

Typical big data applications can be real-time fraud detection, complex competitive analysis, call center optimization, consumer sentiment analysis, intelligent traffic management, and management of smart power grids [3].

Big data is defined as datasets unsuitable to be processed using conventional database methods. Big data has been characterized by the dimensions volume, velocity, variety, veracity, variability, complexity, and value. Volume refers to the size of the data; velocity refers to the speed at which data needs to be processed, e.g., real time; variety refers to heterogeneity related to range and the types of structured and unstructured data; veracity means that also data sources with some inherent unreliability or uncertainty—for example, social media—can be treated as credible and honest; variability refers to variations in dataflow rates; complexity refers to a huge number of different sources from which data is generated; and value is a data-related measure which increases in processing of large volumes of original data [4].

When derived from trusted sources and with the right analytics, big data can deliver richer insight, because multiple sources and transactions to uncover hidden patterns and relationships are in use. Big data is ideal for information security.

Identifying an unauthorized user from an unknown IP address and a denial of service (DOS) attack are examples of data pattern recognition. The same techniques allow you to analyze network traffic to reveal anomalies that point to a data breach.

8.2 Analytics Taxonomy

A methodology for taxonomy development is proposed in [5] and applied in [6] to a taxonomy for analytics on the cloud. Kim [7] proposes a minimal big data analytics taxonomy. We extend the analytics taxonomy proposal in [8] with diagnostic and visual analytics:

- *Descriptive analytics* or *data mining* uncovers from dataset patterns that offer insight. For example, a company can assess credit risks and categorize customers by their likely product preferences in the sales cycle [9]. Descriptive business analytics can also be called business intelligence [8].
- *Diagnostic analytics* is used for discovery or to determine why something happened. For example, for a marketing campaign in social media, a company can assess the number of posts, mentions, followers, fans, page views, reviews, pins, etc. The online mentions can be distilled into a single view to see what worked in the past campaign and what didn't work.
- *Visual analytics* is described by Thomas et al. [10] as "the science of analytical reasoning facilitated by interactive visual interfaces." It supports perception of patterns, trends, structures, and exceptions also in very big and complex data sources [11]. Visual analytics is particularly valuable in the abstraction of multivariate temporal data [12]. Kleim et al. [13] describe formally the visual analytics process:
  - The visual analytics input consists of heterogeneous data sources such as the Internet, newspapers, books, scientific experiments, expert systems, etc.
  - Datasets are chosen from these rich sources.
  - Each dataset consists of a set of attributes.
  - The goal or output is an insight, which is either directly obtained from the set of created visualizations or through confirmation of some hypotheses.
- *Predictive analytics* uses datasets to identify past patterns to predict the future. Using past financial performance of a company to predict future financial performance is a typical example. A company can use predictive analytics for the entire sales process, analyzing lead source, number of communications, types of communications, social media, documents, CRM data, etc. to support sales, marketing, or other types of complex forecasts [3].
- *Prescriptive analytics* focuses to answer specific questions to propose the best action alternative using optimization, simulation, and heuristics in modeling decisions. For example, in the health-care industry, you can measure the number of patients who are clinically obese, add filters for factors like diabetes and LDL,

and determine where to focus treatment. The same procedure can be applied to almost any industry target group or problem. Prescriptive analytics is not much in use. Currently, 13 % of organizations are using predictive analytics, but only 3 % are using prescriptive analytics [3, 8].

8.3 Intrusion Prevention Analytics

The best defense against intrusion is prevention. Forensic information about sources of intrusion attacks can be used for intrusion prevention, for example, by blacklisting corresponding network addresses or by reconfiguring network firewalls to deny access for further communication from these sources.

An intrusion prevention feature of an IDS is prevention of malicious data communication from entering a network or a network host. This feature has however the drawback that a false positive can prevent legitimate data communication from reaching a network or a network host [14]. Problems of knowledge representation, environmental perception, and self-learning associated with a cognitive intrusion prevention system are discussed in [15].

Implementation of intrusion prevention services for Infrastructure as a Service clouds using hypervisor-based network traffic monitoring is surveyed in [16].

In [17], a reverse proxy is used to prevent SQL injection and other intrusion attacks from reaching a web server.

In [18], a web server intrusion prevention system is designed consisting of a set of secondary nodes connected to the web server and a primary node connected to each secondary node. The primary node receives all requests to the web server and applies load balancing to distribute them to the secondary nodes, which detect and block malicious requests using rules created by a rule-based expert library.

8.4 Intrusion Detection Analytics

In the following section, we assess the main intrusion detection types that exist. These detection types are signature-based detection, anomaly-based detection, and stateful protocol analysis. Intrusion detection analytics can be considered sensitive to a malicious adversary which misleads a classifier by "poisoning" its training data with carefully designed attacks [19]. Therefore strict measures should be taken in both design of models and use of datasets.

8.4.1 Signature-Based Detection

Signature-based intrusion detection is based on the identification of known patterns of data communication and network/network host activity in earlier recognized intrusion incidents/attempts. This intrusion detection method is straightforward and fast since it is based on matching of predefined patterns and on applying predefined rules. However, earlier unknown intrusion attempts and slightly modified known intrusion attempts cannot be detected with signature-based methods.

8.4.2 Detection of Anomalies

Anomaly-based intrusion detection identifies deviations from normal and typical user activity, data communication patterns, application behavior, etc. Different users can have different profiles of normal activity. Normal behavior in a network or in a network host must be learned from a training dataset or from monitoring the data communication and activity in a network or in a network host. The benefit of anomaly-based detection is the ability to detect previously unknown intrusion attempt types.

A botnet is a set of malware-infected network hosts called bots, which are controlled by an attacker network host called the botmaster. The BotCloud research project [20] has developed big data analytics tools for identification of malware-infected hosts participating in a botnet. Botnet detection consists of the following steps (see Fig. 8.1):

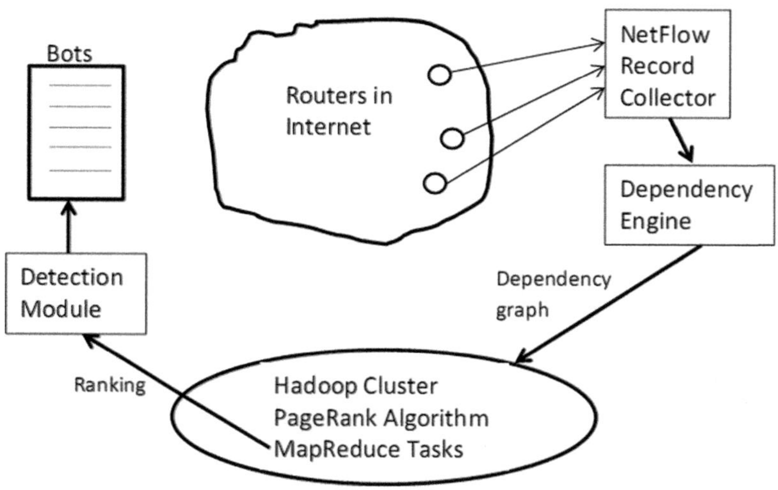

Fig. 8.1 Block scheme of BotCloud framework (Adapted from [20])

- A collector host gathers NetFlow [21] records from routers.
- A dependency engine analyzes the interactions between network hosts and uses the analysis results to create a dependency graph.
- The dependency graph is processed with the PageRank [22] algorithm in order to find out network hosts interconnecting themselves as within a P2P network.
- The adjacency matrix of the dependency graph is distributed among all data nodes of a Hadoop cluster for PageRank execution on Hadoop, in which MapReduce tasks are executed in all Hadoop cluster nodes in order to create a ranking of network hosts.
- A Detection Module analyzes the ranking and identifies bot nodes as highly ranked.

Experimental evaluation of the BotCloud framework has been carried out on NetFlow data delivered by a major ISP in Luxemburg. This NetFlow dataset consisted of 760×10^6 NetFlow records involving 16×10^6 network hosts, altogether a 77 GB dataset. The calculated dependency graph contained 57×10^6 links. The Hadoop cluster consisted of a master node and 11 slave nodes (5 Intel Core 2 Duo 2.13 GHz with 4 GB of memory and 6 Intel Pentium 4 3 GHz with 2 GB of memory). PageRank used three different P2P protocols in creating a ranking of all network hosts. In all evaluations a bot detection rate of nearly 100 % was achieved for a false-positive rate of less than 15 % [20].

An anomaly related to security breaches of data communication or stored data protected by encryption is unauthorized use of secret and private encryption keys. For PKI-based encryption such as TLS/SSL encrypted data communication and S/MIME and PGP, this anomaly can be detected by including certification checks in security analytics applications.

8.4.3 *Stateful Protocol Analysis*

In stateful protocol analysis, the use of a network protocol is compared in each protocol state to predefined specifications on how the protocol should/should not be used. For example, the same protocol activity can be determined as benign for an authenticated user and suspicious for an unauthenticated user. Stateful protocol analysis is quite complex and therefore resource intensive. Another drawback of this detection type is that intrusion attempts based on acceptable protocol behavior are not detected [23].

8.5 Management of Intrusion Events

Handling exceptional events plays a central role in network management. Alarms indicate exceptional states or behaviors. Such events may include intrusion events, congestion, errors, and component failures. A problem is often manifested through a large number of alarms. A severe problem is an event storm (event burst). Fault effects are simultaneously detected, and the management tries to notify symptoms of the problem. The problem is addressed by event correlation tools. Event correlators try to condense many events each with little information to few meaningful events. They passively investigate events to perform a broad first-level localization for a wide range of faults. Classical filtering based on filter patterns is widely used but is not sophisticated enough.

A security event manager (SEM) handles collection, combination, and correlation of logs (events) and data in real time. Essential parts of SEM are intrusion detection system (IDS), intrusion prevention system (IPS), vulnerability scanning, and network and application/user firewalls. A security information manager (SIM) delivers more historical analysis and reports security event data. Event and data collection/correlation, indexed repository for log data, and flexible query/reporting capabilities are supported but not in real time. A combination of SEM and SIM is a security information and event manager (SIEM) [24].

The functionality of a general SIEM can be expressed by the "five Cs" Collection, Consolidation, Correlation, Communication, and Control [25]:

- **Collection**. Log data is collected from a large amount of different kinds of devices. Data communication from a log source to a SIEM needs to be confidential, authenticated, and reliable.
- **Consolidation**. Different types of log formats make it desirable to normalize data to a given format. Thereafter, the aggregation process starts. Different types of events that are of the same type are put together.
- **Correlation**. Different log events are put together to form an attack. The process is processing intensive since threat information from online databases must be downloaded and analyzed to understand an attack.
- **Communication**. Administrators are informed by the SIEM in three ways. An alert is set to the administrator when something is wrong, a report is sent at a predetermined time, or the administrator is actively monitoring the SIEM.
- **Control**. During analysis data is normally stored online. When not in use, the data can be stored in normalized, compressed, and/or encrypted form.

An abstract view of a SIEM is shown in Fig. 8.2.

The underlying principle of a SIEM system is that relevant data about an enterprise's security is produced in multiple locations. It is possible to look at all the data from a single point of view. This approach makes it easier to spot trends and see patterns that are out of the ordinary. The system centralizes the storage and interpretation of logs and allows near real-time analysis. Security personnel may then quickly take defensive actions. Data is collected into a central repository for

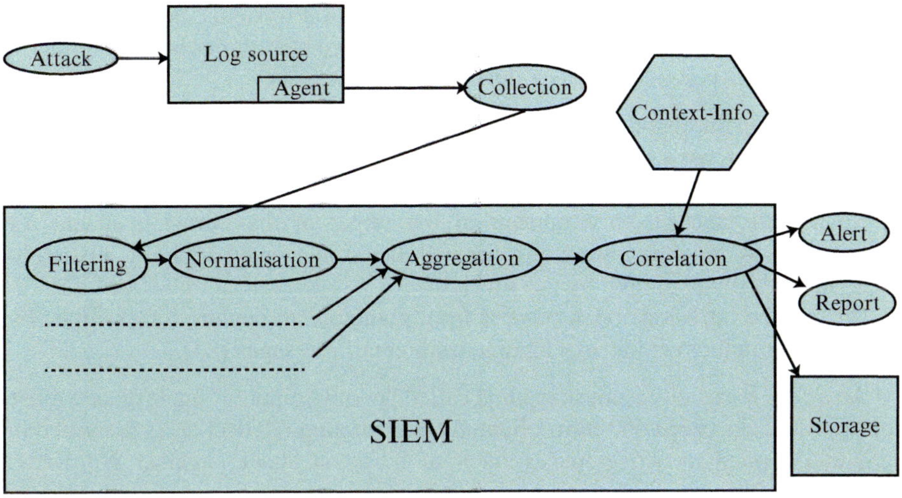

Fig. 8.2 An abstract view of a typical SIEM (Reproduced from [25])

trend analysis and provides automated reporting for compliance and centralized reporting. SIEM systems provide quicker identification, analysis, and recovery of security events. The system can also provide confirmation that an organization's legal compliance requirements are fulfilled.

A SIEM system collects logs and security-related documentation for analysis. Multiple collection agents are deployed in a hierarchical manner to gather security-related events from end user devices, servers, and network equipment. Also specialized security equipment like firewalls, antivirus, or intrusion prevention systems are in use. Events are forwarded by the collectors to a centralized management console, where inspections are performed. For the identification of anomalous events, a profile of the system under normal event conditions must be created.

A basic SIEM system is rule-based or employs a statistical correlation engine to establish relationships between event log entries. Preprocessing may happen at edge collectors, with only certain events being passed. Thus, the volume of information being communicated and stored is reduced, but relevant events may be filtered out too soon.

A SIEM system is typically expensive and complex. The Payment Card Industry Data Security Standard (PCI DSS) compliance has traditionally driven SIEM adoption in large enterprises. A managed security service (MSS) provider can offer smaller organizations a cost-efficient alternative to the implementation of an own SIEM system.

An active intrusion response is automatically generated. In [26], intelligent decision-making agents invoke response executables and scripts for some intrusion attack types.

In a guide to intrusion detection and prevention systems, active intrusion responses are characterized by "can respond to a detected threat by attempting to prevent it from succeeding." Such intrusion responses can:

- Interrupt an intrusion attack by terminating the network connection or user session being used by the attack or by blocking all access to the target of the attack.
- Change the security environment of the target of the intrusion attack, for example, by reconfiguration of a firewall or a router or by applying a patch to a vulnerability exploited by the attack.
- Change the intrusion attack process from malicious to benign, for example, by removing malicious file attachments from email messages [23].

An active intrusion response method called "booby trapping" against code reuse intrusion attacks based on return-oriented programming (ROP) [27] is presented in [28] and applied to active management of intrusion attacks against WordPress websites in [29].

8.6 Forensic Analytics

Many automated tools are available for forensic investigation, but none of these tools can deliver 100 % reliable result information [30]. Forensic investigations after intrusion/intrusion attempts must manage at least the same dataset that was processed before an intrusion alert. Thus, information analytics tools are also needed for forensics.

Network forensics is a subset of both information security and big data. By capturing, recording, and analyzing network events, the source of an attack can be found. Prevention of future attacks may be achieved. A Gartner analyst gives a broader definition of forensics [31]:

- Full packet capture as part of digital evidence gathering.
- Data retention for a period of time.
- Access to captured data using search and other tools.
- Packet header analysis.
- Packet content analysis, including session viewing, application protocol analysis, file extraction, etc.

Forensics may reveal network attacks by using big data tools and techniques for real-time analysis. A faster identification of network threats will cause less damage. A study by the Ponemon Institute has shown that a security incident detected within 60 s can reduce remediation costs by as much as 40 %. Network forensics can be treated as a big data application. The volume and variety dimensions of big data are represented by the necessity to include all network traffic in forensic investigations. The veracity dimension of big data is represented by the necessity to validate data sources in forensic investigations. The velocity dimension of big data is represented by the necessity to carry out forensic investigations immediately to reveal potential threat patterns when forensics is a part of a security analytics application. Network forensics can be used to uncover security issues and as big data analytics to detect security problems and automatically address them. Examples are isolating a server, rerouting data traffic, or a host of other preprogrammed solutions [9].

Nigrini [32] describes thoroughly forensic analytics methods. Forensic capabilities are also included in the commercial security analytics solutions *FlowTraq* [33], *RSA Security Analytics* [34], and *Security Intelligence with Big Data* [35].

8.7 Recovery After Intrusion

The state of a network or a network host is corrupt after an occurred intrusion event. The best possible state to restore is the state immediately before the time of an occurred intrusion event. However, restoration usually uses the last backup before the detected intrusion event. All data stored in the network/network host between the time of this last backup and the time of its rollout is therefore lost. Moreover, a backup is usually very large and recovery is therefore highly time consuming. However, results from forensic investigations can be used for fast recovery after detected intrusion.

Intrusion recovery research is presented in [36–44]. Since most of this research focuses on a single layer of abstraction, operating system (OS) level, application level, or business workflow level, an intrusion recovery framework combining workflow- and OS-level recovery is proposed in Yoon et al. [45]. Bacs et al. [46] present a recent network host disk memory recovery solution DiskDuster, which restores a disk memory to a pre-intrusion state with a minimal loss of data stored before the intrusion incident.

8.8 Analytical Methods for Network Security

The learning methods for alert decision support in anomaly-based intrusion detection must be scalable for unstructured datasets of arbitrary size. Learning methods based on support vector machines (SVM) and on neural networks have previously been used [47]. Recent neural network-based learning methods for intrusion detection are extreme learning machine (ELM) [48–50] and self-organizing feature map (SOM)-based methods [51].

8.9 Design of Data Analytics for Network Security

General security analytics software development requirements are:

- Developer skills to use.
- Programming paradigms such as object-oriented and/or functional programming and distributed/local programming for clouds/local computers.
- Algorithms like Hadoop, neural networks, machine learning, etc.
- A real network such as the Internet, a separate test network, or a simulation environment as a test environment.
- A suitable test dataset.

Available technologies which extend the MapReduce programming model [52] are the open-source tools Hadoop [53], Hive [54], and Pig [55]. MapReduce decomposes data into smaller pieces, which are processed on the network hosts in which they reside, instead of moving the data pieces to other network nodes for processing. Hadoop can process big volumes of stored data but is not designed to stream out answers to continuous queries directed to incoming data streams. For this purpose, a security information and event management (SIEM) solution has traditionally been used [56]. However a SIEM solution is not suitable for aggregating and correlating massive amounts of data [57, 58]. For big data stream processing are available the open-source, distributed, scalable, and fault-tolerant real-time processing system Storm [59], the commercial Splunk software [60], and the commercial IBM Streams platform [61].

For massive data volumes, a scalable and commercially available solution for anomaly detection is FlowTraq Behavioral Fingerprint Generator [62], which learns normal user traffic profiles called "behavioral fingerprints" and creates alerts on deviations from learned user profiles. Jubatus [63] is a scalable open-source platform for online distributed machine learning on big data streams. Apache Mahout [64] is a machine learning software library, which is implemented on the top of Hadoop and uses the MapReduce programming model. Machine learning with Mahout and Hadoop usually consists of the following steps:

- Store data in Hadoop.
- Run data preprocessing in order to vectorize input data (applying filters and/or feature extraction methods).
- Start training jobs, learning at least one model per input vector (only applied during the training phase).
- Execute models—a model classifies an input vector as a threat or not as a threat (the decision is not necessarily a binary decision).

Most of the learning methods described above try to find a nonlinear mapping of the input vector into a high-dimensional feature space. Connecting the high-dimensional feature space to an output layer can then be performed through linear mapping. The decision support system can be implemented as a recommender system, either requiring manual interaction or as a fully automated expert system [65].

An overview of technologies for big data analytics is presented in [66]. The Cloud Security Alliance (CSA) started in 2012 a Big Data Working Group consisting of representatives from industry and academia. This working group has issued a report focusing on the role of big data in network and information security [2]. The report presents examples on big data analytics for security and the Worldwide Intelligence Network Environment (WINE) platform [67] for security experimentation with big data analytics. The international network and information security company RSA has introduced the concept of security analytics for describing the use of information analytics to manage network and information security threats [34]. A similar concept is Security Intelligence with Big Data, introduced by IBM [35, 68].

A recently designed open-source malware analytics tool BinaryPig, based on Hadoop and Pig, is available for malware identification in huge datasets [69]. In [70], BinaryPig is characterized as:

> *... [hoping] to provide a solution to the scalability problem represented by this influx of malware. It promotes rapid iteration and exploration through processing non trivial amounts of malware at scale using easily developed scripts. It was also built to take advantage of preexisting malware analysis scripts.*

In a demo, BinaryPig has detected about 20 MB of malware samples inserted in a 9.5 TB binary dataset. This demo also showed that combining thousands of small binary files into one or more large Hadoop SequenceFile enables Hadoop to distribute processing more efficiently on many nodes. A SequenceFile in this case is comprised of a key/value collection, the key representing the checksum of a certain (malware) binary file and the value the content in raw bytes. Machine learning techniques can then be applied to classify files from the recorded traffic in order to identify potential files with the highest probability of being malware [70].

A practical problem related to testing security analytics applications in a real network like the Internet or in a separate test network is the difficulty to find or to create a suitable test dataset [1]. The first real attempt to create a standardized dataset for research on and development of intrusion detection systems is the KDDcup99 dataset [71], which has been available since 1999. It has been used in much research on network security and in development of several network security solutions [72]. A more recent available dataset, NSL-KDD, consists of selected records from the original KDDcup99 dataset in order to solve some inherent problems of KDDcup99 [73]. The KDDcup99 dataset from 1999 can therefore not be representative for all current types of network security attacks and threats. More novel datasets for research on and solution development for network security are available in [74, 75]. The up-to-date dataset consisting of hundreds of millions of records in Symantec's Worldwide Intelligence Network Environment (WINE) includes occurrences of practically all known host- and network-based attacks from network devices worldwide using Symantec's anti-malware products, from millions of spamming and phishing email accounts, and from Symantec's global honeypot network. However, this dataset is available only for network security researchers funded by the National Science Foundation (NSF) in the USA [76].

Testing a security analytics application prototype in a real network like the Internet is not always possible and is a security risk when testing is possible because of deficiencies, which testing is expected to reveal. To deploy a sufficiently large separate test network containing multiple networked computers, routers, and data links is a very expensive solution even as an Infrastructure as a Service from a cloud service provider. A network simulator can save a lot of money and time. Some available network simulators are assessed in [77].

8.9.1 Detection of Advanced Persistent Threats (APTs)

An advanced persistent threat (APT) is a targeted attack, which operates with a low profile during a long time. The purpose of this type of intrusion attack is to keep a target network or network host unaware of the ongoing intrusion. APT attackers often use stolen user credentials and/or zero-day exploits to avoid triggering IDS alerts. The Verizon data breach investigation for 2010 reports that in 86 % of the cases, data breach evidence was registered in the log files, but no security alerts were triggered by the installed IDS [78]. Big data analytics tools are particularly suitable for APT detection. To detect APT attacks, collection and correlation of large quantities of diverse data (including internal data sources and external shared intelligence data) is a necessity, and long-term historical correlation to incorporate a posteriori attack information in the history of a network or a network host must be performed [2, 57].

8.10 Taxonomy for Network Security Analytics

Taxonomies have been proposed for intrusion detection systems [79] and intrusion response systems [80]. Our proposal applied to network security analytics results in the following taxonomy:

- *Descriptive security analytics* processes data such as network activity logs, security information and events (SIEM), and anomalies in network behavior. It is therefore closely related to security intelligence.
- *Diagnostic security analytics* identifies security attacks, detects anomalies in network behavior of anomalies, and carries out forensic investigations.
- *Visual security analytics* uses data sources such as:
  - Raw packets and data records in network activity traces.
  - Security events in intrusion detection/prevention systems, firewalls, virtual private networks, and anti-malware software.
  - Output from vulnerability scanners.
  - Network events in switches, routers, servers, and network hosts.
  - Logs from network applications.

Visualization techniques are node links, glyphs, scatter plots, histograms, color maps, and tree maps. Visualization examples are communication patterns such as node links between an internal network and external sources, histograms showing port access activity in an internal network, and color maps showing the spread of victim computers in a botnet [11].

- *Predictive security analytics* estimates targets, sources, and methods of possible security attacks and events.
- *Prescriptive security analytics* makes use of diagnostic and predictive outputs to compare and optimize solutions for achieving an improved posterior state. Examples of such solutions are:
  - Protection against malware.
  - Firewall configuration.
  - Protection of network communication.
  - Recovery after security incidents.

8.11 Conclusions

Current networking is rapidly developing towards increased wireless connectivity and mobility in a pervasive environment. A myriad of sensors, actuators, and other smart devices are being connected to the Internet, and at the same time, the amount of unstructured data collected in network log files, etc. is increasing rapidly. Intelligent and effective methods for analyzing the data for finding security attacks are needed, and therefore research on and development of network and information security tools, solutions, and services with big data capacities is currently a hot topic. A taxonomy is presented for network security analytics based on descriptive, diagnostic, visual, predictive, and prescriptive techniques. The survey reveals limited current research into prescriptive methods. Prescriptive methods may offer great potential for dealing with network security threats in a holistic way and warrant more research.

References

1. Verma, R., Kantarcioglu, M., Marchette, D., Leiss, E., Solorio, T.: Security analytics: essential data analytics knowledge for cybersecurity professionals and students. IEEE Security Privacy **13**(6), 60–65 (2015)
2. Big Data Working Group.: Big data analytics for security intelligence, CSA Cloud Security Alliance. https://cloudsecurityalliance.org/download/big-data-analytics-for-security-intelligence/ (2013). Accessed 3 Feb 2016

3. Ingram Micro Advisor.: Four types of big data analytics and examples of their use. http://www.ingrammicroadvisor.com/data-center/four-types-of-big-data-analytics-and-examples-of-their-use (2016a). Accessed 3 Feb 2016
4. Gandomi, A., Haider, M.: Beyond the hype: big data concepts, methods, and analytics. Int. J. Inform. Manag. **35**, 137–144 (2015). doi:10.1016/j.ijinfomgt.2014.10.007
5. Nickerson, R.C., Varshney, U., Muntermann, J.: A method for taxonomy development and its application in information systems. Eur. J. Inf. Syst. **22**(3), 336–359 (2013)
6. Mwilu, O.S., Prat, N., Comyn-Wattiau, I.: Taxonomy development for complex emerging technologies—the case of business intelligence and analytics on the cloud. In: Proceedings of the 19th Pacific Asia Conference on Information Systems (PACIS 2015) (2015)
7. Kim, H.J.: Taxonomy of big data and business analytics. Int. J. Recent Dev. Eng. Tech. **3**(5), 27–30 (2014). http://www.ijrdet.com/files/Volume3Issue5/IJRDET_1114_04.pdf
8. Delen, D.: Real-world data mining: applied business analytics and decision making. Pearson FT Press, Upper Saddle River, NJ, USA (2014)
9. Ingram Micro Advisor.: Understanding big data, forensics, and information security. http://www.ingrammicroadvisor.com/data-center/understanding-big-data-forensics-and-information-security (2016b). Accessed 5 Feb 2016
10. Thomas, J., Cook, K. (eds.): Illuminating the path: research and development agenda for visual analytics. IEEE Press (2005)
11. Shiravi, H., Shiravi, A., Ghorbani, A.A.: A survey of visualization systems for network security. IEEE Trans. Vis. Comput. Graph. **18**(8), 1313–1329 (2012)
12. Sarlin, P.: Self-organizing time map: An abstraction of temporal multivariate patterns. Neurocomputing **99**, 496–508 (2013)
13. Kleim, D.A., Mansmann, F., Schneidewind, J., Thomas, J., Ziegler, H.: Visual analytics: scope and challenges. Lecture Notes in Computer Science, vol 4404, pp. 76–90. Springer, Heidelberg (2008)
14. Wu, T.M.: Intrusion detection systems. Information assurance tools report, 6th edn. Information Assurance Technology Analysis Center IATAC http://nopasara.com/wp-content/uploads/2013/04/intrusion_detection.pdf (2009). Accessed 4 Feb 2016
15. Xu, H., Chen, X., Zhou, J., Wang, Z., Xu, H.: Research on basic problems of cognitive network intrusion prevention. In: Proceedings of the Ninth International Conference on Computational Intelligence and Security (CIS), pp. 514–517 (2013)
16. Laniepce, S, Lacoste, M, Kassi-Lahlou, M, Bignon, F, Lazri, K, Wailly, A.: Engineering intrusion prevention services for IaaS clouds: the way of the hypervisor. In: Proceedings of the 7th International Symposium on Service Oriented System Engineering (SOSE), pp. 25–36, IEEE Press (2013)
17. Mantoro, T., Aziz, A., Yusoff, M., Talib, A.: Log visualization of intrusion and prevention reverse proxy server against Web attacks. In: Proceedings of International Conference on Informatics and Creative Multimedia (ICICM), pp 325–329, IEEE Press (2013)
18. Fang, W., Qian, H., Yong, W., Linlin, Y.: A P2P and rule-based Web application intrusion prevention system. In: Proceedings of the 8th International Conference on Communications and Networking in China (CHINACOM), pp. 410–414. IEEE Press (2013)
19. Biggio, B., Corona, I., Fumera, G., Giacinto, G., Roli, F.: Bagging classifiers for fighting poisoning attacks in adversarial classification tasks. In: Multiple classifier systems, pp. 350–359. Springer, Heidelberg (2011)
20. Francois, J., Wang, S., Bronzi, W., State, R., Engel, T.: BotCloud: detecting Botnets using MapReduce. In: Proceedings of the IEEE International Workshop on Information Forensics and Security. IEEE Press (2011)
21. Cisco IOS Netflow.: http://www.cisco.com/go/netflow (2016). Accessed 5 Feb 2016
22. Page, L., Brin, S., Motwani, R., Winograd, T.: The PageRank citation ranking: bringing order to the Web. Technical Report, Stanford InfoLab, Stanford University, USA. http://ilpubs.stanford.edu:8090/422/ (1999). Accessed 5 Feb 2016

23. Scarfone, K., Mell, P.: Guide to intrusion detection and prevention systems (IDPS) (Draft). Special Publication 800-94 Revision 1 (Draft). NIST National Institute of Standards and Technology, U.S. Department of Commerce. http://csrc.nist.gov/publications/drafts/800-94-rev1/draft_sp800-94-rev1.pdf (2012). Accessed 5 Feb 2016
24. Kakareka A.: Detecting system intrusions. In: Vacca J.R., (ed.) Computer and Information Security Handbook, 2nd edn., pp 47–62, Elsevier, USA (2013)
25. Karlzén, H.: An analysis of security information and event management systems. MSc Thesis, Chalmers University of Technology, Gothenburg, Sweden. http://publications.lib.chalmers.se/records/fulltext/89572.pdf (2009). Accessed 5 Feb 2016
26. Carver, C.A., Hill, J.M.D., Surdu, J.R., Pooch, U.W.: A methodology for using intelligent agents to provide automated intrusion response. In: Proceedings of the IEEE System, Man, and Cybernetics Information Assurance and Security Workshop, pp. 110–116, IEEE Press, West Point (2000)
27. Prandini, M., Ramilli, M.: Return-oriented programming. IEEE Security Privacy **10**(6), 84–87 (2012)
28. Crane, S., Larsen, P., Brunthaler, S., Franz, M.: Booby trapping software. In: Proceedings of New Security Paradigms Workshop (NSPW'13), pp. 95–106, ACM (2013)
29. Paarnio, P., Stenvall, S., Westelund, M., Pulkkis, P.: Active intrusion management for Web server software: Case WordPress. In: Proceedings of the Tenth International Multi-Conference on Computing in the Global Information Technology, IARIA, Malta, pp. 6–12. https://www.thinkmind.org/download.php?articleid=iccgi_2015_1_20_10058 (2015). Accessed 19 Feb 2016
30. EC-Council Press.: Computer forensics investigating network intrusions & cyber crime. Course Technology. Cengage Learning Inc. USA. http://news.asis.io/sites/default/files/Investigating_Intrusions_Network_CyberCrime.pdf (2010). Accessed 8 Feb 2016
31. Chuvakin, A.: 2016Network forensics defined? Gartner Blog Network. http://blogs.gartner.com/anton-chuvakin/2013/01/29/network-forensics-defined/ (2013). Accessed 5 Feb 2016
32. Nigrini, M.: Forensic analytics: methods and techniques for forensic accounting investigations. Wiley Corporate F&A, New York, NY, USA (2011)
33. Full Fidelity Network Forensics.: FlowTraq. http://www.flowtraq.com/corporate/full-fidelity-network-forensics (2015). Accessed 6 Feb 2016
34. RSA Security Analytics.: EMC Advanced security operations center. http://finland.emc.com/security/security-analytics/security-analytics.htm (2016). Accessed 6 Feb 2016
35. Extending Security Intelligence with Big Data Solutions.: White Paper, IBM. http://insight.q1labs.com/ExtendingSecurityIntelligencewithBigData.html (2016). Accessed 6 Feb 2016
36. Ammann, P., Jajodia, S., Liu, P.: Recovery from malicious transactions. IEEE Trans. Knowl. Data Eng. **14**(5), 1167–1185 (2002)
37. Goel, A., Po, K., Farhadi, K., Li, Z., de Lara, E.: The taser intrusion recovery system. In: Proceedings of the Twentieth ACM Symposium on Operating Systems Principles, pp. 163–176, ACM, New York (2005)
38. Hsu, F., Chen, H., Ristenpart, T., Li, J., Su, Z.: Back to the future: a framework for automatic malware removal and system repair. In: Proceedings of the 22nd Annual Computer Security Applications Conference, pp. 257–268, IEEE Press (2006)
39. Jain, S., Shafique, F., Djeric, V., Goel, A.: Application-level isolation and recovery with solitude. In: Proceedings of the 3rd ACM SIGOPS/EuroSys European Conference on Computer Systems, pp 95–107. ACM, New York (2008)
40. Kim, T., Wang, X., Zeldovich, N., Kaashoek, M.F.: Intrusion recovery using selective re-execution. In: Proceedings of the 9th USENIX Conference on Operating Systems Design and Implementation, pp 1–9. USENIX Association, Berkeley (2010)
41. Mahajan, P., Kotla, R., Marshall, C.C., Ramasubramanian, V., Rodeheffer, T.L., Terry, D.B., Wobber, T.: Effective and efficient compromise recovery for weakly consistent replication. In: Proceedings of the 4th ACM European Conference on Computer Systems, pp. 131–144. ACM, New York (2009)

42. Paleari, R., Martignoni, L., Passerini, E., Davidson, D., Fredrikson, M., Giffin, J., Jha, S.: Automatic generation of remediation procedures for malware infections. In: Proceedings of the 19th USENIX Conference on Security, p. 27. USENIX Association, Berkeley (2010)
43. Xiong, X., Jia, X., Liu, P.: Shelf: preserving business continuity and availability in an intrusion recovery system. In: Proceedings of the 2009 Annual Computer Security Applications Conference, pp. 484–493. IEEE Press (2009)
44. Yu, M., Liu, P., Zang, W.: Self-healing workflow systems under attacks. In: Proceedings of the 24th International Conference on Distributed Computing Systems, pp. 418–425. IEEE Press (2004)
45. Yoon, E., Liu, P.: XLRF: A cross-layer intrusion recovery framework for damage assessment and recovery plan generation. In: Sihan Qing, S., Zhou, J., Liu, D. (eds.) Proceedings of the 15th International Conference on Information and Communications Security. Lecture Notes in Computer Science, vol 8233, pp. 194–212. Springer, Switzerland (2013)
46. Bacs, A., Vermeulen, R., Slowinska, A., Bos, H.: System-level support for intrusion recovery. In: Proceedings of the 9th Conference on Detection of Intrusions and Malware & Vulnerability Assessment (DIMVA). Heraklion, Greece. http://www.syssec-project.eu/m/page-media/3/diskduster-dimva12.pdf (2012). Accessed 6 Feb 2016
47. Mukkamala, S., Janoski, G., Sung, A.: Intrusion detection using neural networks and support vector machines. Proc. Int. Joint Conf. Neural Network. **2**, 1702–1707 (2002)
48. Cheng, C., Tay, W.P., Huang, G.-B.: Extreme learning machines for intrusion detection. In: Proceedings of IEEE World Congress on Computational Intelligence (WCCI). IEEE Press, Brisbane, Australia (2012)
49. Jaiganesh, V., Sumathi, P.: Kernelized extreme learning machine with levenberg-marquardt learning approach towards intrusion detection. Int J Comp App **54**(14), 38–44 (2012)
50. Xiang, J., Westerlund, M., Sovilj, D., Pulkkis, G.: Using extreme learning machine for intrusion detection in a big data environment. In: Proceedings of the 2014 Workshop on Artificial Intelligent and Security Workshop, ACM, pp. 73–82. doi: 10.1145/2666652.2666664 (2014)
51. Kayacik, H.G., Zincir-Heywood, A.N., Heywood, M.I.: A hierarchical SOM based intrusion detection system. Eng. App. Artif. Intell. **20**(4), 439–451 (2007)
52. Dean, J., Ghemawat, S.: MapReduce: simplified data processing on large clusters. In: Proceedings of the Sixth Symposium on Operating System Design and Implementation (OSDI'04). San Francisco, CA, USA (2004)
53. Apache Software Foundation.: Welcome to Apache Hadoop. http://hadoop.apache.org (2016). Accessed 6 Feb 2016
54. Apache Software Foundation.: Apache Hive. http://hive.apache.org (2014a). Accessed 6 Feb 2016
55. Apache Software Foundation.: Welcome to Apache Pig! http://pig.apache.org (2015a). Accessed 6 Feb 2016
56. Miller, D.R., Harris, S., Harper, A.A., VanDyke, S., Blask, C.: Security information and event management (SIEM) implementation. McGraw-Hill, New York, NY, USA (2011)
57. Cárdenas, A.A., Manadhata, P.K., Rajan, S.P.: Big data analytics for security. IEEE Security Privacy **11**(6), 74–76 (2013)
58. Chickowski, E.: Moving beyond SIEM for strong security analytics. InformationWeek Dark Reading. http://www.darkreading.com/moving-beyond-siem-for-strong-security-analytics/d/d-id/1141069 (2013). Accessed 7 Feb 2016
59. Apache Software Foundation.: Apache storm. http://storm.apache.org/ (2015b). Accessed 7 Feb 2016
60. Using Splunk Software as a SIEM.: Tech Brief. Splunk Inc. http://www.splunk.com/web_assets/pdfs/secure/Splunk_as_a_SIEM_Tech_Brief.pdf (2013). Accessed 7 Feb 2007
61. IBM Streams.: http://www-03.ibm.com/software/products/en/ibm-streams (2016a). Accessed 7 Feb 2016

62. NBAD Behavioral Fingerprinting.: FlowTraq. http://www.flowtraq.com/corporate/nbad-behavioral-fingerprinting (2016). Accessed 7 Feb 2016
63. Jubatus Overview.: http://jubat.us/en/overview.html (2016). Accessed 7 Feb 2016
64. Apache Software Foundation.: What is Apache Mahout? http://mahout.apache.org (2014b). Accessed 7 Feb 2016
65. Haykin, S.O.: Neural networks and learning machines, 3rd edn. Pearson Prentice Hall, Upper Saddle River, NJ, USA (2009)
66. Think Big, A Teradata Company.: Technologies we use https://thinkbiganalytics.com/leading_big_data_technologies/ (2016). Accessed 7 Feb 2016
67. Dumitras, T., Shou, D.: Toward a standard benchmark for computer security research: the worldwide intelligence network environment (WINE). In: Proceedings of the EuroSys BADGERS Workshop. Salzburg, Austria (2011)
68. IBM Security Intelligence with Big Data.: White Paper, IBM. http://www-03.ibm.com/security/solution/intelligence-big-data (2016b). Accessed 7 Feb 2016
69. Trost, J., Calhoun, T., Hanif, Z.: ENDGAME. BinaryPig: scalable binary data extraction in Hadoop. Presentation slides. BlackHat, USA. https://media.blackhat.com/us-13/US-13-Hanif-Binarypig-Scalable-Malware-Analytics-in-Hadoop-Slides.pdf (2013). Accessed 7 Feb 2016
70. Hanif, Z., Calhoun, T., Trost, J.: BinaryPig: scalable static binary analysis over Hadoop. White Paper, BlackHat, USA. https://media.blackhat.com/us-13/US-13-Hanif-Binarypig-Scalable-Malware-Analytics-in-Hadoop-WP.pdf (2013). Accessed 7 Feb 2013
71. KDD Cup 1999 Data.: The UCI KDD Archive, information and computer science, University of California. http://kdd.ics.uci.edu/databases/kddcup99/kddcup99.html (1999). Accessed 12 Feb 2016
72. Tavallaee, M., Bagheri, E., Wei, L., Ghorbani, A.A.: A detailed analysis of the KDD CUP 99 data set. In: Proceedings of the Second IEEE Symposium on Computational Intelligence for Security and Defence Applications. IEEE Press (2009)
73. UNB ISCX NSL-KDD DataSet.: Information security centre of excellence (ISCX), University of New Brunswick. http://www.unb.ca/research/iscx/dataset/iscx-NSL-KDD-dataset.html (2015). Accessed 13 Feb 2013
74. UNB ISCX Intrusion Detection Evaluation DataSet.: Information security centre of excellence (ISCX), University of New Brunswick. http://www.unb.ca/research/iscx/dataset/iscx-IDS-dataset.html (2012). Accessed 13 Feb 2016
75. ADFA Intrusion Detection Datasets.: https://www.unsw.adfa.edu.au/australian-centre-for-cyber-security/cybersecurity/ADFA-IDS-Datasets/ (2013). Accessed 13 Feb 2016
76. Symantec Corporation.: Symantec University Research. https://www.symantec.com/about/corporate-profile/technology/university-research (2016). Accessed 13 Feb 2016
77. Pan, J.: A survey of network simulation tools: current status and future developments. http://www.cse.wustl.edu/~jain/cse567-08/ftp/simtools/ (2008). Accessed 13 Feb 2016
78. Verizon Inc.: 2010 Data breach investigation report. http://www.verizonenterprise.com/resources/reports/rp_2010-data-breach-report_en_xg.pdf (2010). Accessed 7 Feb 2016
79. Axelsson, S.: Intrusion detection systems: a survey and taxonomy. Technical report 99–15. Department of Computer Engineering, Chalmers University of Technology, Gothenburg, Sweden (2000)
80. Stakhanova, N., Basu, S., Wong, J.: A taxonomy of intrusion response systems. Int. J. Inform. Comput. Secur. **1**(1/2), 169–184 (2007)

Chapter 9
Security in Wireless Sensor Networks (WSNs) and Their Applications

C.V. Anchugam and K. Thangadurai

Abstract As wireless sensor networks continue to grow, so does the need for effective security mechanisms. Because sensor networks may interact with sensitive data and/or operate in hostile unattended environments, it is imperative that these security concerns be addressed from the beginning of the system design. However, due to inherent resource and computing constraints, security in sensor networks poses different challenges than traditional network/computer security. There is currently enormous research potential in the field of wireless sensor network security. Thus, familiarity with the current research in this field will benefit researchers greatly. With this in mind, we survey the major topics in wireless sensor network security, present the obstacles and the requirements in the sensor security, classify many of the current attacks, and finally list their corresponding defensive measures.

9.1 Introduction

Wireless sensor networks have now arrived into prominence because they hold the potential to develop several segments of our economy and life, from environmental observation and conservation, to production and business quality management, to automation within the transportation and healthcare industries. The design, implementation, and operation of a sensor network need the confluence of the many disciplines, together with signal process, networking and protocols, embedded systems, data management, and distributed algorithms. Such networks are usually deployed in resource-constrained environments, for example, with battery-operated nodes running unbound. These constraints dictate that sensor network issues are best approached in a very hostile manner, by collectively considering the physical, networking, and application layers and creating main design trade-offs across the layers.

C.V. Anchugam (✉) • K. Thangadurai
P.G. and Research Department of Computer Science, Government Arts College (Autonomous), Karur, Tamil Nadu 639 005, India
e-mail: anchugam.mca@gmail.com

I.M. Alsmadi et al. (eds.), *Information Fusion for Cyber-Security Analytics*, Studies in Computational Intelligence 691, DOI 10.1007/978-3-319-44257-0_9

Advances in wireless networking, micro-fabrication and integration (e.g., sensors and actuators manufactured using Micro Electro-Mechanical System (MEMS) technology), and embedded microprocessors have enabled a new generation of massive-scale sensor networks appropriate for a range of commercial and military applications. The technology guarantees to revolutionize the method we have a tendency to live, work, and interact with the physical environment. In a very typical sensor network, each sensor node operates unbound and incorporates a microchip and a little quantity of memory for signal process and task scheduling. Every node is supplied with one or more sensing devices such as acoustic microphone arrays, video or still cameras, and infrared (IR), seismic, or magnetic sensors. Every sensor node communicates wirelessly with some different local nodes among its radio communication that vary.

Sensor networks extend the existing Internet cavernous into the physical setting. The resulting new network is orders of magnitude a lot of expansive and dynamic than the present TCP/IP networks and is making entirely new types of traffic that are quite totally different from what one finds on the Internet currently. Data collected by and transmitted on a sensor network describes conditions of physical environments, for example, temperature, humidity, or vibration, and requires advanced query interfaces and search engines to effectively support user-level functions. Sensor networks could inter-network with an IP core network via a number of gateways. A gateway routes user queries or commands to acceptable nodes in a sensor network. It also routes sensor data, sometimes collective and reviewed, to users who have requested it or are expected to utilize the information. A data repository or storage service may be present at the gateway, in addition to data logging at each sensor. The repository may serve as an intermediary between users and sensors, providing a relentless data storage. It is well known that communicating 1 bit over the wireless medium at short ranges consumes way more energy than process that bit.

The information management and networking for sensor networks will require more than just building faster routers, switchers, and browsers. A sensor network is designed to collect information from a physical environment. In several applications, it's a lot of acceptable to address nodes in a sensor network by physical properties, like node locations or proximity, than by IP addresses. However and wherever data is generated by sensors and consumed by users can have an effect on the way data is compressed, routed, and collected. Because of the peer-to-peer connectivity and therefore the lack of a worldwide infrastructure support, the sensors ought to rely on discovery protocols to construct local models about the network and environment.

Wireless sensor networks are a trend of the past few years and that they involve deploying a large number of small nodes. The nodes then sense environmental changes and report them to different nodes over flexible network architecture. Sensor nodes are great for deployment in hostile environments or more outsized geographical areas. The sensor nodes influence the strength of collaborative efforts to provide higher-quality sensing in time and space as compared to traditional stationary sensors, which are deployed within the following two ways:

- Sensors can be positioned far from the actual phenomenon, i.e., something known by sense perception. During this approach, large sensors that use some complex techniques to distinguish the targets from environmental noise are required.
- Several sensors that perform solely sensing are deployed. The position of the sensors and communication topology is fastidiously designed. They transmit time series of the sensed phenomenon to central nodes wherever computations are performed and data are combined.

A wireless sensor network could be assortment of nodes organized into a cooperative network. Every node consists of process capability (one or a lot of microcontrollers, CPUs, or DSP chips), could contain multiple types of memory (program, data, and flash memories), have an RF transceiver (usually with a single omnidirectional antenna), have an influence supply (e.g., batteries and solar cells), and accommodate various sensors and actuators. The nodes communicate wirelessly and sometimes self-organize when being deployed in an ad hoc fashion. Currently, wireless sensor networks are getting down to be deployed at an accelerated pace. It is not unreasonable to expect that in 10–15 years the world will be covered with wireless sensor networks with access to them via the Internet. This could be thought of because the Internet is turning into a physical network. Wireless sensor network is widely used in electronics. This new technology is exciting with unlimited potential for various application areas together with environmental, medical, military, transportation, diversion, home automation and traffic control crisis management, homeland defense, and smart spaces.

9.2 A Brief History of Sensor Networks

The wireless sensor network paradigm that has been introduced in the previous section is not novel. In fact, there have been "sensor networks" before "wireless sensor networks" came into being [1]. The first known deployment of a "sensor network" occurred in the US military, during the Cold War. For example, a system of acoustic sensors (hydrophones) was deployed at strategic locations on the ocean bottom to detect and track quiet Soviet submarines. This system was called the sound surveillance system (SOSUS) and has evolved into civil system that monitors events in the ocean, such as seismic and animal activity [2]. Other "sensor network" systems included networks of air defense radars that defended the continental United States and Canada. This system has conjointly evolved over the years to incorporate aerostats as sensors and airborne warning and control system (AWACS) planes and is additionally used for drug interdiction. These sensor networks generally adopted a hierarchical processing structure, and in many cases, human operators played a key role in the system.

Modern analysis on sensor networks started on the late 1970s with the Distributed Sensor Network (DSN) program at the US Defense Advanced Research

Projects Agency (DARPA). Technology components for a DSN were identified in a Distributed Sensor Network workshop in 1978 [3]. These included sensor HW, communication capabilities, processing algorithms including self-location, and distributed software. Artificial intelligence was also being considered at the time, in particular, for understanding signals and assessing situations, as well as various distributed problem-solving techniques. The resulting DSN program had to address distributed computing support, signal processing, tracking, and test beds.

There were various interesting results derived from the DSN program in the 1980s. Researchers at Carnegie Mellon University (CMU) developed an indoor test bed with signal sources, acoustic sensors, and VAX computers. This test bed focused on providing a network operating system that allowed flexible, transparent access to distributed resources needed for a fault-tolerant DSN [4]. Researchers at the Massachusetts Institute of Technology (MIT), Cambridge, targeted on knowledge-based signal process techniques [5] for following low-flying aircrafts and helicopters using a distributed array of acoustic microphones. Other applications included distributed vehicle monitoring using a functionally accurate, cooperative architecture [6] and tracking multiple targets in a distributed environment by using multiple-hypothesis tracking algorithms [7]. Still, the size of the devices that were part of these systems was significant, and also communication was done by Ethernet and (in extreme cases) microwave radio.

It was the advances in microelectromechanical system (MEMS) technology, wireless communications, and digital electronics that allowed the existence of wireless sensor networks. While DARPA continued its military-oriented research with the SensIT program (1999), whose goal was to form sensor networks to be utilized in battlefield surveillance, reconnaissance, and targeting [8], new research programs without military-oriented goals started to appear. For example, *The Disappearing Computer*, a FET proactive initiative in the EU Fifth Framework Programme influenced by the "pervasive computing" vision of Mark Weiser and other visionaries, started funding sensor network-related projects, such as "Smart-Its" [9]. Also, civil applications that took advantage of the sensing capabilities of sensor networks, such as health applications and home applications, were envisioned [10].

9.3 Wireless Sensor Network vs. Ad Hoc Network

A mobile ad hoc network (MANET), typically referred to as a mobile mesh network, could be a self-configuring network of mobile devices connected by wireless links. Every device in a MANET is liberal to move independently in any direction and so can modify its links to different devices frequently. The distinction between wireless sensor networks and ad hoc networks is outlined below:

– The number of sensor nodes in a sensor network can be several orders of magnitude beyond the nodes in an ad hoc network.

- Sensor nodes are heavily deployed.
- Sensor nodes are prone to failures.
- The topology of a sensor network changes terribly frequently.
- Sensor nodes mainly use broadcast communication paradigm, whereas most ad hoc networks are supported by point-to-point communication.
- Sensor nodes are restricted in power, computational capacities, and memory.
- Sensor nodes might not have universal identification (ID) because of the large quantity of overheads and huge number of sensors.
- Sensor networks are deployed with a selected sensing application in mind, whereas ad hoc networks are regularly created for communication purpose.

9.4 Characteristics of WSN

Characteristics preventing the utilization of conventional security protocols in WSNs and only belonging to WSN are summarized below. Taking into account the mentioned characteristics throughout design and development of protocols will increase the usability of them [6].

- Large Scale
  General applications of WSNs need geographical coverage of large areas [10]. The number of nodes in WSNs may exceed tens of thousands [11].
- Limited Resources
  The requirement that WSNs must be with low installation and operation cost necessitates that sensor node should have simple hardware. For this reason, operation and communication resources in WSNs are limited. For example, one in every of the generic sensor varieties, TelosB, has 16-bit 8 MHz processor, 48 KB main memory, and 1024 KB nonvolatile storage. Each protocol should be designed taking into consideration limitations in processor capability, memory, and radio communication [10].
- Redundancy
  Because of node redundancy, each event is detected by the multiple sensor nodes on the network and therefore increases the amount of data to be transferred over it. In other words, redundancy will increase the number of data sent to the base station and reduce the lifetime of the network [10]. To get rid of data redundancy, clustering protocols are used.
- Security
  WSN applications, like military systems and medical monitoring systems, are terribly sensitive in terms of security. As a result of the restricted resources of the sensor nodes, traditional security mechanisms can't be utilized in WSNs. Therefore, the security mechanisms of WSNs should be designed considering limited resources and malicious sensors [10].

9.5 Structure of a Wireless Sensor Network

The structure of a WSN includes different topologies for radio communication networks. A short discussion of the network topologies that apply to wireless sensor networks is outlined below.

9.5.1 *Star Network (Single Point to Multipoint) [11]*

A star network could be a communication topology wherever a single base station will send and/or receive a message to a variety of remote nodes as shown in Fig. 9.1. The remote nodes aren't permissible to send messages to each other. The advantage of this type of network for wireless sensor networks includes simplicity, ability to stay the remote node's power consumption to a minimum. It conjointly permits low-latency communications between the remote node and therefore the base station. The disadvantage of a network is that the base station should be among radio transmissions which vary in all the individual nodes and it isn't as strong as other networks as a result of its dependency on a single node to manage the network.

9.5.2 *Mesh Network [11]*

A mesh network permits transmitting data from one node to the other node within the network that is among its radio transmission ranges as shown in Fig. 9.2. This permits for what is known as multi-hop communications, that is, if a node desires to send a message to a different node that is out of the radio communication range, it will use an intermediate node to forward the message to the specified node. This configuration has the advantage of redundancy and scalability. If an individual node fails, a remote node still will communicate to the other node in its range that successively will forward the message to the specified location. Additionally, the range of the network isn't essentially restricted by the range in between single nodes; it will simply be extended by adding a lot of nodes to the system.

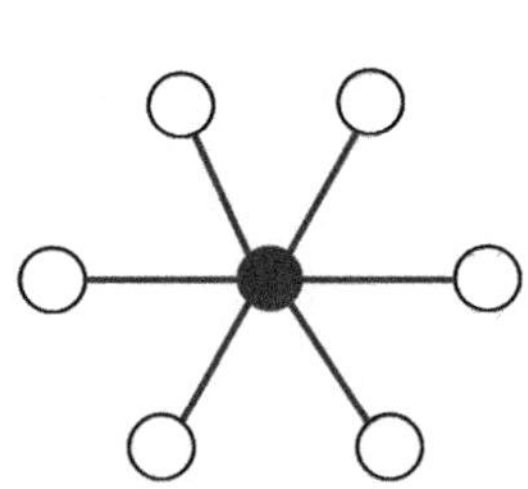

Fig. 9.1 A star network topology

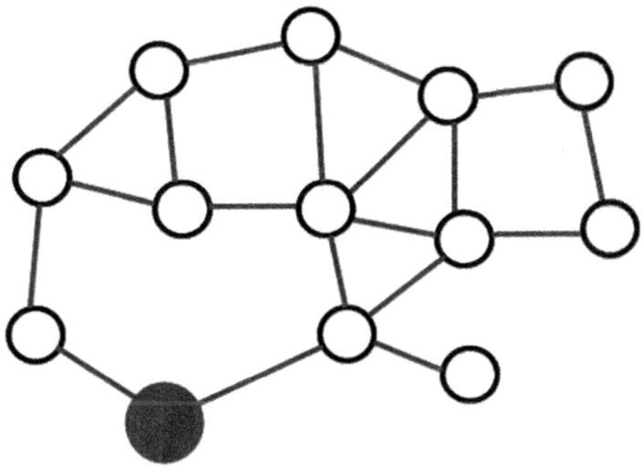

Fig. 9.2 A mesh network topology

The disadvantage of this type of network is in power consumption for the nodes that implement the multi-hop communications that are typically beyond for the nodes that don't have this capability, often limiting the battery life. Additionally, as the numbers of communication hops to a destination will increase, the time to deliver the message also increases, particularly if low-power operation of the nodes could be a demand.

9.5.3 Hybrid Star–Mesh Network [11]

A hybrid between the star and mesh network provides a robust and versatile communication network while maintaining the flexibility to stay the wireless sensor node's power consumption to a minimum. During this configuration, the sensor nodes with the lowest power aren't enabled with the flexibility to forward messages. This permits for token power consumption to be maintained. However, different nodes on the network are enabled with multi-hop capability, permitting them to forward messages from the low-power nodes to different nodes on the network. Generally, the nodes with the multi-hop capability have higher power and, if attainable, are usually blocked into the electrical main line. This is the topology enforced by the up and returning mesh networking standard referred to as ZigBee. Figure 9.3 shows the hybrid star–mesh network.

9.6 Structure of a Wireless Sensor Node

A sensor node is created from four basic elements, sensing unit, process unit, transceiver unit, and power unit, which is shown in Fig. 9.4. It also has application-dependent extra elements: a location finding system, an influence generator, and a mobilizer. Sensing units are typically composed of two subunits: sensors and analog-to-digital converters (ADCs) [12]. The analog signals made by the sensors are converted to digital signals by the ADC and so fed into the process

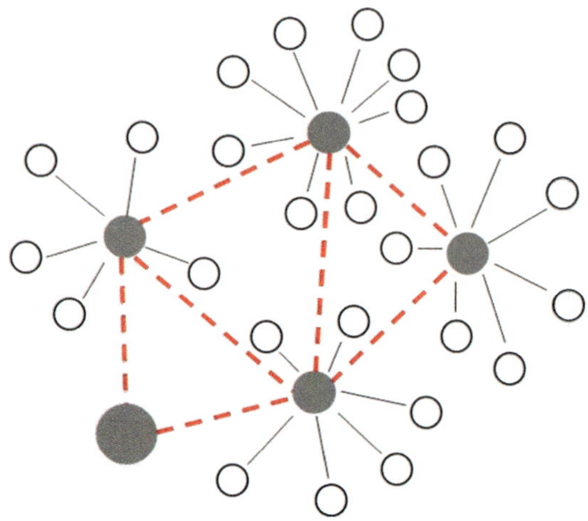

Fig. 9.3 A hybrid star–mesh network topology

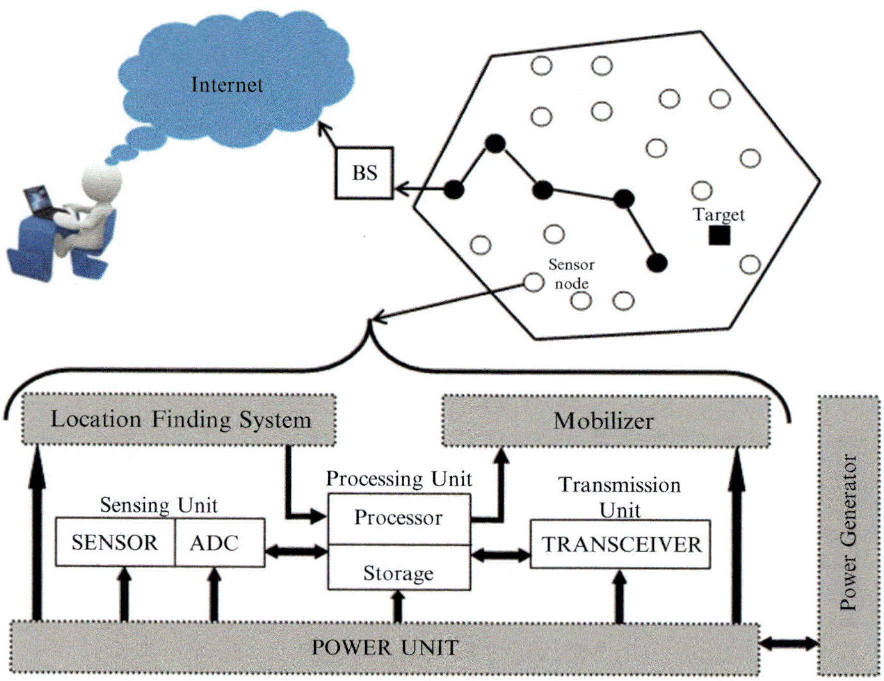

Fig. 9.4 The components of a sensor node

unit. The process unit is usually related to a little storage unit, and it will manage the procedures that build the sensor node which collaborates with the other nodes to hold out the assigned sensing tasks. A transceiver unit connects the node to the network. One of the most important components of a sensor node is the power unit.

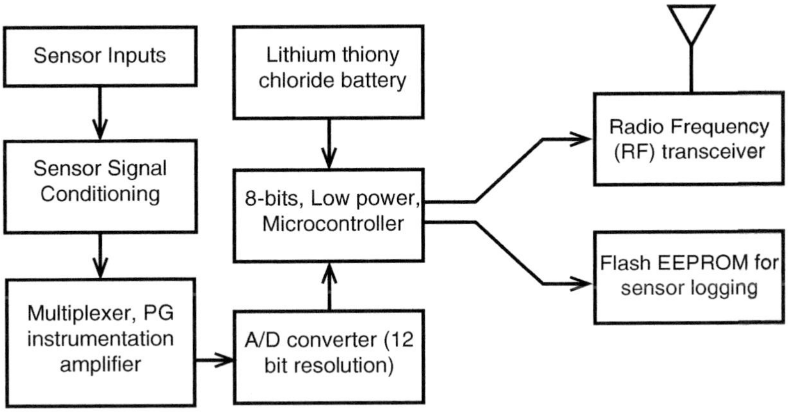

Fig. 9.5 Functional block diagram of a sensor network

A power unit is supported by an influence of a scavenging unit like solar cells. The other subunits of the node are application dependent.

A functional block diagram of a flexible wireless sensing node is provided in Fig. 9.5. A modular design approach provides a flexible and versatile platform to handle the requirements of a good sort of applications. For example, betting on the sensors to be deployed, the signal acquisition block is reprogrammed or replaced. This permits for a good sort of totally different sensors to be used with the wireless sensing node. Similarly, the radio link is also swapped out as needed for a given applications' wireless demand varies and therefore the need for two-way communications.

Using a nonvolatile storage, the remote nodes acquire data on command from a base station or by an occasion detected by one or more inputs to the node. Moreover, the embedded code upgraded through the wireless network within the field. The microprocessor incorporates a number of functions including:

- Managing data assortment from the sensors.
- Performing power management functions.
- The sensor data can be interfacing to the physical radio layer.
- Managing the radio network protocol.

A key facet of any wireless sensing node is to reduce the power consumed by the system. Usually, the radio subsystem needs the most important quantity of power. Therefore, data is distributed over the radio network only if it's needed. An algorithm is to be loaded into the node to determine when to send data supported the detected event. Moreover, it is vital to minimize the power consumed by the sensor itself. Therefore, the hardware ought to be designed to permit the microprocessor to judiciously manage power to the radio, sensor, and sensor signal conditioner [10].

9.7 Components of a Sensor Node (Hardware)

Figure 9.6 shows a schematic of a typical sensor node hardware hierarchy.

As seen in Fig. 9.6, there are several hardware components that make up a typical sensor node:

- *Low-power embedded processor:* The computational tasks on a WSN device represent the process of both locally sensed information and information communicated by other sensors. Currently, primarily as a result of economic reasons, the embedded processors are usually well restricted in terms of computational power (small MHz area). As a result of the constraints of such processors, devices normally run specialized component-based embedded operating systems, such as TinyOS. They incorporate advanced low-power design techniques, such as sleep modes and dynamic voltage scaling, to provide energy savings [3].
- *Memory/storage:* In the storage, both program memory (memory for the instruction set of the processor) and data memory (for storing measured data and other local information, e.g., the location of the node) are included. The range of the memory is usually restricted as a result of economic reasons. With the continuing price reduction of memory devices, the quantities of storage and memory used on sensor nodes increase over time.
- *Radio transceiver:* WSN devices contain a low-rate, short-range wireless radio (10–100 kbps, <100 m). Whereas presently quite restricted in capability too, these radios are likely to enhance in sophistication over time—together with enhancements in the value, spectral potency, immunity to noise, fading, and interference. Radio communication is usually the foremost power-intensive operation in a very WSN device, and therefore the radio should incorporate energy-efficient sleep and wake-up modes.

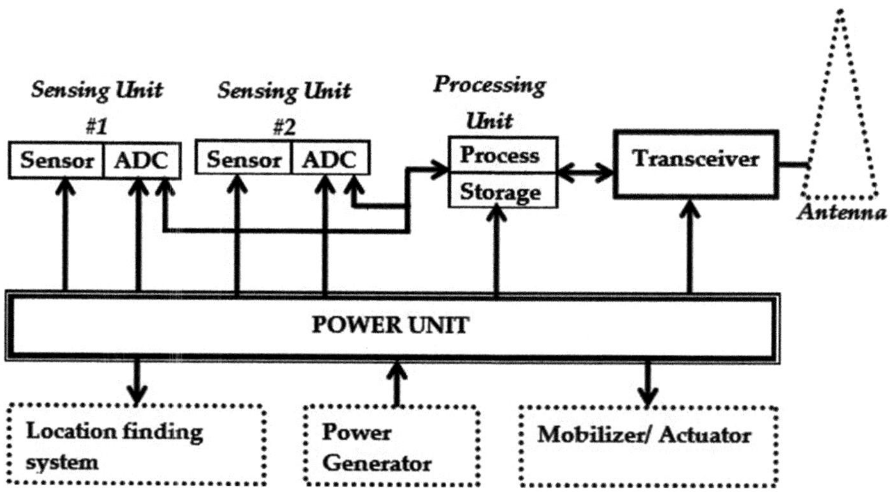

Fig. 9.6 Typical sensor node hardware hierarchy

- *Sensors with ADC unit:* WSN devices typically support solely low-data-rate sensing, owing to the constraints of energy and bandwidth. In several applications, multimodal sensing is important leading to the very fact that each device may have multiple sensors implemented. Which specific sensors are used is application based. The ADC unit interprets the analog signals, provided by the sensors, to digital signals, which will be processed by the processor unit.
- *Location finding system:* To investigate the measured data, in several WSNs, it's vital to know in which location the data was monitored. However sadly, only a few applications enable the designer to pre-configure the location of the sensor nodes. Significantly, for random deployed WSNs that are used, for example, for outdoor operations, location finding systems, normally based on satellite GPS, have to be implemented [4].
- *Power source:* Typically, the power source could be a little battery. The finite battery power is probably going to be the bottleneck in most WSN applications. However, in some applications, a handful of nodes are also wired to continuous power source, or energy harvest techniques could give a little quantity of renewed energy.

9.8 Components of a Sensor Node (Software)

Figure 9.7 shows a schematic of a typical sensor node software hierarchy. The software applications of typical sensor device nodes will generally be separated into five subsystems [5]:

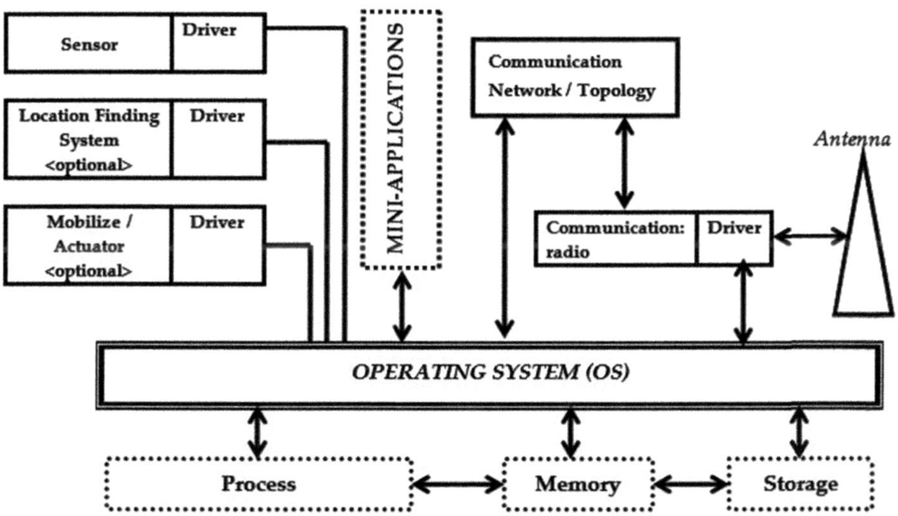

Fig. 9.7 Typical sensor node software hierarchy

- *Operating System Microcode:* It is also known as middleware. It represents the code that is utilized by the high-level software modules to support an assortment of functions. The middleware also covers the software from the machine-level functionality of the microprocessor. A celebrated example for a normally used operating system for sensor networks is TinyOS.
- *Sensor Drivers:* The sensor drivers are software modules that manage basic functions of the sensor transceivers. Depending on the kind of the sensor, they handle to transfer the proper configuration and settings onto it.
- *Communication Processors:* The communication processors manage the communication functions, including routing, packet buffering and forwarding, topology maintenance, medium access control (e.g., contention mechanisms, direct-sequence spread-spectrum mechanisms), and encryption.
- *Communication Drivers:* These software modules operate the details of the radio channel transmission link, including clocking and synchronization, signal encoding, bit recovery, bit counting, signal levels, and modulation.
- *Data-Processing Mini-Applications:* Basic applications, e.g., data processing, signal-value storage and manipulations, etc. They are supported at the node level for in-network processing.

9.9 Communication Structure of a Wireless Sensor Network

The sensor nodes are usually scattered in a sensor field. Each of these scattered sensor nodes has the capabilities to collect data and route data back to the sink and the end users. Data are routed back to the end user by a multi-hop infrastructure-less architecture through the sink. The sink may communicate with the task manager node via the Internet or satellite.

The protocol stack utilized by the sink and the sensor nodes is given in Fig. 9.8. This protocol stack combines power and routing awareness. It integrates data with networking protocols, communicates power efficiently through the wireless medium, and promotes cooperative efforts of sensor nodes. The protocol stack consists of the application layer, transport layer, network layer, data link layer, physical layer, power management plane, mobility management plane, and task management plane [12]. Different kinds of application software can be built and used on the application layer depending on the sensing tasks. This layer creates hardware and software of the lowest layer transparent to the end user. The transport layer helps to take care of the flow of data if the sensor network application requires it. The network layer takes care of routing the data delivered by the transport layer, specific multi-hop wireless routing protocols between sensor nodes and sink. The data link layer is liable for multiplexing of data streams, frame detection, media access control (MAC), and error control. Since the setting is very noisy and sensor nodes are often mobile, the MAC protocol should be power aware and ready to

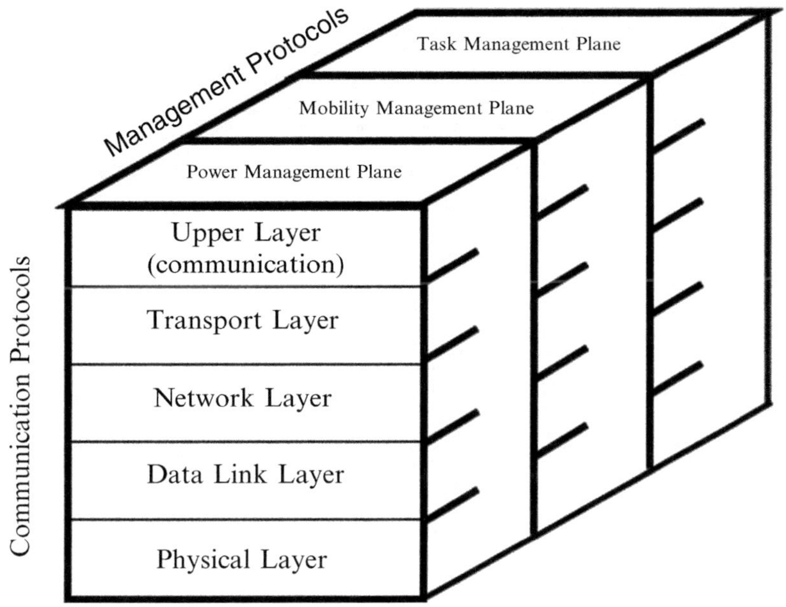

Fig. 9.8 Wireless sensor network protocol stack

reduce collision with neighbors' broadcast. The physical layer addresses the requirements of a straightforward however strong modulation, frequency selection, data encryption, transmission and receiving techniques.

In addition, the power, mobility, and task management planes monitor the power, movement, and task distribution among the sensor nodes. These planes facilitate the sensor nodes which coordinate the sensing task and lower the overall energy consumption.

Figure 9.8 depicts a generic protocol stack model that can be utilized to describe the communication equipment (also see Table 9.1). Table 9.2 shows some typical lower-layer protocols that are in principle applicable to WSNs; overall, a light-weight protocol stack is sought for WSNs. The problems here relate to the following:

9.10 Security Goals for Sensor Networks

The security goals are categorized as primary and secondary. The primary goals are called as standard security goals such as confidentiality, integrity, authentication, and availability (CIAA). The secondary goals are data freshness, self-organization, time synchronization, and secure localization.

Table 9.1 Possible WSN protocol stack

Upper layer	In-network applications, including application processing, data aggregation, external querying query processing, and external database
Layer 4	Transport, including data dissemination and accumulation, caching, and storage
Layer 3	Networking, including adaptive topology management and topological routing
Layer 2	Link layer (contention): channel sharing (MAC), timing, and locality
Layer 1	Physical medium: communication channel, sensing, actuation, and signal processing

Table 9.2 Possible lower-layer WSN protocols

	GPRS/GSM 1xRTT/CDMA	IEEE 802.11b/g	IEEE 802.15.1	IEEE 802.15.4
Market name for standard	2.5G/3G	Wi-Fi	Bluetooth	ZigBee
Network agent	WAN/MAN	WLAN and hotspot	PAN and desk area network (DAN)	WSN
Application focus	Wide area voice and data	Enterprise applications (data and VoIP)	Cable replacement	Monitoring and control
Bandwidth (mbps)	0.064–0.128+	11–54	0.7	0.020–0.25
Transmission range	3000+	1–300+	1–30+	1–300+
Design factors	Range and transmission quality	Enterprise support, scalability, and cost	Cost, ease of use	Reliability, power, and cost

- *Data Confidentiality*

  Confidentiality is the ability to hide messages from a passive attacker in order that any message communicated via the sensor network remains confidential. This can be the most main problem in network security. A sensor node should not reveal its data to the neighbors.
- *Data Authentication*

  Authentication guarantees the consistency of the message by recognizing its origin.
- *Data Integrity*

  Data integrity in sensor networks is required to confirm the reliability of data and refers to the ability to confirm that the message has not been tempered with, altered, or modified. While the network has confidentiality measures, there is still a clear stage that the data integrity has been cooperated by alterations. The integrity of the network is going to be in trouble when:

  – Any one node acts as malicious node in the network and also injects false data.
  – Unbalanced conditions due to wireless channel basis damage or loss of data.

- *Data Availability*
  Availability determines whether a node has the ability to use the resources and whether the network is available for the messages to communicate. However, failure of the base station or cluster leader's accessibility will finally threaten the entire sensor network. Thus, availability is of main significance for maintaining an operational network.

Secondary Goals

- *Data Freshness*
  Even if data confidentiality and data integrity are assured, there's a need to confirm the freshness of every message. Informally, data freshness suggests that the data is recent, and it ensures that no previous messages are replayed. To resolve this issue, a nonce or another time-related counter is often extra into the packet to confirm data freshness.
- *Self-Organization*
  A wireless sensor network is usually an ad hoc network that requires every sensor node to be autonomous and flexible and sufficient and to be self-healing and self-organizing consistent with completely different situations. There is no fixed infrastructure available for the aim of network management in a sensor network. This inherent feature brings an excellent challenge to wireless sensor network security. If self-organization is lacking in a sensor network, the injury resulting from an attack or even the risky environment may be devastating.
- *Time Synchronization*
  Most sensor network applications believe some kind of time synchronization. Sensors might need to calculate the end-to-end delay of a packet as it travels between two pairwise sensors. An additional collaborative sensor network might need cluster synchronization for tracking applications.
- *Secure Localization*
  Often, the utility of a sensor network can believe its ability to accurately mechanically find every sensor within the network. A sensor network designed to find faults can correct location information so as to pinpoint the location of a fault. Regrettably, an attacker will simply manipulate nonsecured location information by reporting false signal strengths, replaying signals.
- *Security Mechanisms*
  In WSN the security mechanisms may literally want to observe, prevent, and get over the security attacks. A large form of security schemes is often invented to counter malicious attacks, and these are often classified as high level and low level. Low-level security primitives for securing sensor networks include:

  – *Key Establishment and Trust Setup*
    The primary constraint of setting up the sensor network is the establishment of cryptographic keys. Generally the sensor devices have restricted process power, and also the public key cryptographic primitives are too expensive to follow. Key establishment techniques need to scale to networks with hundreds or thousands of nodes.

- *Secrecy and Authentication*
  Most of the sensor network applications need protection against eavesdropping, injection, and modification of packets. The earliest sensor networks are probably to use link-layer cryptography, as a result of this approach that provides the greatest ease of deployment among currently available network cryptographic approaches.
- *Privacy*
  Like other traditional networks, the sensor networks have also force privacy consideration. At the start, the sensor networks are deployed for legitimate purpose which might subsequently be used in unforeseen ways. Providing awareness of the presence of sensor nodes and data acquisition is particularly important.
- *Robustness to Communication Denial of Service*
  An adversary attempts to interrupt the network's operation.
- *Secure Routing*
  Routing and data forwarding may be a crucial service for enabling communication in sensor networks. Unfortunately, current routing protocols suffer from many security vulnerabilities.

9.11 Security Threats and Attacks in WSN

Attackers may devise different types of security threats to make the WSN system unstable. Wireless networks are vulnerable to security attacks due to the broadcast nature of the transmission medium.

9.11.1 Security Threats

According to the capability of attacker [13], threats in WSNs can be classified into the following categories:

- *External Versus Internal Attacks*
  External attacks return from nodes that don't belong to a WSN. Outsider has no access to most cryptographic materials in sensor network. External attacks might cause passive eavesdropping on data transmissions moreover as it will reach injected false data into the network to consume network resources and lift denial-of-service (DoS) attack. On the contrary, the internal attacks occur once legitimate nodes of a WSN behave in unplanned or unauthorized ways. An internal attacker or insider is a licensed participant within the sensor network who seeks to disrupt operations or exploit organizational assets.

- *Passive Versus Active Attacks*
  Passive attacks include eavesdropping or observing packets exchanged within a WSN, whereas active attacks involve some modifications of the data stream or the creation of a false stream.
- *Mote-Class Versus Laptop-Class Attacks*
  In mote-class (sensor-class) attacks, an adversary attacks a WSN by using a few nodes with related capabilities as that of network nodes. In laptop-class attacks, an adversary can use more powerful devices like laptop, etc. and may do far more damage to a network than a malicious sensor node. These devices have a processing power, greater transmission range, and energy reserve than the network nodes.
  In a sensor network, sensors monitor the changes of specific parameters and report back to the sink according to the need. Whereas sending the report, the information in transit could also be attacked to supply wrong information to the base stations or sinks. The weakness in a system security design, implementation, configuration, or limitations that would be exploited by attackers is thought as vulnerability or flaw.

9.11.2 Security Attacks

Attacks on the computer system or network are often broadly classified as interruption, interception, modification, and fabrication [14].

- *Interruption:* Interruption is an attack on the availability of the network, for example, insertion of malicious code, message corruption, physical capturing of the nodes, etc. The main purpose is to launch DoS attacks.
- *Interception:* Interception is an attack on confidentiality. The sensor network can be compromised by an adversary to achieve unauthorized access to sensor node or data stored within it.
- *Modification:* Modification is an attack on integrity. Modification means an unauthorized party not only accesses the data but also tampers it, for example, by modifying the data packets being transmitted or causing a DoS attack such as flooding the network with bogus data. The major purpose is to confuse or mislead the parties concerned within the communication protocol. This can be typically aimed toward the network layer and also the application layer, as a result of the richer semantics of those layers.
- *Fabrication:* Fabrication is an attack on authentication. In fabrication, an adversary injects false data and compromises the trustworthiness of the information relayed. This threatens the message faithfulness. This operation also can facilitate DoS attacks by flooding the network.

Attacks also can be classified as host-based and network-based attacks:

- Host-based attacks

  It is further divided into three types: software compromise, hardware compromise, and user compromise:

  - Software compromise involves breaking the software running on the sensor nodes (buffer overflows).
  - Hardware compromise involves tampering with the hardware to extract the program code, data, and keys stored within a sensor node.
  - User compromise involves compromising the users of a WSN, e.g., by cheating the users into revealing information like passwords or keys concerning the sensor nodes.

- Network-based attacks

  It has two perspectives: layer-specific compromises and protocol-specific compromises. This includes all the attacks on information in transit. Aside from that it additionally includes deviating from protocols. Attacker gains an unfair advantage for itself in the usage of the network. In addition, the attacker manifests selfish behaviors, i.e., behaviors that deviate from the intended functioning of the protocol.

9.11.3 Layering-Based Attacks

Though there is no such standard layered architecture of the communication protocol for WSN, here, we have summarized possible attacks and their security resolution approaches in different layers with respect to ISO–OSI layer in Table 9.3 [15, 16].

Most of the routing protocols proposed for ad hoc and sensor network don't seem to be designed to handle security connected problems. Therefore, there's plenty of scope for attacks on them. Completely different probable attacks on the flow of data and control information are often classified as in [17]. Two types of attacks in physical layer are (1) jamming and (2) tampering:

- *Jamming:* This is one of the DoS attacks in which the adversary attempts to disrupt the operation of the network by distribution a high-energy signal. To defense against this attack, use spread-spectrum techniques for radio communication.
- *Tampering:* Sensor networks generally operate in outside environments. Because of unattended and distributed nature, the nodes in an exceedingly WSN are extremely at risk of physical attacks [18]. The physical attacks might cause irreversible injury to the nodes. The challenger can extract cryptographic keys from the captured node, tamper with its circuitry, change the program codes, or even replace it with a malicious sensor.

Table 9.3 Layering-based attacks and possible security approach

Layer	Attacks	Security approach
Physical layer	Jamming and tampering	Use spread-spectrum techniques and Medium Access Control (MAC) layer admission control mechanism
Data link layer	Jamming and collision	Use error-correcting codes and spread-spectrum techniques
Network layer	Sinkhole	Redundancy checking
	Sybil	Authentication, monitoring
	Wormhole	Authentication, probing
	Hello flood	Authentication, packet leashes by geographical and temporal info.
	ACK flooding	Authentication, bidirectional link authentication, verification
Transport layer	Injects false messages and energy drain attacks	Authentication
Application layer	Flooding	Client puzzles
	Desynchronization	Authentication
	Attacks on reliability	Cryptographic approach

The link layer is responsible for MAC, data frame detection, error control, and multiplexing of data streams [19]. Attacks at this layer include purposefully unfairness in allocation, resource exhaustion, and created collisions:

- *Continuous Channel Access (Exhaustion):* A malicious node disrupts the MAC protocol by always requesting or transmitting over the channel. This eventually leads a starvation for other nodes in the network with respect to channel access. One of the countermeasures to such an attack is rate limiting to the MAC admission control such that the network can ignore excessive requests. The second technique is to use time-division multiplexing.
- *Collision:* This is very much similar to the continuous channel attack. A collision occurs when two nodes attempt to transmit on the same frequency simultaneously. When packets collide, a change will likely occur in the data portion, causing a checksum mismatch at the receiving end. The packet will then be discarded as invalid.
- *Unfairness:* Repeated application of these exhaustion- or collision-based MAC layer attacks, or an abusive use of cooperative MAC layer priority mechanisms, can lead into unfairness. This kind of attack is a partial DOS attack, but results in marginal performance degradation. One major defensive measure against such attacks is the usage of small frames, so that any individual node seizes the channel for a smaller duration only.
- *Denial of Service (DoS):* Denial of service (DoS) [20, 21] is produced by the unintentional failure of nodes or malicious action. This attack is a pervasive

threat to most networks. Sensor networks being very energy sensitive and resource limitations are very vulnerable to DoS attacks. Wood and Stankovic [22] explored various DoS attacks that may happen in every network layer of sensor networks. The simplest DoS attack tries to exhaust the resources available to the victim node by sending extra unnecessary packets, thus preventing legitimate network users from accessing services or resources to which they are entitled. DoS attack is meant not only for the adversary's attempt to subvert, disrupt, or destroy a network but also for any event that diminishes a network's capability to provide a service. In WSNs, several types of DoS attacks in different layers might be performed. At the physical layer, DoS attacks could be jamming and tampering. At the link layer, DoS attacks are collision, exhaustion, and unfairness, whereas at the network layer, attacks are neglect and greed, homing, misdirection, and black holes. Moreover, at the transport layer, this attack could be performed by malicious flooding and desynchronization.

The network layer of WSNs is vulnerable to the different types of attacks such as: spoofed routing information, selective packet forwarding, sinkhole, Sybil, wormhole, hello flood, acknowledgment spoofing, etc.:

- *Spoofed, altered, or replayed routing information:* This is the most common direct attack against a routing protocol. This attack targets the routing information exchanged between the nodes. Adversaries may be able to create routing loops, attract or repel network traffic, extend or shorten source routes, generate false error messages, partition the network, and increase end-to-end latency. The standard solution for this attack is authentication, i.e., routers will only accept routing information from valid routers.
- *Selective forwarding:* In a multi-hop network like a WSN, all the nodes need to forward messages accurately. An attacker may compromise a node in such a way that it selectively forwards some messages and drops others [23].
- *Sybil attack:* In this attack, a single node presents multiple identities to all other nodes in the WSN. This may mislead other nodes. Hence, routes believed to be used by disjoint nodes with respect to node that can have the same adversary node. A countermeasure to Sybil attack is the use of a unique shared symmetric key for each node with the base station. Sybil attack is defined as a malicious device illegitimately taking on multiple identities. In Sybil attack [24], an adversary can appear to be in multiple places at the same time. In other words, a single node presents multiple identities to other nodes in the sensor network either by fabricating or stealing the identities of legitimate nodes.
- *Sinkhole attack:* By sinkhole attack, the adversary tries to attract nearly all the traffic from a particular area through a compromised node. A compromised node which is placed at the center of some area creates a large "sphere of influence," attracting all traffic destined for a base station from the sensor nodes. The attacker targets a place to create a sinkhole where it can attract the most traffic, possibly closer to the base station so that the malicious node could be perceived as a base station.

- *Hello flood attack:* Many protocols require nodes to broadcast HELLO packets for neighbor discovery, and a node receiving such a packet may assume that it is within the (normal) radio range of the sender. A laptop-class attacker with large transmission power could convince every node in the network that the adversary is its neighbor, so that all the nodes will respond to the HELLO message and waste their energy. We can prevent this attack by verifying the bidirectionality of local. Another way to prevent the HELLO flood attack is the use of authenticated broadcast protocols.
- *Wormhole:* A wormhole is a low-latency link between two portions of a network over which an attacker replays network messages [23]. This link may be established either by a single node forwarding messages between two adjacent but otherwise non-neighboring nodes or by a pair of nodes in different parts of the network communicating with each other. The latter case is closely related to a sinkhole attack as an attacking node near the base station can provide a one-hop link to that base station via the other attacking node in a distant part of the network.
- *Acknowledgment spoofing:* Several sensor network routing algorithms rely on implicit or explicit link-layer acknowledgments. Due to the inherent broadcast medium, an adversary can spoof link-layer acknowledgments for "overheard" packets addressed to neighboring nodes. Protocols that choose the next hop based on reliability issues are susceptible to acknowledgment spoofing. This results in packets being lost when traveling along such links.
- *Sniffing attack:* Sniffing attack is a good example of interception or listen-in channel attack. In this attack an adversary node is placed in the proximity of the sensor grid to capture data. The collected data is transferred to the intruder by some means for further processing. Sniffing attacks can be prevented by using proper encryption techniques for communication.

The attacks that can be launched on the transport layer in a WSN are flooding attack and desynchronization attack:

- *Flooding:* An attacker may repeatedly make new connection requests until the resources required by each connection are exhausted or reach a maximum limit. It produces severe resource constraints for legitimate nodes. One proposed solution to this problem is to require that each connecting client demonstrates its commitment to the connection by solving a puzzle. As a defense against this class of attack, a limit can be put on the number of connections from a particular node.
- *Desynchronization:* Desynchronization refers to the disruption of an existing connection [22]. An attacker may, for example, repeatedly spoof messages to an end host causing the host to request the retransmission of missed frames.

There are also other attacks like energy drain attack, black hole attack, homing, and node replication attacks:

- *Energy-drain attacks:* This is battery powered and dynamically organized. It is difficult or impossible to replace/recharge sensor node batteries. Because there is a limited amount of energy available, attackers may use compromised nodes to inject fabricated reports into the network or generate a large amount of traffic in the network.

- *Black hole attack:* The black hole attack positions a node in range of the sink and attracts the entire traffic to be routed through it by advertising itself as the shortest route.
- *Homing:* Another interesting type of attack is homing. In a homing attack, the attacker looks at the network traffic to deduce the geographic location of critical nodes, such as cluster heads or neighbors of the base station. The attacker can then physically disable these nodes.
- *Node replication attacks:* This is an attack where the attacker tries to mount several nodes with the same identity at different places of the existing network. There are two methods for mounting this attack. In the first method, the attacker captures one node from the network, creates clones of a captured node, and mounts in different places of the network. In the second method, an attacker may generate a false identification of a node, then makes a clone out of this node, and mounts in different places of the network.

Depending on the network architecture and information used while taking routing decision, routing protocol in WSNs can be classified into flat-based routing, hierarchical-based routing, location-based routing, and network flow or quality of service (QoS)-aware routing.

9.12 Security Protocols

In this section, TinySec, MiniSec, IEEE 802.15.4, SPINS, LSec, LLSP, LISA, and LISP are described.

9.12.1 TinySec

TinySec [25] developed by the University of Berkeley is a link-layer security architecture that has been included in the TinyOS version. Its design is based on ease of use and minimal load brought on sensor network. TinySec supports two different security options: encryption with identity authentication and only authentication. In identity authentication encryption, data is encrypted and an identity authentication code (MAC) is added to the package. However, in only authentication method, data is not encrypted but only authentication of the package is realized with a MAC. As it is understood from this, in TinySec, the identity authentication is a must for each package, but encrypting the data is an option that can be decided according to the application. In encryption of messages, Skipjack block encryption, 8-bit initialization vector (IV), and code block chaining (CBC) are used. There is no restriction on keying method; in practice, a single key pair (one for the encryption of data and the other for the calculation of MACs) is selected for the whole network according to the desired level of security. TinySec at the tightest security level where identity authentication encryption is used brings 10 % extra load on energy, delay, and bandwidth. However, in cases where only authentication is used, this ratio drops to 3 %.

9.12.2 SPINS

SPINS [26], developed by Berkeley University, consists of μTESLA protocol used in identity authentication broadcasting, SNEP protocol providing confidentiality, identity authentication between two nodes and data freshness, and a routing protocol based on these. SNEP offers the below possibilities:

1. Semantic security: semantic security, meaning an attacker listening to the network cannot obtain any information about the plain text even if more than one encrypted copy of the same plain text is received, is realized by a counter shared between the receiver and the sender and incremented in each message exchange.
2. Identity authentication: the receiving node verifies the identity of the sender with the MAC used.
3. Recursion protection: the counter in MAC prevents old messages to be sent again.
4. Weak freshness: the counter used between the receiver and sender for semantic security ensures the message received is sent after the previous one.
5. Low communication overhead: keeping the counter on the receiver and sender, not placing it in the message, reduces communication overhead.

In conventional approaches, identity authentication is done by asymmetrical methods. However, hardware restrictions of sensors are highly insufficient for the quite expensive asymmetrical methods. μTESLA gives the logic of asymmetry to identity authentication with symmetric methods. The sender creates a MAC for the message packages to be broadcasted by using a key known by only itself and by using a one-way function. It broadcasts the key of the message a certain time after the message is sent. Thus, the possibility of changing the contents of the package is removed. At the receiver end, the package kept in a buffer memory is authenticated by using this key. RC5 is used in encryption. For all this identity authentication process, μTESLA needs synchronization between the receiver and the sender even if it is loose.

9.12.3 LISP

LISP aims security solutions in large-scale wireless networks consisting of a large number of nodes with limited resources. To scale networks consisting of a large number of nodes, Park and Shin divide them into clusters, select a head for each cluster, and create a key server. LISP [27] (lightweight security protocol) has a new switching mechanism. It uses switching mechanism by using head cluster and key servers. The advantages are: (1) it uses an effective key broadcast which does not need ACKs to be sent, (2) it uses check bits created without adding them to the data message, (3) it might recover the lost keys, (4) it refreshes key without data encryption or decryption.

The benefits of LISP in protecting critical information against attacks can be summarized as follows: (1) data integrity prevents tampering of data that is sent. (2) Access control is achieved by controlling the inputs to the network. (3) Key refreshing provides protection against nodes that may jeopardize the network.

LISP protocol may combine together with security the other services (routing, data distribution, and location). LISP is a flexible and energy-sensitive protocol. In addition, because it does not need ACK and other control packages, it is quite strong against DoS [28] attacks.

9.12.4 IEEE 802.15.4

IEEE 802.15.4 [29, 30] defines medium access and physical layers for wireless private area networks (WPANs). Although this protocol was not developed for WSN, it is used in WSNs because of its low power consumption, low cost, and flexibility. Currently, this protocol works on Micaz, TelosB nodes produced by the company CrossBow. ZigBee strong encryption AES-128 is used. ZigBee provides freshness. Controlling freshness prevents repeated attacks. Counter is reset when a new key is created. ZigBee provides integrity and prevents an attacker from changing the message. Integrity options are 0, 32, 64, and 128 bit, by default 64 bit. ZigBee provides authentication. Authentication tests whether the right person is reached or not and prevents the attacker showing the device like another one. Authentication is possible at the network and device levels. Authentication at the network level is achieved by using a public network key. Authentication at device level is achieved by using the unique link key between devices. ZigBee provides encryption and prevents an attacker from intercepting and listening. ZigBee uses 128-bit AES encryption. Encryption security is provided at the network, and at the device level, a public key is used at the network-level encryption. It prevents attacks because of a very low memory usage. The device-level encryption uses a common link key. ZigBee uses three types of keys. The master key provides long-term security between two devices. The link key provides security between two devices. The network key provides security on the network.

9.12.5 LSec

LSec [31] provides authentication and authorization with simple key exchange scheme. Furthermore, it has protection mechanisms against data confidentiality, breaches, and illegal events. There is a variety of security attacks on sensor networks. As examples of DoS, eavesdropping, replay attacks, tempering the message, and malicious nodes can be mentioned. To defend against these types of attacks, LSec uses data confidentiality, identity authentication, data integrity, defense against intruders, and some security mechanisms. These problems can be solved partially when the communication among the nodes is encrypted, but a complete solution requires a strong key exchange and distribution scheme. LSec provides identity authentication and authorization, simple secure key exchange, defense mechanism against breaches, data privacy, and usage of asymmetrical and symmetrical encryption together. LSec protocol is simulated on sensor network simulator and emulator (SENSE). There is no application of it.

9.12.6 LISA

LISA [32] includes security solutions listed below:

1. Semantic security: the same data is encrypted in different ways by increasing the value of the counter after each data.
2. Identity authentication: it ensures that the data is from the right node.
3. Protection against replay attacks: it prevents old messages from being repeated.
4. Weak freshness: base station verifies that the message generated is after the previous one.

9.12.7 MiniSec

MiniSec [33] is implemented on Telos platform. While TinySec provides low security at low power consumption, ZigBee [34] provides high security at high power consumption. According to the authors, MiniSec provides high security at low power consumption. Three techniques are used to achieve this. First, block encryption method is used to provide privacy and authentication. But there is only one pass over the data. Second, initialization vector (IV) is used as a very few bits. Third, basic gaps are used during unicast and broadcast communication. In the unicast mode, the power consumption of the radio is reduced by making extra computations and using synchronized counters. In the broadcast mode, bloom filter mechanism is used. Skipjack is used as the encryption algorithm and OCB as the encryption mode. It is defenseless against DoS attacks.

9.12.8 LLSP

LLSP [35] provides minimum cost identity authentication, data integrity, and semantic security by using only symmetric security algorithms. The key mechanism determines key management issues in WSNs. It includes the questions of how the cryptograph keys are distributed, shared, and updated. An appropriate keying mechanism depends on the factors such as the target hazard model, the network communication in practice, security requirements, and ease of use.

9.13 Security Mechanism

The security mechanisms are actually used to detect, prevent, and recover from the security attacks. A wide variety of security schemes can be invented to counter malicious attacks, and these can be categorized as high level and low level. Figure 9.9 shows the order of security mechanisms.

9.13.1 Low-Level Mechanism

Low-level security primitives for securing sensor networks include:

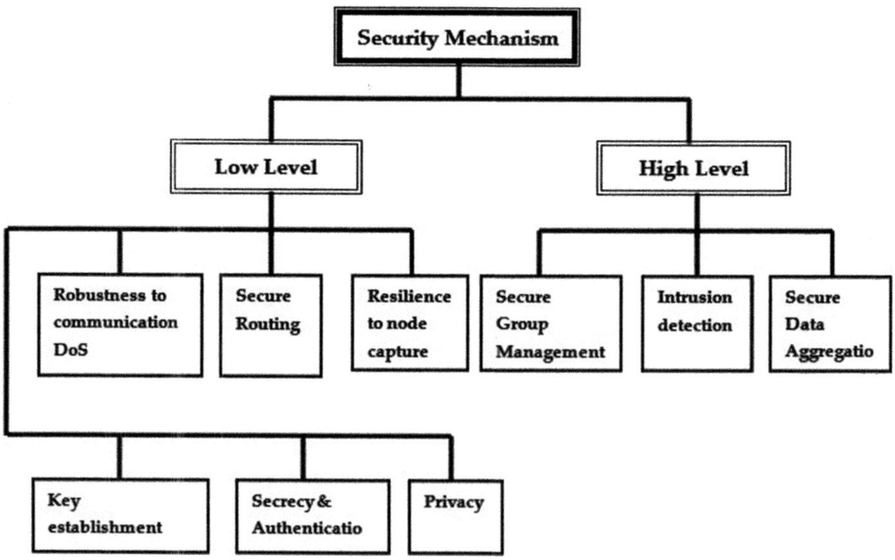

Fig. 9.9 Security mechanism

1. Key establishment and trust setup
2. Secrecy and authentication
3. Privacy
4. Robustness to communication denial of service
5. Secure routing
6. Resilience to node capture

1. Key establishment and trust setup
   The primary requirement of setting up the sensor network is the establishment of cryptographic keys. Generally the sensor devices have limited computational power, and the public key cryptographic primitives are too expensive to follow. Key establishment techniques need to scale to networks with hundreds or thousands of nodes. In addition, the communication patterns of sensor networks differ from traditional networks; sensor nodes may need to set up keys with their neighbors and with data aggregation nodes. The disadvantage of this approach is that attackers who compromised sufficiently and many nodes could also reconstruct the complete key pool and break the scheme [36].

2. Secrecy and authentication
   Most of the sensor network applications require protection against eavesdropping, injection, and modification of packets. Cryptography is the standard defense. Remarkable system trade-offs arise when incorporating cryptography into sensor networks. For point-to-point communication [37], end-to-end cryptography achieves a high level of security but requires that keys be set up among all end points and be incompatible with passive participation and local broadcast. Link-layer cryptography with a network-wide shared key simplifies key setup and supports passive participation and local broadcast, but intermediate nodes might eavesdrop or alter messages. The earliest sensor networks are likely to use link-layer cryptography, because this approach provides the greatest ease of deployment among currently available network cryptographic approaches [38].

3. Privacy
   Like other traditional networks, the sensor networks have also force privacy concerns. Initially the sensor networks are deployed for legitimate purpose which might subsequently be used in unanticipated ways. Providing awareness of the presence of sensor nodes and data acquisition is particularly important [36].

4. Robustness to communication denial of service
   An adversary attempts to disrupt the network's operation by broadcasting a high-energy signal. If the transmission is powerful enough, the entire system's communication could be jammed. More sophisticated attacks are also possible; the adversary might inhibit communication by violating the 802.11 medium access control (MAC) protocol by, say, transmitting while a neighbor is also transmitting or by continuously requesting channel access with a request-to-send signal [36].

5. Secure routing
   Routing and data forwarding is a crucial service for enabling communication in sensor networks. Unfortunately, current routing protocols suffer from many security vulnerabilities. For example, an attacker might launch denial-of-service attacks on the routing protocol, preventing communication. The simplest attacks involve injecting malicious routing information into the network, resulting in routing inconsistencies. Simple authentication might guard against injection attacks, but some routing protocols are susceptible to replay by the attacker of legitimate routing messages [38].
6. Resilience to node capture
   One of the most challenging issues in sensor networks is resiliency against node capture attacks. In most applications, sensor nodes are likely to be placed in locations easily accessible to attackers. Such exposure raises the possibility that an attacker might capture sensor nodes, extract cryptographic secrets, modify their programming, or replace them with malicious nodes under the control of the attacker. Tamper-resistant packaging may be one defense, but it's expensive, since current technology does not provide a high level of security. Algorithmic solutions to the problem of node capture are preferable [36].

9.13.2 High-Level Mechanism

High-level security mechanisms for securing sensor networks include secure group management, intrusion detection, and secure data aggregation:

1. Secure group management
   Each and every node in a wireless sensor network is limited in its computing and communication capabilities. However, interesting in-network data aggregation and analysis can be performed by groups of nodes. For example, a group of nodes might be responsible for jointly tracking a vehicle through the network. The actual nodes comprising the group may change continuously and quickly. Many other key services in wireless sensor networks are also performed by groups. Consequently, secure protocols for group management are required, securely admitting new group members and supporting secure group communication. The outcome of the group key computation is normally transmitted to a base station. The output must be authenticated to ensure it comes from a valid group [36].
2. Intrusion detection
   Wireless sensor networks are susceptible to many forms of intrusion. Wireless sensor networks require a solution that is fully distributed and inexpensive in terms of communication, energy, and memory requirements. The use of secure groups may be a promising approach for decentralized intrusion detection [36].

3. Secure data aggregation
   One advantage of a wireless sensor network is the fine-grain sensing that large and dense sets of nodes can provide. The sensed values must be aggregated to avoid overwhelming amounts of traffic back to the base station. For example, the system may average the temperature of a geographic region, combine sensor values to compute the location and velocity of a moving object, or aggregate data to avoid false alarms in real-world event detection. Depending on the architecture of the wireless sensor network, aggregation may take place in many places in the network. All aggregation locations must be secured [38].

9.14 Sensor Network Applications

The evolution of sensor networks into a generic wireless "sensing layer" for computer systems has opened a wide range of application possibilities. These kinds of networks cannot be considered as the "panacea" for all the sensing needs of computer systems, since they are not specially suitable for very complex applications (such as the Large Hadron Collider [39]) or applications with hard quality of service (QoS) requirements. Nevertheless, because of their particular characteristics, the benefits of using wireless sensor networks for sensing the real world are numerous:

- The network can survive in its deployment area for long periods of time.
- The network can be able to run unattended.
- The overall cost of the network is low, due to the price of the nodes and the use of a wireless interface.
- The network can be deployed by nonexperts and can be easy to maintain.
- The network can react to abnormal events, maintaining the availability of its services.

There are many types of applications that can take advantage of all these benefits. According to Culler et al. [40], those applications can be classified into the following categories:

1. Monitoring space: The sensor network simply monitors the physical features of a certain environment. This category includes applications such as environmental and habitat monitoring, precision agriculture, indoor climate control, surveillance, treaty verification, and intelligent alarms.
2. Monitoring things: The sensor network controls the status of a physical entity. This category includes applications such as structural monitoring, ecophysiology, condition-based equipment maintenance, medical diagnostics, and urban terrain mapping.
3. Monitoring interactions: The sensor network monitors the interactions of things (both inanimate and animate) with each other and the encompassing space. This category includes applications such as wildlife habitats, disaster management,

critical (information) infrastructure systems, emergency response, asset tracking, healthcare, and manufacturing process flow.

While all the application areas presented in the previous classification are mere ideas of where wireless sensor networks could be applied, the research communities have already proven the usefulness of wireless sensor networks in real-world settings. Some examples of prototypes and research areas include: monitoring of aging infrastructures [41], critical (information) infrastructures [42], surveillance [43], detecting equipment vibration [44], control of vineyards [45], water quality analysis [46], control of glacier behavior [47], monitoring of an active volcano [48], habitat monitoring [49], firefighter's assistance [50], monitoring of assisted-living residents [51–53], healthcare [54], smart environments [55], checking availability of washing machines [56], and optimization of HVAC systems [57]. In fact, wireless sensor networks have already jumped from the research laboratories to the commercial world, with applications such as precision agriculture [58], pipeline and freighter monitoring [44], and many others.

Not only wireless sensor networks are useful for providing services to specific applications but also are an important part of the "pervasive computing" paradigm and the "Internet of Things" paradigm due to their "sensing layer" nature. Envisioned by Mark Weiser in 1991 [59] and continued by other visionaries such as Satyanarayanan [60], the major idea behind pervasive computing is that "computers will disappear into the background, weaving themselves into the fabric of everyday life until they are indistinguishable from it." In other words, the environment itself will be saturated with computing and communication capabilities, yet gracefully integrated with human users. The elements of that augmented environment (e.g., urban spaces [61]) will collaborate to process information coming from both the real world (e.g., traffic conditions, room temperatures) and the virtual world (e.g., internal state of traffic lights and air conditioner), providing context-adapted services to themselves, other machines, and humans.

On the other hand, the vision of the "Internet of Things" [62] considers that all the elements of these pervasive environments will be connected to the Internet; thus, objects and locations in the real world (e.g., restaurants, roads) will be linked to information on the web. As a result, it will be possible to convert real-world events into meaningful information that could be analyzed anywhere, providing real entities (e.g., humans) and virtual entities with the capability of making better decisions. In both paradigms, the sensor networks play an important role either by providing an intelligent environment the ability to understand its context or by enhancing the capabilities of a single entity (e.g., a toy) with the ability to "sense."

By using the wireless sensor network paradigm, we can be able to create a heterogeneous set of applications linking the real world with the virtual world. However, a specific configuration of a wireless sensor network may not work for two different applications. In fact, different applications have different requirements, and these requirements will have a significant influence over the actual structure, architecture, and protocols of a wireless sensor network. For example, we can cite the transmission media used in the wireless communications. In underwater sensor networks (UWSN [63]), sensor nodes are deployed below the ocean surface, and it is necessary to use underwater acoustic modems, which have low bandwidth and are prone to communication errors. On the other hand, in underground sensor networks (USN [64]), sensor nodes are deployed completely below the ground, where it is necessary to cope with extra challenges such as path loss due to material absorption. These differences in the transmission channel will influence the behavior of the sensor nodes and the behavior of the sensor network as a whole.

9.15 Conclusion and Future Work

The aim of this chapter is to discuss few important issues of WSNs, from the application, design, and technology points of view. For designing a WSN, we need to consider different factors such as the flexibility, energy efficiency, fault tolerance, high sensing fidelity, low cost, and rapid deployment, above all the application requirements. However, realization of sensor networks needs to satisfy several constraints such as scalability, cost, hardware, topology change, environment, and power consumption. Since these constraints are highly tight and specific for sensor networks, new wireless ad hoc networking protocols are required. To meet the requirements, many researchers are engaged in developing the technologies needed for different layers of the sensor network protocol stack.

Future research on WSN will be directed toward maximizing area throughput in clustered wireless sensor networks designed for temporal or spatial random process estimation; accounting for radio channel, PHY, MAC, and NET protocol layers and data aggregation techniques; simulation and experimental verification of lifetime-aware routing and sensing spatial coverage; and the enhancement of the desired sensing spatial coverage evaluation methods with practical sensor model.

The advances of wireless networking and sensor technology open up an interesting opportunity to manage human activities in a smart home environment. Real-life activities are often more complex than the case studies for both single user and multiuser. Investigating such complex cases can be very challenging while we consider both single and multiuser activities at the same time.

Future work will focus on the fundamental problem of recognizing activities of multiple users using a wireless body sensor network. Wireless sensor networks hold the promise of delivering a smart communication paradigm which enables setting up an intelligent network capable of handling applications that evolve from user requirements. We believe that in the near future, WSN research will put a great impact on our daily life. For example, it will create a system for continual observation of physiological signals while the patients are at their homes. It will lower the cost involved with monitoring patients and increase the efficient exploitation of physiological data, and the patients will have accessWireless sensor networks (WSNs): to the highest-quality medical care in their own homes. Thus, it will avoid the distress and disruption caused by a lengthy inpatient stay.

References

1. Chong, C., Kumar, S.P.: Sensor networks: evolution, opportunities, and challenges. Proc. IEEE **91**(8), 1247–1256 (2003)
2. Nishimura, C.E., Conlon, D.M.: IUSS dual use: monitoring whales and earthquakes using SOSUS. Mar. Technol. Soc. J. **27**(4), 13–21 (1994)
3. Proceedings of the Distributed Sensor Nets Workshop. Carnegie Mellon University, Pittsburgh, USA (1978)
4. Rashid, R., Robertson, G.: Accent: a communication oriented network operating system kernel. ACM SIGOPS Operating Systems Review, vol. 15, no. 5, December 1981
5. Lacoss, R.T.: Distributed mixed sensor aircraft tracking. In: Proceedings of the 6th American Control Conference, pp. 1827–1830, Minneapolis, USA, June 1987
6. Lesser, V.R., Corkill, D.D.: The distributed vehicle monitoring testbed: a tool for investigating distributed problem solving networks. Readings from the AI magazine, pp. 69–85, ISBN: 0-929280-01-6 (1989)
7. Chong, C.Y., Chang, K.C., Mori, S.: Distributed tracking in distributed sensor networks. In: Proceedings of the American Control Conference (1982)
8. DARPA. SensIT—Sensor Information Technology. http://www.sainc.com/sensit/goals.htm
9. Smart-Its Consortium. The Smart-Its Project. http://www.smart-its.org/
10. Akyildiz, I.F., Su, W., Sankarasubramaniam, Y., Cayirci, E.: Wireless sensor networks: a survey. Comput. Netw. **38**(4), 393–422 (2002)
11. Wilson, J.: Technology. Sensor, Handbook. Newnes, Elsevier, Burlington, MA (2005)
12. Akyildiz, I.F., Su, W., Sankarasubramaniam, Y., Cayirci, E.: A survey sensor on I.E.E. networks. IEEE Commun. Mag., **40**(8), pp. 102–114, (2002)
13. Karlof, C., Wagner, D.: Secure routing in sensor networks: attacks and countermeasures. Ad Hoc Netw. J. (Elsevier), Special Issue on Sensor Network (SNPA), pp. 293–315, September 2003
14. Chen, X., Makki, K., Yen, K., Pissinou, N.: Sensor network security: a survey. IEEE Commun. Surv. Tut. **11**(2), 52–73 (2009). Second Quarter
15. Saxena, M.: Security in Wireless Sensor Networks—A Layer Based Classification. Technical Report, Center for Education and Research in Information Assurance and Security-CERIAS, Purdue University. pages.cs.wisc.edu/~msaxena/papers/2007-04-cerias.pdf (2007)
16. Sen, J.: A survey on wireless sensor network security. Int. J. Commun. Netw. Inf. Secur. **1**(2), 59–82 (2009)
17. Mohanty, P., Panigrahi, S., Sarma, N., Satapathy, S.S.: Security issues in wireless data gathering protocols. J. Theor. Appl. Inf. Technol. **13**(1), 14–27 (2010)
18. Wang, X., Gu, W., Schosek, K., Chellappan, S., Xuan, D.: Sensor Network Configuration Under Physical Attacks. Technical Report (OSU-CISRC-7/04-TR45), Department of Computer Science and Engineering, Ohio State University, July 2004
19. Akyildiz, I.F., Su, W., Sankarasubramaniam, Y., Cayirci, E.: A survey on sensor networks. IEEE Commun. Mag. **40**(8), 102–114 (2002)
20. Sharifnejad, M., Shari, M., Ghiasabadi, M., Beheshti, S.: A survey on wireless sensor networks security. SETIT (2007)
21. Wang, B.T., Schulzrinne, H.: An IP traceback mechanism for reflective DoS attacks. In: Canadian Conference on Electrical and Computer Engineering, vol. 2, pp. 901–904, 2–5 May 2004
22. Wood, A.D., Stankovic, J.: Denial of service in sensor network. IEEE Comput. Mag. **35**(10), 54–62 (2002)
23. Karlof, C., Wagner, D.: Secure routing in wireless sensor networks: attacks and countermeasures. In: Proceedings of the 1st IEEE International Workshop on Sensor Network Protocols and Applications, pp. 113–127, May 2003
24. Douceur, J.R.: The Sybil attack. In: First International Workshop on Peer-to-Peer Systems (IPTPS'02), LNCS, vol. 2429, pp. 251–260, March 2002

25. Karlof, C., Sastry, N., Wagner, D.: TinySec: a link layer security architecture for wireless sensor networks. In: Proceedings of the 2nd International Conference on Embedded Networked Sensor Systems (SenSys '04), Baltimore, MD, USA, pp. 162–175, November 2004
26. Perrig, A., Szewczyk, R., Tygar, J.D., Wen, V., Culler, D.E.: SPINS: security protocols for sensor networks. Wirel. Netw. **8**(5), 521–534 (2002)
27. Park, T., Shin, K.G.: LiSP: a lightweight security protocol for wireless sensor networks. ACM Trans. Embed. Comput. Syst. **3**(3), 634–660 (2004)
28. Wood, A.D., Stankovic, J.A.: Denial of service in sensor networks. Computer **35**(10), 54–62 (2002)
29. IEEE-TG15.4, PART 15.4: Wireless Medium Access Control (MAC) and Physical Layer (PHY) Specifications for Low-Rate Wireless Personal Area Networks (LR-WPANs), IEEE Standard for Information Technology (2003)
30. Koubaa, A., Alves, M., Tovar, E.: IEEE 802.15.4: a federating communication protocol for time-sensitive wireless sensor networks. In: Sensor Networks and Configurations: Fundamentals, Techniques, Platforms and Experiments, pp. 19–49. Springer, Berlin, Germany (2007)
31. Shaikh, R.A., Lee, S., Khan, M.A., Song, Y.C.: LSec: lightweight security protocol for distributed wireless sensor network. In: Personal Wireless Communications. Lecture Notes in Computer Science, vol. 4217, pp. 367–377 (2006)
32. Tripathy, S.: LISA: lightweight security algorithm for wireless sensor networks. In: Proceeding of Distributed Computing and Internet Technology, 4th International Conference, ICDCIT 2007, Bangalore, India, December 17–20. Lecture Notes in Computer Science, vol. 4882, pp. 129–134 (2007)
33. Luk, M., Mezzour, G., Perrig, A., Gligor, V.: MiniSec: a secure sensor network communication architecture. In: Proceedings of the 6th International Conference on Information Processing in Sensor Networks (IPSN '07), pp. 479–488, April 2007
34. ZigBee Alliance. Zigbee Specification. Technical Report Document, Version 1.0, ZigBee Alliance, 2005
35. Lighfoot, L.E., Ren, J., Li, T.: An energy efficient link-layer security protocol for wireless sensor networks. In: Proceedings of the IEEE International Conference on Electro/Information Technology (EIT '07), Chicago, IL, USA, pp. 233–238, May 2007
36. Perrig, A., Stankovic, J., Wagner, D.: Security in wireless sensor networks. Commun. ACM **47**(6), 53–57 (2004)
37. Djenouri, D., Khelladi, L., Badache, N.: A survey of security issues in mobile ad hoc and sensor networks. IEEE Commun. Surv. Tut. **7**, 2–28 (2005)
38. Pathan, A.S.K., Lee, H.-W., Hong, C.S.: Security in wireless sensor networks: issues and challenges. In: Advanced Communication Technology (ICACT), 6p (2006)
39. European Organization for Nuclear Research (CERN). LHC—The Large Hadron Collider. http://lhc.web.cern.ch (2008)
40. Culler, D., Estrin, D., Srivastava, M.: Overview of sensor networks. IEEE Comput. **37**(8), 41–49 (2004)
41. Wired and Wireless Intelligent Networked Systems (WINES)—Smart Infrastructure Project. http://www.winesinfrastructure.org/
42. Lopez, J., Montenegro, J.A., Roman, R.: Service-oriented security architecture for CII based on sensor networks. In: Second International Workshop on Security, Privacy and Trust in Pervasive and Ubiquitous Computing (SecPerU 2006), Lyon, France, pp. 1–6, June 2006
43. He, T., Krishnamurthy, S., Luo, L., Yan, T., Gu, L., Stoleru, R., Zhou, G., Cao, Q., Vicaire, P., Stankovic, J.A., Abdelzaher, T.F., Hui, J., Krogh, B.: VigilNet: an integrated sensor network system for energy-efficient surveillance. ACM Trans. Sens. Netw. **2**(1), 1–38 (2006)
44. Sensor Nets/RFID. Intel Corporation. http://www.intel.com/research/exploratory/wireless_sensors.htm
45. Millman, G.J.: Accenture. Virtual Vineyard. http://www.accenture.com/Global/Research_and_Insights/Outlook/By_Alphabet/VirtualVineyard.htm

46. Ramanathan, N., Balzano, L., Estrin, D., Hansen, M., Harmon, T., Jay, J., Kaiser, W.J., Sukhatme, G.: Designing wireless sensor networks as a shared resource for sustainable development. In: Proceedings of the International Conference on Information and Communication Technologies and Development (ICTD 2006), Berkeley, USA, pp. 256–265, May 2006
47. Martinez, K., Padhy, P., Elsaify, A., Zou, G., Riddoch, A., Hart, J.K., Ong, H.L.R.: Deploying a sensor network in an extreme environment. In: Proceedings of Sensor Networks, Ubiquitous and Trustworthy Computing (SUTC 2006), Taichung, Taiwan, pp. 186–193, June 2006
48. Lees, J.M., Johnson, J.B., Ruiz, M., Troncoso, L., Welsh, M.: Reventador Volcano 2005: eruptive activity inferred from seismo-acoustic observation. J. Volcanol. Geotherm. Res. **176** (1), 179–190 (2008)
49. Szewczyk, R., Polastre, J., Mainwaring, A., Culler, D.: Lessons from a sensor network expedition. In: Proceedings of the 1st European Workshop on Wireless Sensor Networks (EWSN 2004), Berlin, Germany, pp. 307–322, January 2004
50. University of California, Berkeley. FIRE Project. http://fire.me.berkeley.edu/
51. Harvard University. The Code Blue Project. http://www.eecs.harvard.edu/~mdw/proj/codeblue
52. Wood, A., Virone, G., Doan, T., Cao, Q., Selavo, L., Wu, Y., Fang, L., He, Z., Lin, S., Stankovic, J.: ALARM-NET: Wireless Sensor Networks for Assisted-Living and Residential Monitoring. Technical Report CS-2006-11, Department of Computer Science, University of Virginia (2006)
53. Cadi Scientific Pte Ltd. CADI SmartSenseTM. http://www.cadi.com.sg
54. Nordic Collaboration Project. Biomedical Wireless Sensor Network Project. http://www.bwsn.net
55. Minder, D., Marr´on, P.J., Lachenmann, A., Rothermel, K.: Experimental construction of a meeting model for smart office environments. In: Proceedings of the First Workshop on Real-World Wireless Sensor Networks (REALWSN 2005), Stockholm, Sweden, June 2005
56. Krishnamurthy, A., Kamasi, B., McCoy, C., Mallela, K.: Project Laund-re-mote
57. Watts, W., Koplow, M., Redfern, A., Wright, P.: Application of Multizone HVAC Control Using Wireless Sensor Networks and Actuating Vent Registers. Texas University A&M Repository (2007)
58. Crossbow Technology, Inc. eKo Pro—Precision Agriculture. http://www.xbow.com/eko/ (2008)
59. Weiser, M.: The computer for the twenty-first century. Sci. Am. **265**(3), 94–104 (1991)
60. Satyanarayanan, M.: Pervasive computing: vision and challenges. IEEE Pers. Commun. **8**(4), 10–17 (2001)
61. Calabrese, F., Kloeckl, K., Ratti, C., Bilandzic, M., Foth, M., Button, A., Klaebe, H., Forlano, L., White, S., Morozov, P., Feiner, S., Girardin, F., Blat, J., Nova, N., Pieniazek, M.P., Tieben, R., van Boerdonk, K., Klooster, S., van den Hoven, E., Martın Serrano, J., Serrat, J., Michelis, D., Kabisch, E.: Urban computing and mobile devices. IEEE Pervasive Comput. **6**(3), 52–57 (2007)
62. International Telecommunication Union. The Internet of Things. ITU Internet Reports (2005)
63. Akyildiz, I., Pompili, D., Melodia, T.: Underwater acoustic sensor networks: research challenges. Ad Hoc Netw. J. (Elsevier) **3**(3), 257279 (2005)
64. Akyildiz, I.F., Stuntebeck, E.P.: Wireless underground sensor networks: research challenges. Ad Hoc Netw. J. **4**(6), 669–686 (2006)

Chapter 10
Emerging Trends in Security System Design Using the Concept of Social Behavioural Biometrics

M.L. Gavrilova, F. Ahmed, S. Azam, P.P. Paul, W. Rahman, M. Sultana, and F.T. Zohra

Abstract This chapter investigates how existing biometric research can be advanced by integrating it with the social behavioural information. Analytical discussions on how social behavioural biometrics can be extracted and applied in various security, and authentication applications will be presented. This chapter also provides some insights onto current and emerging research in the multimodal biometric domain, formulates open questions and investigates future directions. Answers to those questions will assist not only in establishment of the new methods in the biometric security domain but also provide insights into the future emerging topics in the big data analytics and social networking research.

10.1 Introduction

We live in a deeply interconnected society, where aspects of someone's personal and social life, professional affiliations, shopping profiles, hobbies and interests become more and more intervened. One special place where various facets of life become even more prominent is our cyberworld profiles, or so-called online identities. Not surprising, such areas as big data analytics, decision-making, decision fusion, artificial intelligence, pattern recognition and biometrics now have a permanent presence in the online domain. For instance, there is an extensive interest in research based on users of social networks (Facebook, Twitter, etc.) into user preferences and interactions based on the spheres of their interests or their affiliation [1]. Many human-computer interaction studies now include such aspects as computational aesthetics—understanding someone's preferences and interests with the goal of building aesthetic profiles [2]. In big data analytics domain, web browsing histories and social network activity patterns of millions of users are being collected and analysed on a daily basis. On the other hand, in security

M.L. Gavrilova (✉) • F. Ahmed • S. Azam • P.P. Paul • W. Rahman • M. Sultana • F.T. Zohra
University of Calgary, Calgary, AB, Canada
e-mail: marina@cpsc.ucalgary.ca

I.M. Alsmadi et al. (eds.), *Information Fusion for Cyber-Security Analytics*, Studies in Computational Intelligence 691, DOI 10.1007/978-3-319-44257-0_10

research, social behavioural biometric authentication as well as artificial biometrics emerged as powerful research tools, based on the multitude of features exhibited by users in their online interactions [3, 4].

Traditional definition of biometric research scope usually includes recognising someone's identity from collected biometric data [5], which now includes physiological, behavioural, soft and recently introduced social traits [6]. Physiological features (facial, ear, iris, fingerprint) can be often collected through some specialised image or video-capturing devices, such as infrared sensors, remote temperature-measuring devices, Kinect and so on. Behavioural characteristics, which traditionally include the way a person walks (gait), the way a person talks (voice), the way a person writes (typing patterns, keystroke pressure) or the way a person authenticates documents (signature), can be obtained from variety of sensors. Soft biometrics include easily collected but not as unique as the previous two biometric data, i.e. age, gender, height, weight, eye colour or hair colour of a person. Even the way a person socially interacts or expressed their preferences can become a rich source of authentication information. Very recently, a social behavioural biometrics (or social biometrics) was proposed as a new way to obtain supplementary but sometimes crucial for authentication information. Under this umbrella, the social behavioural features are extracted from the way users interact through various social online and offline networks. This includes user's online presence patterns (time, day, month), the nature of interaction (tweets, blogs, chats), the content of interaction (topics, likes, opinions), online game-playing strategies, virtual world avatar preferences, etc. [3, 7]. One of the main features, which is crucial for research on the social behavioural biometrics, is the communication patterns in the networks of users and the composition of such networks themselves. It is generating a lot of interest and getting traction in biometric research, as well as in related fields looking into human interaction, physiological studies, user profiling, pattern recognition, authorship identification and collective intelligence [3, 8]. The idea can be transferred to the real world as well. For instance, in a given social context, some social behavioural patterns such as friends and acquaintances, daily routine, psychological states, style of conversation, gestures and emotions during conversations, preferences, spatial information and activity logs can play important role in recognising a person. Such patterns can also provide a unique set of tools to promote better understanding of collaborative processes, workflow dynamics and risk analysis in collaborative/social environments in real and virtual worlds.

This chapter investigates how existing biometric multimodal systems can be advanced by integrating social behavioural information. Analytical discussions on how social behavioural biometrics can be extracted and applied in various security and authentication applications will be presented. This chapter also provides some insights onto current and emerging research in biometric domain, formulates open questions and investigates future directions in this vibrant research field. The answers to those questions will assist not only in the establishment of new methods in the biometric security domain but also provide insights into the future emerging research topics in our deeply interconnected world.

10.2 Literature Review

The realm of biometric research is experiencing advancement in both breadth and depth directions. The depth advancement includes enhancing the performances of well-established biometrics such as fingerprint, signature, face, etc., by proposing new feature extraction, fusion and classification algorithms. We observe that the most exciting developments are occurring in the breadth direction, where a wide range of physiological and behavioural traits is emerging as new biometrics. In order to take this ongoing advancement one step further, we propose to mine the emerging social behavioural traits from human behaviour such as online networks, aesthetic preference, facial expression/emotions/social signals, brain signals, gait, activities, etc. Although some of these candidates such as expressions, brain signals and activities are being studied in different domains including biometrics, they have not been studied in collaborative/social environment as social behavioural biometrics. In the following sections, we present analytical discussions on the existing works on the proposed social behavioural biometric candidates, i.e. aesthetic preference, emotions/social signals, brain signals and gait and activities.

Social behavioural pattern analysis is the new emerging domain in biometric recognition, led by the team of researchers in the Biometric Technologies Laboratory at the University of Calgary. The idea behind this is to infer behavioural patterns from the day-to-day social interaction. In this domain, the investigation of facial expression or collaborative activity as a social biometrics remains unexplored. Similar to facial expressions, gait- and activity-related soft biometrics have great potential for continuous and unobtrusive authentication. Gait recognition involves identifying a person by analysing his/her walking pattern, which is particularly useful in scenarios where explicit user interaction is not desired.

Activity-related biometric authentication provides an unobtrusive and natural alternative for physiological biometrics that can exploit everyday life activities involving interaction with objects or people for extracting biometric signature. For example, Drosoua et al. [9] proposed a novel activity biometrics based on the motion patterns of prehension movement. Later, Bazazian and Gavrilova [10] utilised behavioural and social contextual metadata to increase the recognition performance of gait recognition based on multimodal fusion. Their proposed method exploited the daily routine activities of users to extract important behavioural patterns and contextual information using a context extractor and matcher.

In recent years, the mass growth of online social networks has introduced a completely new platform of analysing human behaviour. Recent studies showed that behavioural footprints exist in online interactions of users via online social media [3, 6, 11]. Another recent research showed that editing behaviour of users in collaborative environment such as Wikipedia can help to predict the identity of authors [12]. In these works, analysis of social data over a period of time explores underlying behavioural pattern and demonstrated encouraging performance in user identification. The authors identified online social behavioural interactions as the potential candidates of behavioural biometrics. The applications of the proposed social behavioural biometric features are as diverse as person authentication, access control, anomaly detection, customer profiling, behaviour analysis, situation awareness, risk analysis, friend recommendation systems and so on.

10.3 Social Interactions as Biometric Features

Social interactions and behaviour have been the focus of interest for researchers' spanning domains of social science, psychology, neuroscience, organisational behaviour and targeted marketing for many years. However, analysis of social behaviour has not been carried over from the perspective of user authentication. Research in this direction can assist finding the answers to the following open questions:

1. Are there some features extracted from social behavioural information that are permanent and unique enough to be used for individual authentication?
2. What benefits the combination of social behavioural biometrics with traditional biometrics (physiological/behavioural/soft) can provide?
3. Is it possible to use social behavioural activities for risk assessment and security threat prevention?

Recent research in this domain partially addressed the above questions. It postulated that social network analysis can provide an additional insight on how social behaviour of human beings can be utilised as identifying information for use in security and access control systems [3, 7]. This is the first step towards a new generation of biometric systems that are capable of understanding and interpreting social actions/behaviours of humans and utilising them for authentication purposes. This process is analogous to the recognition processes of human brain, where social interactions, emotional responses and contextual information play significant role along with visual perception during recognition of familiar faces [13]. The more knowledge about a person is available, the better recognition results can be achieved. For instance, if a person is very familiar (i.e. a family member, a close friend), one can make a confident guess on his/her identity even in a less than ideal conditions (distant location, insufficient light) based on a negligible amount of information such as appearance, gesture, posture, voice tones, style, content or context of conversation, etc. Thus, we envision the next-generation biometric systems as intelligent machines capable of utilising all aspects of information (physiological, behavioural and social) in order to make a confident human-like decision in unfavourable conditions, where the traditional biometrics would perform poorly. In other words, the next-generation biometric systems should be equipped with the mechanisms of extracting social features from the everyday social interactions of individuals, adaptively expanding behavioural patterns as well as learning new knowledge by analysing those features over time. Thus,

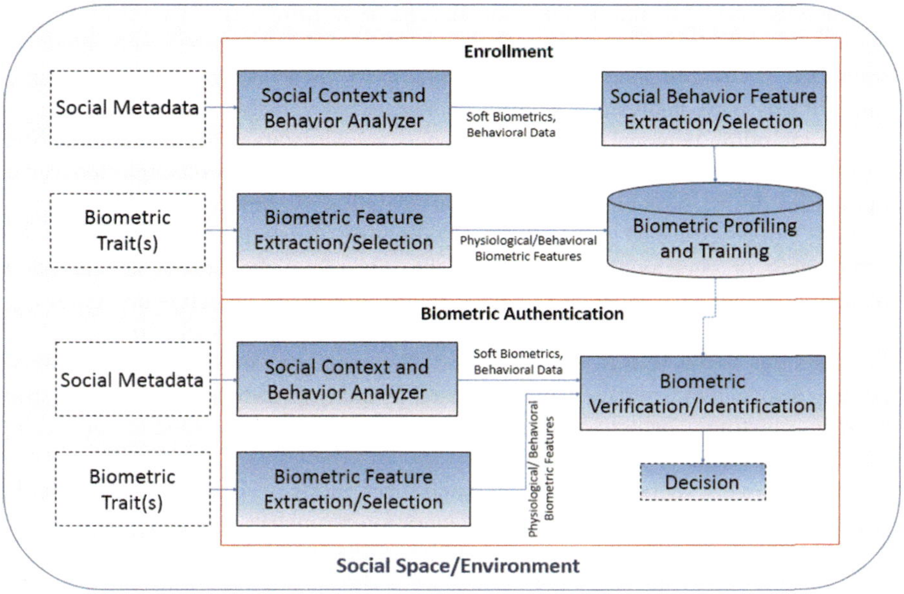

Fig. 10.1 A generalised flow diagram of the enrolment and authentication system, using fusion of social, contextual, behavioural and physiological data

those behavioural and social patterns will be utilised to authenticate individuals with higher confidence even in the absence of other traditional biometric traits.

Knowledge extracted from social behaviour and contextual information can improve accuracy of existing multimodal biometric decision-making systems. It is possible to fuse such social information with existing physiological or behavioural biometrics as part of a multimodal system to make a confident decision on the person's identity. Figure 10.1 presents a generalised framework of the proposed social biometric system. The framework in Fig. 10.1 shows that auxiliary features (e.g. social behavioural features, contextual information, etc.) and soft biometrics (e.g. aesthetic preference, age, gender, profession, interest, community, etc.) can be obtained by analysing everyday social activities of a person in a social environment (e.g. meeting room, family gathering, office space, etc.). During an enrolment phase, features are extracted from physiological and behavioural traits as well as from social data in a given social space. All features are then stored in the training database as biometric profiles of enrolled individuals. During an authentication phase, physiological, behavioural and social features can be extracted depending upon the availability of data; then the test profile will be created using available features. Finally, identification or verification decision is made based on the matching score of the test and training profiles. The major advantage of the proposed fusion of social data and soft biometrics is that a reliable authentication decision can be obtained from the biometric systems regardless of the distortion or missing features of the physiological or behavioural biometrics.

Knowledge about individuals' social behaviour can be discovered by mining their social data in a particular social space [3, 7]. For instance, a person's close friend list can be obtained by keeping a record of his accompanying persons over time. In this way, analysing social data may reveal valuable information about a person including personal choice, preference, social acquaintances and contextual and spatio-temporal information. This information can be directly exploited as soft biometrics during biometric authentication. Alternately, unique social behavioural patterns can be exploited as a modality of biometric trait and fused at feature, match score, rank or decision levels. One of the intriguing phenomena is that social behavioural biometrics can be extracted by observing the known behavioural biometrics (e.g. expression, interactions, gestures, voice, activities, etc.) of individuals in a specific social setting over a period. For instance, an idiosyncratic way of starting a speech of a person can be revealed by analysing voice data acquired from regular meetings, which can act as social behavioural biometrics during authentication [3].

The following sections contain detailed discussions on how social data can be used as biometric features in the next-generation biometric systems.

10.3.1 Online Communication as Social Biometrics

10.3.1.1 Overview

During this era of social media, many of everyday communications have been extended into cyberworld. Social communications can be broadly classified into two categories: online and offline [3]. Online interactions can take place as blogs, posts, discussion forums, online games, virtual worlds or social networks. Nowadays, communication via online social networks such as Facebook, Twitter, Myspace, LinkedIn, Flickr, Instagram, etc., has become an integral part of our daily routine, which offers new opportunities of studying human social interactions in cyber domain. Figure 10.2 shows some examples of online domains where social communication occurs and social biometric features can be extracted from. Extraction of social biometrics from online communications and their subsequent uses in biometric applications is a very promising new direction of research. Some of the challenges and open issues are discussed in the following subsection.

10.3.1.2 Open Questions and Future Research

The security in cyberworld is as important as security in the real world [3, 6]. In some cases, it is more crucial since breaching the security in cyberspace may jeopardise our life beyond monetary loss. The application of online social biometrics would open interesting possibilities to enhance person authentication, access control, anomaly detection, customer profiling, behaviour analysis, situation

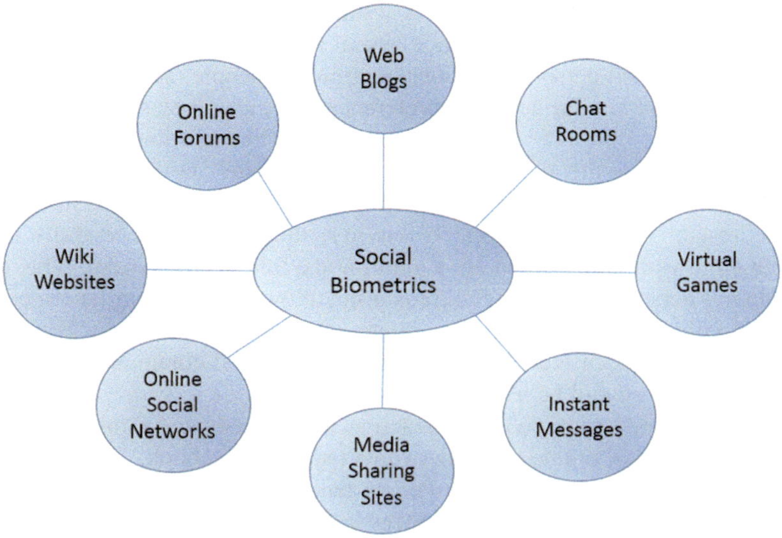

Fig. 10.2 Some online application domains of social biometrics [3, 6]

awareness and risk analysis [3, 8]. For instance, the traditional way of online user verification is to use username, password and security questions. Memorising security questions and their answers is quite painstaking for users. Social biometrics obtained from social media can aid auto-generation of security questions and verification. This would increase the security to the users' account in online domain as well as lessen the pain of setting security questions and memorising the answers. However, unlike other biometric traits, the social biometric features are not readily available. Moreover, social data acquired from web is considered as big data, which are not very informative in raw form. For such reasons, data sets are needed to be crawled from social media for a longer period of time and require exhaustive analysis to explore idiosyncratic behavioural features or demographic information. In addition to this, online social biometrics for one application domain would be very different from other domains. Therefore, each application domain needs to be investigated individually to explore consistent domain-specific social biometrics. All of those questions must be addressed in order to be able to integrate social biometrics into the traditional authentication systems.

10.3.2 Aesthetic Preferences as Social Biometrics

10.3.2.1 Overview

Every human being has his own aesthetic preference or evaluation system that guides him to choose his favourite item over a list of options. Different people have different taste, favour, likeness and judgement of beauty, which can distinguish a person from others. However, person's aesthetic preferences are not fully unique. They depend on one's life experience, social environment, race and influence of their upbringing and personal relations. A group of people (such as members of a family) may have similar aesthetic preferences, because they share similar social environment and cultural and life experiences. The question is whether these aspects can make aesthetic preference or "personal taste" into a social soft biometric trait. While social behavioural biometrics typically yield higher error rates than the well-established physiological biometrics, they can be successfully used in combination with other traits in a multimodal biometric system. The unique behavioural perspective that social traits bring allows to reduce instances of incorrect user authentication, in the cases where physiological biometric data is noisy, erroneous or compromised. Moreover, patterns of social communication and aesthetic preferences could be harder to imitate or spoof.

Over the recent years, people interaction with friends and family increasingly involves a variety of online social networks (such as Facebook, Twitter, Pinterest, Flickr, YouTube and many more). People tend to share their feelings, thoughts, judgments as well as aesthetic preferences in the form of texts, emoticons, images and videos. In Flickr ("Flickr"), a person can share his personal photographs and favourite images to everyone or to a specific community [14]. Figure 10.3 shows several examples of images which are collected from a Flickr user's favourite image tab. Images are taken from the Flickr data set used in aesthetic preferences research [15] at the University of Verona, Italy, in collaboration with some other institutions. This research bridges the gap between artistic preferences and human's intrinsic behaviour and poses a question whether visual aesthetic preferences can be used as behavioural biometric features [2, 15]. To answer that question, a bag of heterogeneous image features is analysed to find the visual preference model of a person that can distinguish them from others in a unique way (Csurka [16]). A similar image-based online social network is [17] ("Pinterest"), where a person can attach (or "Pin") images that look interesting to him during web browsing. Through these online social networks, people's visual preferences are now available to a

Fig. 10.3 Examples of images from the Flickr social network ("Flickr")

broad community. The most challenging part of using this data as a biometric trait is that aesthetic preference (such as visual preference) of a person is highly subjective. There is no comprehensive scientific explanation on what dictates one's visual preferences in an objective way [18]. A majority of research have been done to find the good features of images for the purpose of automatic aesthetic evaluation of photographic images [19, 20], for aesthetic image classification [21, 22] and for finding the answer of what makes an image memorable and popular [23, 24].

10.3.2.2 Social Aesthetic Biometric System Concept

Similarly to visual aesthetic images, we postulate now that the same idea can be applied to other types of data, visual or audio or textile. We call this concept a social aesthetic biometrics (SAB). For instance, favourite movies, cartoons and animations can be used as a new type of biometrics. Similarly, audio preferences, such as favourite type of music, video clips or instruments, can be another soft social biometrics. Moreover, even favourite fabrics or spaces can be used as spatial textile soft biometrics. Figure 10.4 illustrates the concept of generating N person's biometric templates using SAB features of their favourite visual or audio preferences.

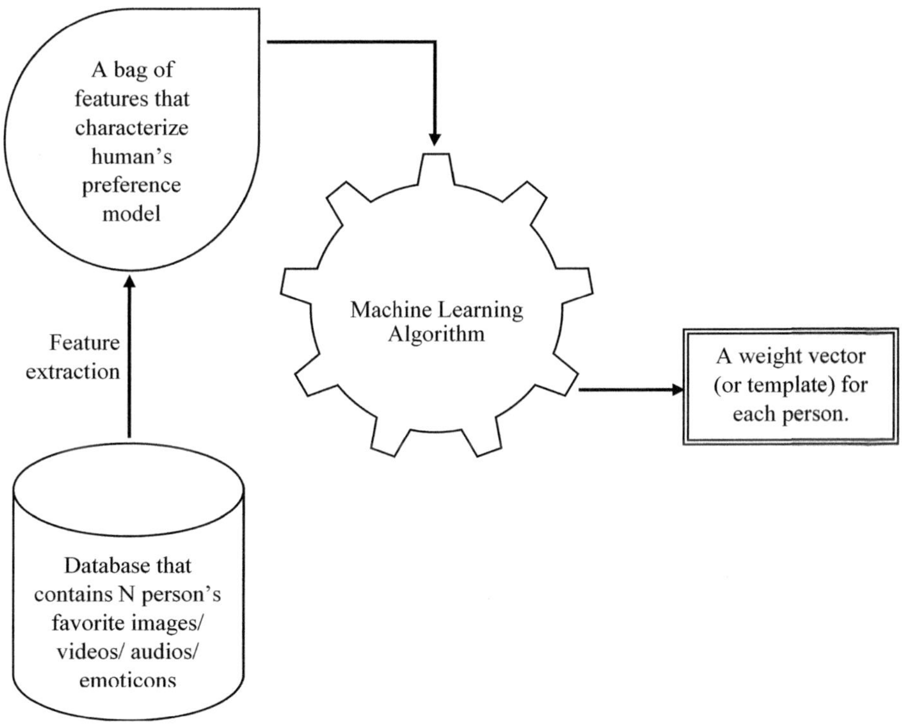

Fig. 10.4 A block diagram that illustrates the concept of generating N person's biometric templates using aesthetic features of their favourite images/audios/videos/fabrics/emoticons

An important factor for successful utilisation of audio or visual aesthetics is the sufficient number of samples (such as images or sounds) used during template generation and matching. Typically, a high recognition rate can be achieved with the increased number of samples.

10.3.2.3 Open Problems and Future Work

There are a number of open research problems related to this newly introduced social aesthetic biometrics trait. A difficult part of this research is the translation of human's aesthetic preferences into a set of computable rules that can characterise a person in a distinct way. Scientific modelling of someone's distinct personal taste is not a trivial task. Frequently, it is impossible for a human to provide reasons behind his/her aesthetic preferences. Extensive psychological research is needed to find the answer of what features can model the aesthetic preference of an individual that can distinguish one person from another in a group.

Only very recently, researchers consider only the image aesthetic preference as a biometrics [15]. Other multimedia contents like videos, emoticons and audios are not explored yet. Recently, these types of multimedia data become popular and easily accessible through different online social networks. Finding distinct preference model for identification of a person using this type of data can be a new research direction in the field of social behavioural biometrics.

We can use a person's aesthetic preference of personal association and friends in online social networks (such as Facebook, Twitter) as another novel social behavioural biometrics. Again, we need to design a friend preference model of an individual by analysing profiles of his friends. Here, the challenging part is finding the appropriate feature set of a profile that can contribute significantly to the person's friend preference model.

In addition to the exciting new research directions outlined above, another promising research would be the fusion of multiple preference models. We can combine a person's visual preferences and friend preferences together to learn a multimodal preference model [25] as a social biometrics. Also, good investigation and analysis are needed to find the appropriate machine learning algorithm for training the multimodal aesthetic preference model.

We can extract other soft biometrics, such as gender, age and ethnicity, from the aesthetic preference of a person. The intuition is that different age groups and ethnic groups of people have different aesthetic preferences. Even socially, male–female have different aesthetic preferences. For example, colour preferences are usually different between male and female ("Digital Synopsis"). So, predicting gender, age and ethnic information from person's aesthetic preference can be a new area of research.

A person's aesthetic preference is a recent addition to the social biometric traits. The intuition is that a person can be characterised in a distinct way using his personal taste, preferences and likes. However, this social biometric trait needs to be investigated thoroughly to consider its uniqueness, stability over time and differentiating features. Computational interpretation of such trait may be very complex. As a future work, it is important to address the aforementioned open problems. More specifically, we can introduce new distinctive image features based on image's contextual meaning to improve the recognition performance. Deep learning algorithms can be applied to learn the visual preference model and to generate the template. Also, we need to conduct investigation to find potential features or psychological aspects that can distinguish gender, different age groups and ethnic groups from visual preferences. Finally, we need to consider the issues of usability and practicability during designing and prototyping of a biometric system which includes social aesthetic trends.

10.3.2.4 Application Domain

One of the main application domains of social aesthetic biometrics is the forensic area. As traditional soft biometrics, such as gender, age, ethnicity and body geometry, forensic experts can use the aesthetic preference to identify the suspects or the criminals. Users of various online social networks continuously leave behind their behavioural footprints in a form of text, blogs, photographs, images, videos and emoticons. So it is possible for the forensic experts to collect samples of someone's aesthetic preferences. Another application domain is the multimodal biometric system where a social biometrics can provide support to other primary biometric traits to improve the recognition rate of the system.

Aside from security and forensic applications, image aesthetic preference model can be used inside an image view recommender system [15]. Marketers may apply the preference model to identify a list of customers (from online social network) and thus target only certain customers during publishing images. Another application can be a customised desktop with images sorted according to viewer's visual preferences. Finally, political campaigns can benefit from understanding aesthetic principles of the general population in their geographical locale.

10.3.3 Facial Expression Analysis as a Social Biometrics

10.3.3.1 Overview

Humans are social beings by nature. We develop and learn about the world around us through social referencing which enables us to interact with appropriate emotions and social actions. Research on social behavioural patterns aims at identifying physical actions and observable emotions associated with individuals and environment and finding a way to incorporate those emotions and actions to bridge the gap between social human beings and "unsocial machines", i.e. computers [26]. Human psychology, interpersonal impression, community, culture and the environment around individuals define the way of their social interactions. Facial expression is a fundamental component in social interaction. Although basic facial expression and emotions are fairly similar across people, certain behavioural expressions, intensity of emotions, signs used in interaction and response to a specific action vary person to person. Thus, similarly to the physical world, it may be possible to identify a person from the facial expression or action patterns at the time of social interactions. Over the past decade, facial actions are extensively studied to extract the emotions behind the expressions [27–33]; however, the investigation of facial expressions as social biometrics remains unexplored. A tremendous progress in modern technologies made it possible to start investigation of the nonverbal communications in a workplace, a classroom, a meeting room or any other collaborative environment. People act deliberately or unconsciously in these scenarios, which may reflect different personalities, emotions, intentions and attitude. This information, if properly utilised, can be useful in extracting characterising features of individuals, which can be considered as social biometric traits.

10.3.3.2 Social Biometric Feature Detection from Facial Expression

Automated facial expression recognition in the context of social biometrics can be broken down into typical components: data acquisition, registration, feature extraction, recognition and extraction of behavioural pattern.

The first step for any automated biometric recognition is the data acquisition. It can be captured in different environment or settings, and various acquisition devices can be used to capture the data. Nowadays, not only specialised image-capturing devices but also webcams, video recorders, surveillance cameras, wearable devices and fully equipped smart meeting rooms can be used to acquire the data. For effective expression recognition, face registration is very important. The registration process is used to reduce the head pose variation and model spontaneous interaction. Face registration can be categorised as the whole face registration, the partial registration and the point registration [34]. The next step is feature extraction. The geometric or appearance-based information can be extracted from the facial image, such as facial points, colour, texture information, etc., and can be consequently used to extract the facial actions or cues. The facial action sets can be classified in different categories. One of the most popular categorisations is the Facial Action Coding System (FACS) [35], which uses action units (AUs) concept to categorise facial expressions in the human face. The facial expressions can be recognised by detecting the action units (AUs) and by incorporating the temporal phase of the action units, as well as the combination of AUs. The last step is extraction of behavioural patterns. The detected facial expressions can be used to extract person-specific social behavioural traits and typical emotional states. They can be also used to model the time-dependent facial actions, analyse the social attitude, personality and its effect and recognise the role a person plays in the conversation or in the meeting. Figure 10.5 shows the block diagram of a proposed system for detecting social biometric features from the facial expressions.

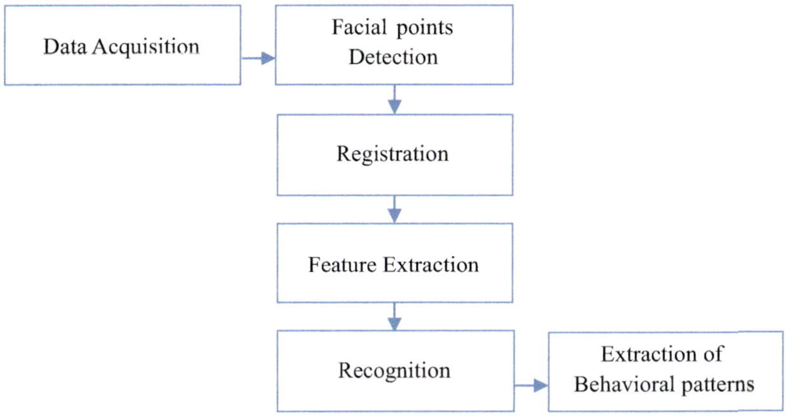

Fig. 10.5 Proposed architecture of a social biometric system based on facial expressions

10.3.3.3 Open Problems and Future Research

Despite decades of research on facial expression authentication, there are still some issues and open problems which are considered only partially in the literature.

(a) In order to extract features from the facial expressions for behaviour analysis, data must be analysed over a long period of time. Also, the situation and environment that produce the key emotions may vary from person to person. Thus, interpretation of the context of actions should be considered while extracting the behavioural patterns from the social interactions.

(b) Expressions are not always associated with the same emotions. Most of the existing systems do not consider different emotions for similarly looking expressions. Again, for spontaneous social interaction, it is very important to consider the head pose variations because it can be part of the individual social behaviour. Moreover, depending on the culture and context, the same facial action can be interpreted as a different social behaviour.

(c) Research on facial expressions lacks study of micro-expressions. Micro-expressions can be defined as very brief facial expressions which last only for a fraction of a second. They occur when a person either intentionally or instinctively try to hide an emotion which can also be counted as a behavioural feature.

(d) Most of the existing systems for facial expression recognition use posed expression data set for their validation which is different from natural expressions and gestures. Social behaviours are part of the unconscious cognition which reflects the true identity of a person. The lack of well-annotated databases of unposed facial behaviour needs to be addressed so that it would not hinder the way of detecting individual personality.

In addition to establishing emotion recognition as social behavioural biometrics and investigating its features, and corresponding open problems identified above, there are a few highly interesting complimentary research directions. The rapid growth of depth imaging technology makes it possible to analyse the 3D model of faces [34]. Depth information can also address the illumination and pose invariance facial actions recognition. Sometimes, it is possible to identify certain facial affects in a 3D face model which would be hardly noticeable in the 2D imaging. Incorporating the high-dimensional information will help to identify more distinguishing features.

In the learning process of social behaviours, infants take cues from other people in the environment and learn which action or behaviour is appropriate towards environmental object, person and situations. A similar learning process can be introduced to a robot or any intelligent interface through synthetic social behaviour [36]. Social robot can respond against the social interaction if it is possible to provide information regarding the environment, emotional states and the outcome of the behaviour. Statistical model can be created from the behavioural patterns extracted from human facial expression, and this synthetic behavioural pattern can be integrated with the artificial agents.

Finally, facial expressions can be interpreted as the behavioural traits in the context of the events that caused that behaviour or expression to occur. For example, long-term observation of the facial expression, gesture and pattern of voice can be applicable in the identification of interpersonal relations and long-term effects of isolation on a space station [37]. Location of occurrence of the emotional events and intensity of such events can also be used as one of the psycho-emotional features of social behavioural biometric research.

10.3.4 Collaborative Environment Activity Recognition as a Social Behavioural Biometrics

10.3.4.1 Overview

While facial expressions can be one way to study emotions, behaviour or actions of an individual in a collaborative environment, similarly, individual's pose, gait and hand gestures can be equally well qualified to provide clues on the emotional states or develop dynamics in a group during social interaction. Recently, gait and gesture analysis has been considered for virtual and augmented reality, motion and video retrieval, 3D human body modelling and animation and healthcare applications [38–41]. However, activity-related behavioural biometrics has been rarely studied from a position of person authentication, especially in the context of collaborative intelligent environment. Some emerging research in this direction focused on activities that express individual physiology and style through multi-joint movements [42], sensing seat-based anthropometric biometrics that models the user weight distribution patterns during sitting action [43] and study movement-based biometrics that models grasping, reaching and object manipulation [9]. Another recent study [10] on fusion of context metadata with gait recognition has shown a significant improvement over gait-only biometric recognition, where user contextual and behavioural patterns are modelled based on his/her daily routine.

10.3.4.2 Gait and Activity-Related Biometric Signature Extraction Using Kinect

While vision-based biometric gait recognition has been a topic of interest for the past 20 years, the recent popularisation of the Kinect sensor introduced by the Microsoft has resulted in an affordable research tool for investigation of gait recognition. Due to the various features of the Kinect data acquisition and processing, it has been a popular addition in various real-world applications, such as home monitoring (Stone and [44]), healthcare [45] and surveillance [46]. Ball et al. [47] were some of the first researchers to conduct initial study on Kinect-based gait recognition, utilising an unsupervised clustering. Another approach proposed in Preis et al. [48] used 13 biometric features, including height, leg lengths, torso length, upper arms, forearms, step length and speed. In 2015, another study [49] presented a feature fusion-based gait recognition method where the raw skeletal data is transformed into scale and view independent feature representations, namely, joint relative distance and joint relative angle. Then, a dynamic time warping (DTW)-based kernel was utilised for a rank level fusion (Fig. 10.6).

Gait- and activity-related features are typically sensitive to changes in clothing and carrying conditions. This affects gait- and activity-related biometric authentication in dynamic environment [49]. Hence, recent gait recognition methods [9–10] incorporate complementary modalities to increase the robustness of the developed system. In this context, activity-related soft biometric signatures and context metadata can be used to increase the overall performance of a gait recognition system. For example, in a typical meeting room set-up, a meeting starts with participants entering the room and sitting on the chair. In this scenario, we can

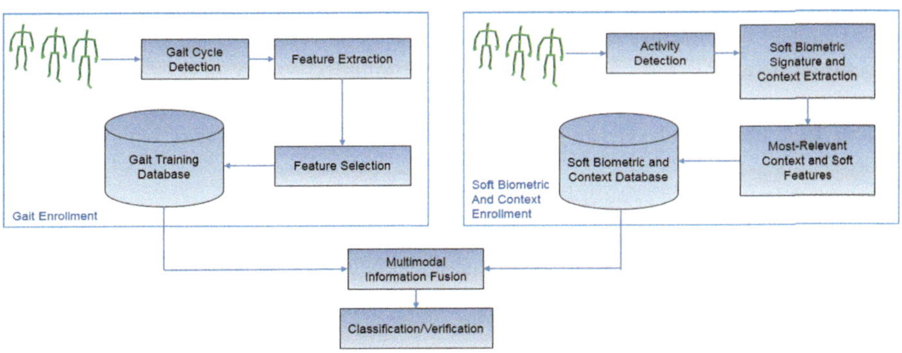

Fig. 10.6 Different data streams that can be obtained from Kinect [49]

couple the sensing seat-based authentication with Kinect-based skeleton data, which can boost the recognition performance.

Different types of behavioural and social context metadata can be also used to boost gait recognition. In the previous study [10], it was found that user information related to the daily routine such as being at a specific location during a specific time (e.g. being in some workplace during the morning) and specific conditions related to the location (e.g. working on a desk in an office, attending a meeting, etc.) provide valuable context metadata that can be used to improve the biometric recognition performance. This information can be treated as social behavioural gait-based biometrics and sued to enhance authentication. Similarly, social behavioural information can be extracted from a collaborative environment [50] where user behaviour related to the manner of communication, situational responses, temporal patterns of user activity, preference of spatial location and interaction among group members and project organiser can be utilised as metadata.

10.3.5 EEG as a Social Biometrics

10.3.5.1 Overview

Many traditional physiological biometrics discussed above have vulnerability—they can be easily replicated or spoofed. On the other hand, bioelectric signals such as EEG (electroencephalography) and ECG (electrocardiography) start to play a vital role as they are non-vulnerable in the case of a spoof attack. ECG is studied in cardiology and relates to the heart as a vital organ of a human body. In a biometric research, we are mostly interested in behavioural patterns, such as human neurological activity, represented by ECG as a brain signal. Brain signal has a certain frequency band [51], such as alpha, beta, theta and gamma, with each band associated with a particular frequency range and activity.

The activity of each band of brain signal is different. They are summarised in Table 10.1. These could be an important feature for establishing brain signal as behavioural biometrics. We can record the brain activity of a person when he/she spends time on his/her social network or community—such as browsing in social

Table 10.1 EEG signal frequency band (compiled from [52, 53], types of brainwave online article)

Band	Bandwidth	Description
Alpha	[8, 12] Hz	Most dominant feature for the subject at resting state with eye close
Beta	[12, 19] Hz	Indicates the alert state specially when the subject is in dynamic thinking, concentration state
Theta	[4, 8] Hz	Indicates the subject is in depression, inattentiveness
Gamma	[19, 21] Hz	Indicates the anxiety, stress, and meditation
Delta	[0.5, 4] Hz	Appears at brain injuries, learning problems, and inability to think

network, reaction of a person when he/she gives or sees comment or like in the post in social network, excitement and stress state of the brain when a person plays game on virtual network, activity of the brain in an official meeting, exam hall and social cultural functions, etc.

There are various types of techniques to extract the features from the EEG signal [54, 55]. These can be classified into three categories: time domain feature extraction, frequency domain feature extraction and time-frequency domain feature extraction. The statistical features are known as time domain features and include mean absolute value, median, variance, zero crossing, slop sign changes and waveform length. Fourier transform is typically used to analyse the frequency domain features. Wavelet transform is used to represent the signal in a time-frequency domain. The capability of a time-frequency domain analysis of a wavelet transform can be useful to separate the alpha, beta and theta band and consequently to extract important features. For considering EEG as biometric identification, it requires four main characteristics, such as universality, uniqueness, collectability and permanence. Recent research shows that EEG signal has those characteristics [56].

10.3.5.2 Open Problems

As EEG is a relatively new type of biometrics, there are many open problems related to using EEG as biometric identifier. The brain activity of a person varies time to time. It also depends on the social environment of a person who is living or doing his/her activity. In the morning, a person can be in a jolly mood, at afternoon in stressful condition. These states will vary person to person. Moreover, it is difficult to control the person's social environment. So, whether a person's brain signals can be used as social biometrics, extensive research is needed to find some particular features that can be utilised for using EEG for person identification.

EEG can provide some unique information which can be considered in a context of social behaviour biometrics. Table 10.1 shows that each band of EEG provides some specific characteristics. EEG can be collected during various social settings, such as conversation, problem solving, physical activity, group presentation, photograph observation, etc. Similar to other social biometrics, EEG can be fused with other biometric characteristics as part of a multimodal system. Moreover, EEG signal has genetic characteristics. It can be used to identify the genetic history of population group—such as birth defects, diseases, etc. Some of the challenges originate from the fact that EEG signal is difficult to collect, is subject to incorrect electrode placement and is highly dependent on the mood and emotional state of a subject. EEG-based biometric system is an emerging research topic and may open new research directions and applications in the future. It is impossible to control the person's social environment or network. But some of the possible future developments may include by utilising the EEG signal people under same mental condition such as when they are solving a particular math problem that will increase EEG acceptability as biometric identifier. Moreover, some of the possible future developments may include by minimising the other factors for using brain signal as person identification. A system should be designed so that we can minimise the noise effect in recording EEG signal. So, for using brain signal as social biometrics, firstly ensuring the quality of EEG data by developing a proper system of EEG recording and, then, other soft biometric identifiers such as age and gender can be combined with the EEG signal that will increase the acceptability of EEG signal as social biometric identifier as social networking activity varies age to age and gender to gender. Young generation social networking area will not be the same as the old generation.

10.3.5.3 Conclusion as Future Direction

Brain signal as social biometrics is one of the new research directions. However, for using brain signal as social biometric identifier needs to be analysed perfectly to consider its acceptability, uniqueness and stability over time. As a future work, it is essential to address the mentioned open problem. For that, some of the possible future developments may include by developing a system model which will be noise sensitive during EEG recording, recording the EEG signal at a long period of time, finding the particular features which can be used in various mental state and higher level algorithm to combine the EEG signal with other soft biometrics that will increase the system stability, usability and acceptability.

10.4 Social Behavioural Biometric Application Domains

The social behavioural biometrics listed above can be utilised in a wide variety of application domains. The list below provides some examples of emerging technologies and application domains where social biometric traits can play a vital role.

- *Efficient person recognition*: Despite extensive research in the field of individual biometrics, wide variety of distortions during data acquisition and variable quality samples make use of single biometrics almost obsolete in a current world. The rise of multimodal biometrics and numerous studies demonstrated benefits of incorporating more than one biometric type. In this chapter, we argue that inclusion of social behavioural biometrics in a multimodal system provides unique advantages over traditional biometrics. For example, addition of social biometric features extracted from the facial expression with the partial physiological features extracted from the face can improve the performance of the face-based biometric system. Features extracted from facial actions, such as interpersonal impressions and emotional states, can be integrated with other nonverbal communications such as patterns of rhythm and sound, gait, gesture and posture to filter the dominant characteristic of a person [10].

- *Automated measurement of engagement in collaborative environment*: Student engagement in the class, active meeting involvement and detecting group activity, all these can be automatically extracted from the environment by processing nonverbal activities of the participant [57]. For launching effective education system, finding the student engagement can help in certain ways, such as improving grade, involving in educational games and introducing intellectual tutoring system and online-based courses. In the workplace, identifying the rising leaders, meeting room activities and hiring candidate for specific job requirement can be achieved by analysing nonverbal behaviours [50]. Also group happiness or dissatisfaction can be identified through the analysis of social interactions, activities and/or facial expressions.
- *Healthcare*: Facial image or expression can be analysed for detecting the presence of pain, depression and other psycho-emotional patterns [26, 58]. For example, researchers observed some facial actions, such as brow lowering, cheek raising, eyes closure and nose wrinkling, can indicate the intensity of pain. A depressed person seems to be less communicative, more reserved and responds slower which can be reflected in their facial expressions, actions and other nonverbal communications. Automatic detection of these kinds of behavioural patterns can be used in the treatment of psychological disorder as well as in general healthcare.
- *Security and forensic*: Early detection of certain physiological patterns, such as anxiety and nervousness, can prevent certain situations from occurring. This type of situation awareness system can be used in security for the preliminary screening or in surveillance areas [59]. In interviewing and interrogation of criminal activities, analysis of facial expressions can indicate some behavioural pattern, such as eye movement and facial muscle tightening, which can detect mal-intent and deception.
- *Prediction of ad liking*: With the progress of the Internet and telecommunication, it became much easy and cheaper to promote new products. Many business organisation place their advertisement on the web or various social networks. It is not only cost-effective but also allows to attract specific consumer group. Emotional responses towards an advertisement can determine the acceptance of the new product among the people [60]. Thus, automated methods for capturing the emotional responses to measure the effectiveness of the advertisement and purchase intension among the viewers can be beneficial for both the consumers and producers. In addition, aesthetical preferences of individuals can play an important role in targeted advertisement or in political campaign.
- *Social media interaction*: Facial expression and gestures are widely used for interpreting emotional or psychological traits. It also reveals personality traits or interpersonal impression. In the social media, specifically in social video sites, people upload different types of videos. This type of videos reflects interpersonal impression of the person or speaker because most of the media content contains the upper body part of the speaker where the face appearance is clearly visible. Facial expressions and nonverbal feedback extracted from these videos can reflect emotional states, as well as various personality traits which can be used

to judge the interpersonal characteristics of a person. Moreover, blogging activities such as Twitter or social network associations can provide significant information about user preferences, likes and intentions [3, 6, 7, 12].
- *Interaction with virtual world*: In the virtual world, avatar or virtual agents are graphical representation of a person. They mimic the social behavioural traits of the corresponding person [4, 8]. Avatar or virtual agents are used in gaming environment and also as a counsellor or coach in the virtual world. Virtual agents are expected to generate different emotions through facial expressions, such as smiles, sighs or expressive attitude, during interaction to make it more realistic [26]. This kind of social action needs mutual engagement, use of space and perception of the context and most importantly identifying the behavioural traits of the relevant person. Through detecting behavioural traits from facial expression of the user, such dynamic virtual environment can be constructed. The same ideas can be also applied to other social agents and humanoid robots.

10.5 Conclusions

This chapter introduced a new concept of biometric fusion using physiological, behavioural and social characteristics of humans and avatars. We identified some behavioural biometrics as the potential candidates of the newly emerging social biometrics. We also introduced social aesthetic biometrics as a new subarea of social biometrics. Methodologies on how biometric information can be extracted from social, contextual, temporal and behavioural data of individuals are also presented. This chapter also presents methodical classification of the open problems, challenges, application domains and future research. The discussion presented in this chapter will assist in further establishment of the next generation of intelligent multimodal biometric system. Our future research includes studying human behaviour in a broad range of social environments from biometric perspective.

References

1. Paul, P.P., Gavrilova, M.L., Alhajj, R.: Decision fusion for multimodal biometrics using social network analysis systems. IEEE Trans. Man. Cybern. Syst. **44**(11), 522–1533 (2014)
2. Segalin, C., Perina, A., Cristani, M.: Personal aesthetics for soft biometrics: a generative multi-resolution approach. 16th International conference on multimodal interaction (ICMI '14), pp. 180–187 (2014)
3. Sultana, M., Paul, P.P., Gavrilova, M.: A concept of social behavioral biometrics: motivation, current developments, and future trends. International conference on Cyberworlds, pp. 271–278 (2014)
4. Yampolskiy, R., Gavrilova, M.: Artimetrics: biometrics for artificial entities. IEEE Robot Autom. Mag. **19**(4), 48–58 (2012)

5. Jain, A.K., Ross, A., Prabhakar, S.: An introduction to biometric recognition. IEEE Trans. Circuits Syst. Video Technol. **14**(1), 420 (2004)
6. Sultana, M., Paul, P.P., Gavrilova, M.: Mining social behavioral biometrics in Twitter. International conference on cyberworlds, pp. 293–299 (2014)
7. Sultana, M., Paul, P.P., Gavrilova, M.: Social behavioral biometrics: an emerging trend. Int. J. Pattern Recognit. Artif. Intell. **29**(8), 1556013-1-20 (2015)
8. Gavrilova, M., Yampolskiy, R.: Applying biometric principles to avatar recognition. Trans. Comput. Sci. **XII**, 140–158 (2011)
9. Drosou, A., Ioannidisa, D., Tzovarasa, D., Moustakasb, K., Petroua, M.: Activity related authentication using prehension biometrics. Pattern Recognit. **48**(5), 1743–1759 (2015)
10. Bazazian, S., Gavrilova, M.: A hybrid method for context-based gait recognition based on behavioral and social traits. Trans. Comput. Sci. LNCS **9030**, 115–134 (2015)
11. Sultana, M., Paul, P.P., Gavrilova, M.: Identifying users from online interactions in Twitter. In: Gavrilova, M.L. (ed.) Transactions on Computational Science XXVI, pp. 111–124. Springer, Berlin (2016)
12. Paul, P.P., Sultana, M., Matei, S.A., Gavrilova, M.L.: Editing behavior to recognize authors of crowdsourced content. IEEE international conference on systems, man, and cybernetics (SMC), pp. 1676–1681 (2015)
13. Haxby, J., Hoffman, E., Gobbini, I.: Human neural systems for face recognition and social communication. Biol Psychiatry **51**(1), 59–67 (2002)
14. Flickrdomain. https://www.flickr.com/. Accessed 19 Jan 2016
15. Lovato, P., Bicego, M., Segalin, C., Perina, A., Sebe, N., Cristani, M.: Faved! Biometrics: tell me which image you like and I'll tell you who you are. IEEE Trans. Inf. Forensic Secur. **9**(3), 364–374 (2014)
16. Csurka, G., et al.: Visual categorization with bags of keypoints. Workshop on statistical learning in computer vision, ECCV 1, pp 1–22 (2004)
17. Pinterest domain. https://www.pinterest.com/. Accessed: 19 Jan 2016
18. Leder, H., Belke, B., Oeberst, A., Augustin, D.: A model of aesthetic appreciation and aesthetic judgments. Br. J. Psychol. **95**(4), 489–508 (2004)
19. Aydin, T., Smolic, A., Gross, M.: Automated aesthetic analysis of photographic images. IEEE Trans. Vis. Comput. Graph **21**(1), 31–42 (2015)
20. Jiang, W., Loui, A., Cerosaletti, C. (2010) Automatic aesthetic value assessment in photographic images. 2010 I.E. international conference on multimedia and expo (ICME), pp. 920–925 (2010).
21. Marchesotti, L., Perronnin, F., Larlus, D., Csurka, G.: Assessing the aesthetic quality of photographs using generic image descriptors. IEEE international conference on computer vision (ICCV), pp.1784–1791 (2011)
22. Xiaohui, W., Jia, J., Yin, J., Cai, L.: Interpretable aesthetic features for affective image classification. 20th IEEE international conference on image processing (ICIP), pp. 3230–3234 (2013).
23. Isola, P., Xiao, J., Parikh, D., Torralba, A., Oliva, A.: What makes a photograph memorable? IEEE Trans. Pattern Anal. Mach. Intell. **36**(7), 1469–1482 (2014)
24. Khosla, A., Sarma, A.D., Hamid, R.: What makes an image popular? 23rd international conference on World wide web (WWW '14), pp. 867–876 (2014)
25. Gavrilova, M., Monwar, M.: Multimodal Biometrics and Intelligent Image Processing for Security Systems. IGI book, Hershey, PA (2013)
26. Vinciarelli, A., Pantic, M., Heylen, D., Pelachaud, C., Poggi, I., D'Errico, F., Schroeder, M.: Bridging the gap between social animal and unsocial machine: a survey of social signal processing. IEEE Trans. Affect Comput. **3**(1), 69–87 (2012)
27. Chew, S., Lucey, P., Lucey, S., Saragih, J., Cohn, J., Matthews, I., Sridharan, S.: In the pursuit of effective affective computing: the relationship between features and registration. IEEE Trans. Syst. Man. Cybern. B **42**(4), 1006–1016 (2012)

28. Eleftheriadis, S., Rudovic, O., Pantic, M.: Discriminative shared Gaussian processes for multiview and view-invariant facial expression recognition. IEEE Trans. Image Process **24** (1), 189–204 (2015)
29. Li, Y., Wang, S., Zhao, Y., Ji, Q.: Simultaneous Facial Feature Tracking and Facial Expression Recognition. IEEE Trans. Image Process **22**(7), 2559–2573 (2013)
30. Littlewort, G., Whitehill, J., Wu, T., Fasel, I., Frank, M., Movellan, J., Bartlett, M.: The computer expression recognition toolbox (CERT). IEEE International conference of automatic face and gesture recognition workshops, pp. 298–305 (2011)
31. Lucey, S., Matthews, I., Hu, C., Ambadar, Z., Cohn, J.: AAM derived face representations for robust facial action recognition. IEEE International conference of automatic face and gesture recognition, pp. 155–160 (2006)
32. Tian, Y.L., Kanade, T., Cohn, J.F.: Recognizing action units for facial expression analysis. IEEE Trans. Pattern Anal. Mach. Intell. **23**(2), 97–115 (2001)
33. Wang, S., Liu, Z., Lv, S., Lv, Y., Wu, G., Peng, P., Chen, F., Wang, X.: A natural visible and infrared facial expression data-base for expression recognition and emotion inference. IEEE Trans. Multimedia **12**(7), 682–691 (2010)
34. Sariyanidi, S., Gunes, H., Cavallaro, A.: Automatic analysis of facial affect: a survey of registration, representation, and recognition. IEEE Trans. Pattern Anal. Mach. Intell. **37**(6), 1113–1133 (2015)
35. Ekman, P., Friesen, W.: Facial Action Coding System: A Technique for the Measurement of Facial Movement. Consulting Psy-chologists Press, Palo Alto, CA (1978)
36. Boucenna, S., Gaussier, P., Hafemeister, L.: Development of first social referencing skills: emotional interaction as a way to regulate robot behavior. IEEE Trans. Autonom Mental Dev. **6**(1), 42–55 (2014)
37. Barakova, E., Gorbunov, R., Rauterberg, M.: Automatic interpretation of affective facial expressions in the context of interpersonal interaction. IEEE Trans. Hum. Mach. Syst. **45**(4), 409–418 (2015)
38. Bae, M., Park, I.: Content-based 3d model retrieval using a single depth image from a low-cost 3d camera. Visual Comput. **29**, 555–564 (2013)
39. Barth, J., Klucken, J., Kugler, P., Kammerer, T., Steidl, R., Winkler, J., Hornegger, J., Eskofier, B.: Biometric and mobile gait analysis for early diagnosis and therapy monitoring in Parkinson's disease. Annual international conference of the IEEE engineering in medicine and biology society, EMBC, pp. 868–871 (2011)
40. Zhang, Y., Zheng, J., Magnenat-Thalmann, N.: Example-guided anthropometric human body modeling. Visual Comput. **CGI 2014**, 1–17 (2014)
41. Zhou, L., Zhiwu, L., Leung, H., Shang, L.: Spatial temporal pyramid matching using temporal sparse representation for human motion retrieval. Visual Comput. **30**, 845–854 (2014)
42. Drosou, A.: Activity related biometrics for person authentication. PhD thesis, Imperial College London (2014)
43. Ferro, M., Pioggia, G., Tognetti, A., Carbonaro, N., Rossi, D.D.: A sensing seat for human authentication. IEEE Trans. Inf. Forensic Secur. **4**(3), 451–459 (2009)
44. Stone, E., Skubic, M.: Evaluation of an inexpensive depth camera for passive in-home fall risk assessment. International pervasive computing technologies for healthcare conference, pp. 71–77 (2011)
45. Chang, Y., Chen, S., Huang, J.: A kinect-based system for physical rehabilitation: a pilot study for young adults with motor disabilities. Res. Dev. Disabil. **32**(6), 2566–2570 (2011)
46. Popa, M., Koc, A., Rothkrantz, L., Shan, C., Wiggers, P.: Kinect sensing of shopping related actions. Commun. Comput. Inf. Sci. **277**, 91–100 (2012)
47. Ball, A., Rye, D., Ramos, F., Velonaki, M.: Unsupervised clustering of people from 'skeleton' data. ACM/IEEE international conference on human robot interaction, pp. 225–226 (2012)
48. Preis, J., Kessel, M., Linnhoff-Popien, C., Werner, M.: Gait recognition with kinect. Workshop on kinect in pervasive computing (2012)

49. Ahmed, F., Paul, P., Gavrilova, M.: Dtw-based kernel and rank level fusion for 3d gait recognition using Kinect. Visual Comput. **31**(6-8), 915–924 (2015)
50. Ahmed, F., Gavrilova, M.: Biometric-based user authentication and activity level detection in a collaborative environment. In: Matei, S.A., et al. (eds.) Transparency in Social Media, pp. 166–179. Springer, New York, NY (2015)
51. Sanei, S., Chambers, J.: EEG Signal Processing. John Wiley & Sons Ltd, England (2007)
52. Webster, J.G.: Medical Instrumentation Application and Design. Medical Imaging and Instrumentation Laboratory, Stanford, CA (2009)
53. Types of brain waves online article http://mentalhealthdaily.com/2014/04/15/5-types-of-brain-waves-frequencies-gamma-beta-alpha-theta-delta/. Accessed 17 Jan 2016
54. Nguyen, P., Tran, D., Huang, X., Sharma, D.: A proposed feature extraction method for EEG based person identification. In: International conference artificial intelligence (ICAI) (2012)
55. Smit, D., Posthuma, D., Boomsma, D.I., Geus, E.J.C.: Heritability of background EEG across the power spectrum. Psychophysiology **42**(6), 691–697 (2005)
56. Ruiz Blondet, M., Laszlo, S., Jin, Z.: Assessment of permanence of non-volitional EEG brainwaves as a biometric. International conference on identity, security and behavior analysis (ISBA) (2015)
57. Whitehill, J., Serpell, Z., Lin, Y.C., Foster, A., Movellan, J.R.: The faces of engagement: automatic recognition of student engagement from facial expressions. IEEE Trans. Affect Comput. **5**(1), 86–98 (2014)
58. Cohn, J.F.: Advances in behavioral science using automated facial image analysis and synthesis [social sciences]. Signal Proc. Mag. **27**(6), 128–133 (2010)
59. Poursaberi, A., Vana, J., Mráček, S., Dvora, R., Yanushkevich, S.N., Dra-hansky, M., Shmerko, V.P., Gavrilova, M.L.: Facial biometrics for situational awareness systems. IET Biometrics **2**(2), 35–47 (2013)
60. McDuff, D., El Kaliouby, R., Cohn, J.F., Picard, R.W.: Predicting Ad liking and purchase intent: large-scale analysis of facial responses to Ads. IEEE Trans. Affect Comput. **6**(3), 223–235 (2015)

Chapter 11
Empirical Evidences in Software-Defined Network Security: A Systematic Literature Review

Izzat M. Alsmadi and Mohammad Zarour

Abstract Nowadays, the term software-defined networking (SDN) becomes very popular. It is an approach that decouples the "control plane" and the "data plane" in switches to allow more programmable control of network traffic flows. Currently, several efforts are under way to thoroughly study and deploy SDN, as well as create standards that regulate the use of SDN. Since SDN is considered relatively a new discipline, a very little empirical literature has been aggregated in this field. The objective of this study is to aggregate and synthesize the empirical evidence from literature of SDN security to report the trends, patterns, and current status of the field. A systematic literature review (SLR) has been conducted to synthesize the empirical work in SDN.

11.1 Introduction

Software-defined networking (SDN) is a new evolving networking architecture. The main aspects that can characterize SDN include the following:

- *Separation of control from data*: SDN switches include only forwarding data with no control as in traditional switches. Control is taken from all switches and is centrally allocated in a software-based program called the controller. Communication between the controller and its switches is defined through the OpenFlow protocol.
- *Centralization*: One of the main characteristics of SDN that is getting a lot of concerns is the centralization of control. While this clearly has some advantages,

I.M. Alsmadi (✉)
Department of Computing and Cyber Security, University of Texas A&M, One University Way, San Antonio 78224, TX, USA
e-mail: ialsmadi@tamusa.edu

M. Zarour
College of CS and Information Technologies, Prince Sultan University, P.O. Box 66833, Rafha Street, Riyadh 11586, KSA, USA
e-mail: mzarour@cis.psu.edu.sa

I.M. Alsmadi et al. (eds.), *Information Fusion for Cyber-Security Analytics*, Studies in Computational Intelligence 691, DOI 10.1007/978-3-319-44257-0_11

it has also some serious concerns. From a reliability perspective, a centralized controller can be a single point of failure where the complete network will fail if the controller fails. From a performance perspective, it can be also a bottleneck especially where the controller needs to receive all new packet headers to make decisions about and write those as flow rules in switches. Finally, from security perspective, having control centralized in one place may seduce attackers to focus their attacks on the controller.

- *Programmable*: SDN or OpenFlow networks can be classified as programmable networks as control and management are handled by software programs. This may also introduce a tremendous amount of new opportunities (e.g., automating security services, on-demand security services, etc.), while at the same time it may pose some serious security concerns. The general claims that hardware can be more reliable and secure than software are not new claims from SDN only. However, the flexibility that the software has over the hardware can be also used to continuously and easily test and update this software to improve its security and quality aspects.
- *Dynamic*: Based on programmability, many current services available in networks can be automated based on SDN. For example, access control lists (ACLs) in current or traditional firewalls are added and maintained manually through network administrators. In large complex networks, this is a very tedious task especially when the network grows and changes. In such cases, some rules can be obsolete and should be updated or removed. However, network administrators hesitate to change such rules in fear of underestimating their corresponding impact. It is envisioned through SDN that firewalls and all security controls or services can be automatically or dynamically updated based on the network topology and traffic. However, such ambitious goals are far from achievement currently given that their practical implementations are complex.

Security in SDN networks can be seen from two different perspectives: new security opportunities offered through SDN and security risks and vulnerabilities in SDN.

In terms of security opportunities, it is believed that SDN may work as a stimulator to many network-based applications such as security controls. SDN or "programmable network" as a more generic term can change the way we develop and use security controls. Many companies are recently using the term "next-generation" firewalls, with some of the promises related to programmable networks. For different vendors, in many cases, this term has different meanings. Dynamic and programmable firewalls are two key terms related to "next-generation" firewalls. Current firewalls can be classified as static for several different reasons. Firewalls are static as rules in those firewalls are written manually by network administrators. They write those rules based on their knowledge about the network, its topology, possible threats, and business goals or policies. Classical firewalls are also static as they have to be changed/deleted or updated also manually. If, for example, the network is changed, rules in the firewall should be adjusted manually to accommodate those changes.

As an alternative, SDN promises to build dynamic firewalls that can automatically respond to network changes, threats, etc. It is also dynamic as rules themselves in those firewalls can be written or modified dynamically or programmatically without human intervention. For example, if the firewall detects a new threat, it will try to write new rules that can protect against such new threat. While SDN may have the initial capabilities and promises to develop such dynamic firewalls, however, the path to develop such firewalls will face several practical obstacles or challenges.

In terms of security risks and vulnerabilities in SDN, several publications investigated such issues. It is expected that such new architecture will face new security challenges. One of the most issues that were discussed in relation to vulnerabilities in SDN is the risk that a central network controller can be a serious single point of failure and also target for attacks. Further, when such central controller is a software, fear arises that it can be easier to target than hardware or hybrid software-hardware controller. Initial use cases of SDN controllers targeted several goals including the need for network protocols to be open and unified. This will enable conducting research as an alternative to current closed and vendor-specific network protocols. On the other hand, having those protocols to be open with open-source controllers may facilitate the task of attacking such protocols and controllers.

Another vulnerability issue related to SDN controller is related to the possibility of attacking those networks through flooding or denial of service (DoS). A significant volume of traffic is exchanged between the controller and its switches. In real medium-to-large networks, scalability can be a serious concern of such large volume of traffic. Man-in-the middle (MiM) attacks may try to saturate or flood the controller with traffic which may ultimately cause the connection between the controller and its switches to fail. Attackers may ultimately claim the controller of network switches. Similarly, switches include flow tables that have dynamic rules added in real time based on traffic. Crafted attacks can aim at flooding those flow tables with traffic that ensure all new rules added to the flow table. Those are examples of possible attacks in SDN.

The rest of the chapter is organized as the following. The next section will focus on related work to SDN surveys and systematic mappings. In Sect. 11.3, we showed our research methodology to conduct a systematic mapping for SDN security. In Sect. 11.5, we will elaborate on the results and discussions on collected information. The paper is then concluded in Sect. 11.6.

11.2 Related Work

In this section, we will focus on papers related to SDN with two focuses: surveys or systematic literature review (SLR) and mapping.

While to the best of our knowledge no SLR paper is published in SDN, there are several published survey papers. Existing research conducted surveys in SDN in general or SDN security in particular (e.g., [1–13]). In this section, we will focus on security aspects discussed in those papers.

Scott-Hayward et al. in 2013 surveyed issues related to SDN security. The paper discussed some of the security challenges related to the new SDN architecture. The paper surveyed also some of the approaches and technologies adopted in SDN to deal with security issues and challenges. Authors focused on three major aspects: security analysis, SDN security enhancements, and challenges. They classified SDN security solutions based on SDN layers or interfaces: application, control, data, or a mix between those three components. For example, data leakage, switch flow table flooding, and data modification security problems fall within the data layer. Frauds in rules insertion and policy enforcement can fall in application or control layers. They focused on some of the security concerns that those are expected to be higher in SDN if compared with classical networking. Most of those mentioned are related to the idea of control centralization. For example, denial of service (DoS) and flow table intrusion can have a very serious impact if conducted on SDN. Mechanisms to secure the controller are yet to be thoroughly investigated. The second main security challenge exists due to the nature of SDN that is related to its open architecture. While such open architecture was one of the major goals to make the architecture vendor independent and open source, however, at the same time, it can attract hacking and security threats.

There are other security issues authors pointed to in OpenFlow protocol. For example, the communication between switches and the controller uses an option of transport layer security (TLS) protocol, while such security standard is not tested, adopted by vendors, or even enforced to be used in the communication between the controller and its switches. Security concerns related to SSL/TLS are described also in several other papers. Several investigations showed that it is adopted only by open virtual switch. Most other virtual switches and also physical switches that support OpenFlow have not included TLS.

Middleboxes are applications that are added or developed on top of OpenFlow networks. Middleboxes are used to provide network security functions. Traffic can be redirected, original, or a copy of the original, for those applications to perform security investigations upon. Those middleboxes communicate with the controller back and forth for the necessary information. They can also provide their advice or recommendation to the controller related to security, load balancing, etc. Those can work on the upper, northbound section of the controller. While those middleboxes can provide the important security services and information, they can be also used to attack the controller and illegally access its internal modules. Once a middlebox is compromised, it can be used as a botnet to access or attack the controller.

Hu et al. in 2014 presented a survey paper on SDN. We will point to some security-related issues in the paper. The first issue is related to virtualization and slices' isolation. Authors showed some of the most recent progresses in achieving secure and robust isolation. Those can be different in terms of performance, robustness, and flexibility. In addition, the dynamic allocation of resources among the different slices is a compelling SDN advantage. However, such dynamic allocation should be orchestrated well so that no insiders' threats may occur between the different VMs or slices. Efficient and secure management in the controller is another important security challenge in SDN. Most middleboxes or applications are planned to be developed on the northbound section of the controller. Those applications need to interact with the controller and exchange information with. On the other hand, those can be serious vulnerable points especially as those applications are typically developed by third parties, operate on vulnerable environments, etc.

Conflicts may also occur when different middleboxes try to acquire different information. For example, Jarraya et al. [11] described one example where there may be a case where two middleboxes are requesting flows or information about flows that contradict with each other. In other cases, such requested information itself may cause security exploits. The open architecture of the OpenFlow model is considered an important advantage where the network or its protocols are not tied to a certain vendor. However, from security perspectives, such openness can pose new types of exploits or vulnerabilities.

The SDN controller can be a significant target for security attacks, in addition to traditional networking elements such as routers and switches. Once the controller is attacked, all switches can be out of control and maybe easily attacked or compromised.

Flow loops may occur in some cases in SDN where decisions cannot be made on some flows. The mechanism to detect and prevent occurrences of such loops should be developed and tested. Safety and failure recovery are important subjects in SDN security. The singleton state of the controller can be very risky from those perspectives where the network is relying completely on one component. In addition to techniques such as controller distribution or failure recovery, methods should be put in place to make sure that if failures occur, they will be very short and recoverable.

Jarraya et al. [11] described an SDN research architecture that can be composed based on the different layers (i.e., infrastructure, control, application). They further divided those into sublayers. In the security section, authors included a list of papers discussing security issues in OpenFlow, their goals, and their approaches. Those can be classified as flow table rules, conflicts, etc. Authors focused also on enumerating some of the open security threats in SDN architecture. For example, the communication between the controller and components from the different sides (e.g., south and north bounds) is shown to have some security problems. The controller itself can be another central weak point susceptible to attacks such as flooding, DoS, spoofing, MiM, etc. Switch specifications and many other aspects in the technology are still evolving, and consequently it is normal for such new evolving architecture to have possible bugs and vulnerabilities. The important part is to keep evaluating such security vulnerability and enhance OpenFlow protocol in the next version based on experimental recommendations.

In the new versions of OpenFlow (1.2 and above), flow attributes are increased from 12 to 40 attributes. Many new attributes are covered to describe different details about the traffic beyond the traditional (L2–L3) information. It is hoped in future research that security appliances that require information about traffic from higher layers can show and demonstrate whether added attributes can be very useful or not. In addition, few recent researches evaluated those extra flow attributes and how they are going to impact the network from scalability, security, etc. perspectives.

Kreutz et al. in 2013 and 2014 presented survey papers on security analysis of SDN with challenges and open issues. They described several categories of threats that may target SDN. Those can be either inherently similar to traditional networks or can be unique based on the SDN architecture. They pointed to two major characteristics in SDN that may trigger security threats. Those are first the fact that software controls the network, and the second one is related to control centralization. They also proposed possible solutions to deal with discussed security threats.

While one of the major SDN characteristics was to have an open architecture, however, we think that this is not yet formalized or standardized specially on the northbound communication with the controller. The OpenDayLight Controller project is working on one standard northbound interface (i.e., REST API). There are many serious questions to be handled once such protocol (i.e., northbound protocol) is completed. For example, there are many occasions where implemented northbound APIs have to give instructions and control decisions to the controller. This can be referred to as "controlling the controller." How can this be orchestrated without contradicting the centrality of decision-making in SDN? A proper control delegation method should be proposed to make sure that control delegation, if requested, will be only within very specific boundaries. There are also some security concerns that those applications communicating with the controller may cause security problems, intentionally or unintentionally. Those middleboxes can be seen as insiders to the controller or its network. This is especially true as controller may delegate some control privileges to them. Those middleboxes run in operating systems that can be securely vulnerable. Other applications in the same host can be also botnets and used to access OpenFlow network.

Proposed several recommendations to reduce information disclosure in OpenFlow networks. Authors evaluated the different SDN components and how much they can be vulnerable to different attack types. Authors use the STRIDE attack model (i.e., Spoofing, Tampering, Repudiation, Information disclosure, Denial of service, Elevation of privilege). For example, intelligent rules for time-out randomization can make it difficult for scanners or sniffers to understand network patterns. The same thing can be mentioned about the response time between the controller and switches. A monitoring tool can detect the difference in response time between sending a new and previously sent flows. The existence of such difference in response time is an indicator of an existing OpenFlow network. Countering this type of information sniffing can take several scenarios. In one option, using a fully proactive approach where all flow rules are installed by network administrators is an option. Directed and intelligent countermeasure methods can be also effective in making fake response time based on the nature of the network attack.

Attack tree models proposed and several other researchers can be used to automatically detect the type of network attack. However, such models still seem to be highly semantic and do not include information that can be directly interpreted or related to flow- or packet-level information.

We believe that such attack trees should exist even if the size of the tree is going to be large (assuming that the tree will include all types of network attacks previously discovered with the exact flow- and packet-level information that can identify it). In such types of attack models, attack types will be more specific than the generic models. For example, if this is to implement a programmable firewall, it is not enough to say that we have a DoS attack. We may need to specify the protocol (e.g., ICMP) and the port in addition to all relevant flow and packet attributes. If we want the attack tree model to be automatically interrogated and understood by low-level switches, it should include attributes and metrics related to flows or packets.

11.3 Research Method

Systematic literature review has been conducted to accumulate literature reporting empirical work within the SDN discipline, mainly in security-related issues. Figure 11.1 summarizes the adopted guidelines as reported in Kitchenham's guidelines.

Systematic literature review has three main phases: planning, conducting, and documenting the review. Each one of these steps is described in details in the following sections.

The need for the systematic review (Step 1) is motivated by the hot topic of software-defined networks nowadays. Security in SDN is still an open issue that many researches have been published in the past few years. Such publications include empirical studies containing case studies, experience reports, and experiments. The objective of this research work is to aggregate and synthesize the empirical evidence from literature related to SDN security, trying to explore the trends and categories of work in this field.

Defining research questions (Step 2) is essential to specify what we should extract from the collected publications (Sect. 11.3.5). For instance, we need to know how much research has been conducted since 2005, and then what are the main issues related to SDN security that have been studied in the literature. We are also curious to know who is leading the research in SDN security and which publication channels are publishing SDN research the most. Once we know these issues, we need to explore the possible measures that are used to quantify SDN security. Moreover, we need to identify what factors can affect the SDN security implementation both positively (success factors) and negatively (failure factors). A summary of the research questions is given in Table 11.1

The following Sects. 11.3.1 and 11.3.2 define the review protocol (Step 3). The protocol was evaluated (Step 4) by an independent researcher with experience in conducting systematic reviews. According to his feedback and our own gathered experiences during the process, we iteratively improved the design of the review.

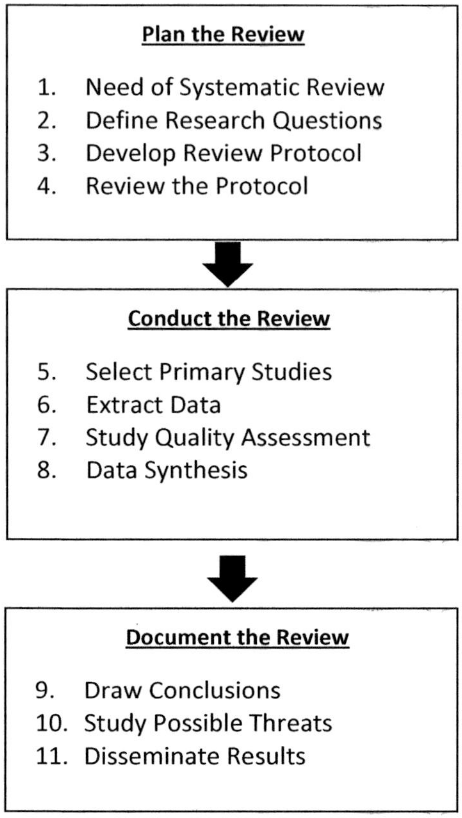

Fig. 11.1 Systematic literature review steps

Table 11.1 Research questions

ID	Question
RQ1	RQ1: How chronically research trends in SDN security evolve?
RQ2	What are the main issues related to SDN security that have been studied in the literature?
RQ3	Who is leading the research in SDN security?
RQ4	Which publication channels are publishing the SDN research the most?
RQ5	What measures are used to quantify SDN security issues?
RQ6	What are the success and failure factors that affect the implementation of SDN security?

11.3.1 Search Process

We have used a general search string to identify all possible publications suitable to answer our research questions. The search string is: "software-defined network and security".

The data sources to be searched include IEEE Xplore, ACM Digital Library, Wiley Online Library, ScienceDirect, and SpringerLink. We used the following

Table 11.2 SDN security identified literature

Resource	No. of studies
IEEE Xplore	122
ACM digital library	64
Wiley online library	30
ScienceDirect	90
SpringerLink	159
Total	465

string to search for candidate papers. The selected data sources are known to publish empirical studies in different fields.

Searching has been conducted manually, and the search string brought 465 studies from all five databases. The search was applied on the databases searching for publications in the period Jan 2005–Jan 2015. The distribution of studies among publication channels is shown in Table 11.2. Each researcher is assigned half of the data sources to conduct the search and apply the inclusion and exclusion criteria (Sect. 11.3.2). Once completed, researchers swap the roles so that each researcher checks the included and excluded papers decided by the other researcher. All disagreements are collected and collated.

11.3.2 Inclusion and Exclusion Criteria

Papers that discuss researchers' experience in SDN security based on empirical evidences and published between Jan 2005 and Jan 2015 are included.

The excluded papers include duplicate publications of the same study where in this case the most comprehensive version is included. We excluded editorials, discussions, tutorials, summaries, theses, and any study that do not present empirical studies.

The study selection criterion has two major phases. In the first phase, the study inclusion/exclusion steps were applied on the title and abstract. In the second phase, the study screening criterion was applied on a full text. In both of the phases, the focus of the study on SDN and existence of empirical work was ensured. As a result of the first screening phase (title and abstract screening), we obtained 465 studies. These studies were screened for full-text inclusion/exclusion in the second phase.

11.4 General Statistics

11.4.1 ACM

A search in ACM retrieved 227 articles. Table 11.3 shows the top events and proceeding series (i.e., workshop, conference).

Table 11.3 Top events and proceeding series

SIGCOMM'14 (32)	SIGCOMM (66)
CoNEXT '14 (7)	CONEXT (12)
HotNets-XIII (6)	AICPS (10)
ANCS '14 (3)	ANCS (8)
CCS'14 (3)	HotNets (8)
IMC '14 (3)	HPDC (6)
EuroSys '15 (3)	CoNEXT (6)
PLDI '14 (2)	HOTNETS-XII (6)

Table 11.4 Top in Springer by journal

Computer science 227	Communication networks 169
Engineering 102	SWE 76
Business and management 46	Database management and information retrieval 71
Materials 12	Signals 69
Mathematics 8	Information systems and applications 60

11.4.2 SpringerLink

Search queries from the SpringerLink return 238 research articles and chapters: 155 chapters and 83 articles. Table 11.4 shows top disciplines and subdisciplines.

11.5 Results and Analysis

We synthesized the extracted data to answer the research questions. This section presents the results obtained from SLR.

11.5.1 RQ1: How Chronically Research Trends in SDN Security Evolve?

Although software-based or programmable networks as a general term for SDN have been there in different contexts for a while, the specific term is coined recently around the year 2007. Since then SDN evolved with focuses on different areas. By looking at early statistical tables, we can see examples of some of the focus areas such as communication networks, software engineering, database management, etc. As we mentioned earlier, SDN security evolved into two directions. The first one is related to security vulnerabilities and risks in SDN. The second one is related to the evolution of security controls and mechanisms as a result of this new architecture.

Table 11.5 SDN security publication per year

ACM	ScienceDirect
2015 (17)	2015 (27)
2014 (107)	2014 (80)
2013 (62)	2013 (24)
2012 (28)	2012 (4)
2011 (6)	2011 (1)
2010 (3)	

Table 11.6 Main research areas in SDN security

Datum center (18)
Future Internet (5)
Computer science (4)
Software (4)
Control plane (3)

Table 11.7 Top five Google Scholar: SDN security

FRESCO: Modular composable security services for software-defined networks
Are we ready for SDN? Implementation challenges for software-defined networks
Towards secure and dependable software-defined networks
OpenFlow random host mutation: transparent moving target defense using software-defined networking
Revisiting traffic anomaly detection using software-defined networking

Table 11.5 shows a very clearly continuous increase in publications in this focus area (for 2015, as of May 2015). The table indicates that SDN security will be one of the major research areas in SDN for the next few years.

11.5.2 RQ2: What Are the Main Issues Related to SDN Security that Have Been Studied in the Literature?

Table 11.6 shows the ScienceDirect focus research areas related to SDN security. Clearly, security in the cloud and data centers is the most important one.

Table 11.7 shows the Google Scholar five top-cited articles related to SDN security

We used top-cited papers in our focused search as one indicator of its research trends. The first paper (FRESCO) showed one of the contributions related to SDN-based security controls. It is expected that future security services will be more modular, customizable, and composable rather than the current generic types of security controls.

Other examples of most cited papers in SDN security include papers that discussed security issues (i.e., challenges and opportunities) in SDN in general without specific contribution or focus. Other papers discussed also some of the vulnerabilities in SDN architecture and how to target such vulnerability. We also identified two major domains in which SDN security publications are trending: the cloud and wireless networks.

11.5.2.1 SDN Security for the Cloud

Both SDN and the cloud share the same security concern. On the other side, they share also the ability to provide flexible and dynamic services. Publications in this particular focus (i.e., SDN cloud security) are also split between security concerns and opportunities.

11.5.2.2 SDN Security for Wireless

Extending SDN and programmable networks to mobile or wireless networks continues to be one of the major research trends in networking and wireless in general. Providing dynamic, customized, and flexible security services is a very promising goal in this scope for researchers, wireless vendors, and service providers. However, the idea of making protocols unified and open still faces some industrial- and business-related obstacles. Flexibility can always have conflicts with security. Making wireless networks or smart device software enabled, controlled, etc. may open new waves of security threats that were less popular in those domains. On the other hand, research trends show that while there are always security-related concerns when software is given more control roles, nonetheless, this never stopped technologies to move toward such directions or trends.

11.5.3 Who Is Leading the Research in SDN Security?

In its current form, SDN started as a research project in Stanford University [14, 15]. Early contributions were also from other popular universities such as the Princeton University, Cornell University, Georgia Institute of Technology, and University of California, Berkeley. Table 11.8 shows ACM-based top authors and institutions.

11.6 Conclusion

Software-defined networking (SDN) is a recently evolving network architecture that aims at enabling users and their application to have better control on the network and traffic. It falls within the general definition of "programmable

Table 11.8 Top authors and institutions

Rexford, Jennifer L (13)	Princeton University (21)
Foster, Nate (12)	Cornell University (15)
Feamster, Nick Greer (9)	Georgia Institute of Technology (12)
Shenker, Scott J (7)	University of California, Berkeley (10)
Walker, David Patrick (6)	University of Southern California (9)
Guha, Arjun (6)	University of Illinois at Urbana-Champaign (9)
Godfrey, Philip Brighten (5)	University of Wisconsin Madison (9)
Koponen, Teemu (5)	Stanford University (6)

networks" in which software is taking more roles in network control and management. This evolution in network architectures is seen as a driver that will change how applications are developed on top of those networks. We focused in security controls and activities in this paper and what impact SDN can cause to them. It is highly likely that this area of network security is going to receive a lot of focus and trends both in research and industry in the very near future.

References

1. Hu, H., Han, W., Ahn, G.-J., Zhao, Z.: FLOWGUARD: building robust firewalls for software-defined networks. In: HotSDN '14: Proceedings of the Third Workshop on Hot Topics in Software Defined Networking, August 2014
2. Hu, C., Liu, B., Zhao, H., Chen, K., Chen, Y., Yu, C., Hao, W.: Discount counting for fast flow statistics on flow size and flow volume. IEEE/ACM Trans. Networking **22**(3), 970–981 (2014)
3. Kreutz, D., Ramos, F.M.V., Verissimo, P.: Towards secure and dependable software-defined networks. In: HotSDN '13: Proceedings of the Second ACM SIGCOMM Workshop on Hot Topics in Software Defined Networking, August 2013
4. Scott-Hayward, S., O'Callaghan, G., Sezer, S.: SDN security: a survey in 2013 I.E. SDN for future networks and services (SDN4FNS), pp. 1–7. Institute of Electrical and Electronics Engineers (IEEE) (2013)
5. Kloeti, R., Kotronis, V., Smith, P.: OpenFlow: a security analysis. In: Proceedings of the 8th Workshop on Secure Network Protocols (NPSec), part of IEEE ICNP, Göttingen, Germany, October 2013
6. Kreutz, D., Ramos, F., Verissimo, P.: Towards secure and dependable software-defined networks. In: Proceedings of the Second ACM SIGCOMM Workshop on Hot Topics in Software Defined Networking, ACM, 2013, pp. 55–60, HotSDN'13, 16 August 2013, Hong Kong, China
7. Kreutz, D., Ramos, F.M.V., Verissimo, P., Rothenberg, C.E., Azodolmolky, S., Uhlig, S.: Software-defined networking: a comprehensive survey, June 2014
8. Hu, F., Hao, Q., Bao, K.: A survey on software-defined network (SDN) and OpenFlow: from concept to implementation. IEEE Commun. Surv. Tutor. **16**(4), 2181–2206 (2014)
9. Nunes, B., Mendonca, M., Nguyen, X.-N., Obraczka, K., Turletti, T.: A survey of software-defined networking: past, present, future of programmable networks. IEEE Commun. Surv. Tutor. **16**(3), 1617–1634 (2014)
10. Xia, W., Wen, Y., Foh, C.H., Niyato, D., Xie, H.: A survey on software-defined networking. IEEE Commun. Surv. Tutor. **99**, 1 (2014)

11. Jarraya, Y., Madi, T., Debbabi, M.: A survey and a layered taxonomy of software-defined networking. IEEE Commun. Surv. Tutor. **16**(1), 1955–1980 (2014)
12. Lara, A., Kolasani, A., Ramamurthy, B.: Network innovation using OpenFlow: a survey. IEEE Commun. Surv. Tutor. **16**(1), 493–512 (2014)
13. Lara, A., Ramamurthy, B., Nagaraja, K., Krishnamoorthy, A., Raychaudhuri, D.: Using OpenFlow to provide cut-through switching in MobilityFirst. Photonic Netw. Commun. **28** (2), 165–177 (2014). doi:10.1007/s11107-014-0461-3. http://link.springer.com/article/10.1007/s11107-014-0461-3
14. Feamster, N., Rexford, J., Zegura, E.: The road to SDN. Queue **11**(12), 20–32 (2013)
15. Feamster, N., Rexford, J., Zegura, E.: The road to SDN: an intellectual history of programmable networks. ACM Queue **11**, 12 (2013)
16. Li, H., Hu, C., Hong, J., Chen, X., Jiang, Y.: Parsing application layer protocol with commodity hardware for SDN. In: ANCS '15: Proceedings of the Eleventh ACM/IEEE Symposium on Architectures for Networking and Communications Systems, May 2015
17. Qu, Y.R., Zhang, H.H., Zhou, S., Prasanna, V.K.: Optimizing many-field packet classification on FPGA, multi-core general purpose processor, and GPU. In: ANCS '15: Proceedings of the Eleventh ACM/IEEE Symposium on Architectures for Networking and Communications Systems, May 2015
18. Garzarella, S., Lettieri, G., Rizzo, L.: Virtual device pass through for high speed VM networking. In: ANCS '15: Proceedings of the Eleventh ACM/IEEE Symposium on Architectures for Networking and Communications Systems, May 2015
19. Hsieh, C.-L., Weng, N.: Scalable many-field packet classification using multidimensional-cutting via selective bit-concatenation. In: ANCS '15: Proceedings of the Eleventh ACM/IEEE Symposium on Architectures for Networking and Communications Systems, May 2015
20. Silva, A.: A short introduction to the coalgebraic method. SIGLOG News. **2**(2) (April 2015)
21. Chetty, M., Kim, H., Sundaresan, S., Burnett, S., Feamster, N., Keith Edwards, W.: uCap: an internet data management tool for the home. In: CHI '15: Proceedings of the 33rd Annual ACM Conference on Human Factors in Computing Systems, April 2015
22. Verma, A., Pedrosa, L., Korupolu, M., Oppenheimer, D., Tune, E., Wilkes, J.: Large-scale cluster management at Google with Borg. In: EuroSys '15: Proceedings of the Tenth European Conference on Computer Systems, April 2015
23. Tobias Distler, Christopher Bahn, Alysson Bessani, Frank Fischer, Flavio Junqueira. Extensible distributed coordination. April 2015 EuroSys '15: Proceedings of the Tenth European Conference on Computer Systems
24. Leners, J.B., Gupta, T., Aguilera, M.K., Walfish, M.: Taming uncertainty in distributed systems with help from the network. In: EuroSys '15: Proceedings of the Tenth European Conference on Computer Systems, April 2015
25. Ambrosin, M., Conti, M., De Gaspari, F., Poovendran, R.: LineSwitch: efficiently managing switch flow in software-defined networking while effectively tackling DoS attacks. In: ASIA CCS '15: Proceedings of the 10th ACM Symposium on Information, Computer and Communications Security, April 2015
26. Dong, X., Lin, H., Tan, R., Iyer, R.K., Kalbarczyk, Z.: Software-defined networking for smart grid resilience: opportunities and challenges. In: CPSS '15: Proceedings of the 1st ACM Workshop on Cyber-Physical System Security, April 2015
27. Ma, J., Sui, X., Sun, N., Li, Y., Yu, Z., Huang, B., Xu, T., Yao, Z., Chen, Y., Wang, H., Zhang, L., Bao, Y.: Supporting differentiated services in computers via programmable architecture for resourcing-on-demand (PARD). In: ASPLOS '15: Proceedings of the Twentieth International Conference on Architectural Support for Programming Languages and Operating Systems, March 2015
28. Rehman, M.S., Boles, J., Hammoud, M., Sakr, M.F.: A cloud computing course: from systems to services. In: SIGCSE '15: Proceedings of the 46th ACM Technical Symposium on Computer Science Education, February 2015

29. Kangarlou, A., Shete, S., Strunk, J.D.: Chronicle: capture and analysis of NFS workloads at line rate. In: FAST'15: Proceedings of the 13th USENIX Conference on File and Storage Technologies, February 2015
30. Chockler, G., Junqueira, F., Rodrigues, R., Vigfusson, Y.: LADIS'14: 8th Workshop on Large-Scale Distributed Systems and Middleware. SIGOPS Operating Systems Review, vol. 49, issue 1, January 2015
31. Chlipala, A.: From network interface to multithreaded web applications: a case study in modular program verification. In: POPL '15: Proceedings of the 42nd Annual ACM SIGPLAN-SIGACT Symposium on Principles of Programming Languages, January 2015
32. Foster, N., Kozen, D., Milano, M., Silva, A., Thompson, L.: A coalgebraic decision procedure for NetKAT. In: POPL '15: Proceedings of the 42nd Annual ACM SIGPLAN-SIGACT Symposium on Principles of Programming Languages, January 2015
33. Alt, L., Beverly, R., Dainotti, A.: Uncovering network tarpits with degreaser. In: CSAC '14: Proceedings of the 30th Annual Computer Security Applications Conference, December 2014
34. Spillner, J., Schill, A.: Algorithms for dispersed processing. In: UCC '14: Proceedings of the 2014 IEEE/ACM 7th International Conference on Utility and Cloud Computing, December 2014
35. Bleikertz, S., Vogel, C., Groß, T.: Cloud radar: near real-time detection of security failures in dynamic virtualized infrastructures. In: ACSAC '14: Proceedings of the 30th Annual Computer Security Applications Conference, December 2014
36. Renner, T., Stanik, A., Körner, M., Kao, O.: Portable SDN applications on the PaaS layer. In: UCC '14: Proceedings of the 2014 IEEE/ACM 7th International Conference on Utility and Cloud Computing, December 2014
37. de Jesus, W.P., da Silva, D.A., de Sousa Júnior, R.T., da Frota, F.V.L.: Analysis of SDN contributions for cloud computing security. In: UCC '14: Proceedings of the 2014 IEEE/ACM 7th International Conference on Utility and Cloud Computing, December 2014
38. Soulé, R., Basu, S., Marandi, P.J., Pedone, F., Kleinberg, R., Sirer, E.G., Foster, N.: Merlin: a language for provisioning network resources. In: CoNEXT '14: Proceedings of the 10th ACM International on Conference on Emerging Networking Experiments and Technologies, December 2014
39. Kuzniar, M., Peresini, P., Kostić, D.: Providing reliable FIB update acknowledgments in SDN. In: CoNEXT '14: Proceedings of the 10th ACM International on Conference on Emerging Networking Experiments and Technologies, December 2014
40. Abaid, Z., Rezvani, M., Jha, S.: MalwareMonitor: an SDN-based framework for securing large networks. In: CoNEXT Student Workshop '14: Proceedings of the 2014 CoNEXT on Student Workshop, December 2014
41. Moradi, M., Wu, W., Li, L.E., Mao, Z.M.: SoftMoW: recursive and reconfigurable cellular WAN architecture. In: CoNEXT '14: Proceedings of the 10th ACM International on Conference on Emerging Networking Experiments and Technologies, December 2014
42. Durairajan, R., Sommers, J., Barford, P.: Controller-agnostic SDN debugging. In: CoNEXT '14: Proceedings of the 10th ACM International on Conference on Emerging Networking Experiments and Technologies, December 2014
43. Castro, I., Cardona, J.C., Gorinsky, S., Francois, P.: Remote peering: more peering without internet flattening. In: CoNEXT '14: Proceedings of the 10th ACM International on Conference on Emerging Networking Experiments and Technologies, December 2014
44. Wang, A., Guo, Y., Hao, F., Lakshman, T.V., Chen, S.: Scotch: elastically scaling up SDN control-plane using vSwitch based overlay. In: CoNEXT '14: Proceedings of the 10th ACM International on Conference on Emerging Networking Experiments and Technologies, December 2014
45. Klingel, D., Khondoker, R., Marx, R., Bayarou, K.: Security analysis of software defined networking architectures: PCE, 4D and SANE. In: AINTEC '14: Proceedings of the AINTEC 2014 on Asian Internet Engineering Conference, November 2014

46. Tasch, M., Khondoker, R., Marx, R., Bayarou, K.: Security analysis of security applications for software defined networks. In: AINTEC '14: Proceedings of the AINTEC 2014 on Asian Internet Engineering Conference, November 2014
47. Mayer, S., Hassan, Y.N., Sörös, G.: A magic lens for revealing device interactions in smart environments. In: SA '14: SIGGRAPH Asia 2014 Mobile Graphics and Interactive Applications, November 2014
48. Muhamedyev, R.I., Kalimoldaev, M.N., Uskenbayeva, R.K.: Semantic network of ICT domains and applications. In: EGOSE '14: Proceedings of the 2014 Conference on Electronic Governance and Open Society: Challenges in Eurasia, November 2014
49. Toso, G., Munaretto, D., Conti, M., Zorzi, M.: Attack resilient underwater networks through software defined networking. In: WUWNET '14: Proceedings of the International Conference on Underwater Networks & Systems, November 2014
50. McDaniel, P., Jaeger, T., La Porta, T.F., Papernot, N., Walls, R.J., Kott, A., Marvel, L., Swami, A., Mohapatra, P., Krishnamurthy, S.V., Neamtiu, I.: Security and science of agility. In: MTD '14: Proceedings of the First ACM Workshop on Moving Target Defense, November 2014
51. Jafarian, J.H.H., Al-Shaer, E., Duan, Q.: Spatio-temporal address mutation for proactive cyber agility against sophisticated attackers. In: MTD '14: Proceedings of the First AM Workshop on Moving Target Defense, November 2014
52. Cuzzocrea, A., Song, I.-Y.: Big graph analytics: the state of the art and future research agenda. In: DOLAP '14: Proceedings of the 17th International Workshop on Data Warehousing and OLAP, November 2014
53. Sommer, R., Vallentin, M., De Carli, L., Paxson, V.: HILTI: an abstract execution environment for deep, stateful network traffic analysis. In: IMC '14: Proceedings of the 2014 Conference on Internet Measurement Conference, November 2014
54. Richter, P., Smaragdakis, G., Feldmann, A., Chatzis, N., Boettger, J., Willinger, W.: Peering at peerings: on the role of IXP route servers. In: IMC '14: Proceedings of the 2014 Conference on Internet Measurement Conference, November 2014
55. Pujol, E., Richter, P., Chandrasekaran, B., Smaragdakis, G., Feldmann, A., Maggs, B.M., Ng, K.-C.: Back-office web traffic on the internet. In: IMC '14: Proceedings of the 2014 Conference on Internet Measurement Conference, November 2014
56. Butt, S., Ganapathy, V., Srivastava, A.: On the control plane of a self-service cloud platform. In: SOCC '14: Proceedings of the ACM Symposium on Cloud Computing, November 2014
57. Shin, S., Song, Y., Lee, T., Lee, S., Chung, J., Porras, P., Yegneswaran, V., Noh, J., Kang, B. B.: Rosemary: a robust, secure, and high-performance network operating system. In: CCS '14: Proceedings of the 2014 ACM SIGSAC Conference on Computer and Communications Security, November 2014
58. Yuan, Y., Alur, R., Loo, B.T.: NetEgg: programming network policies by examples. In: HotNets-XIII: Proceedings of the 13th ACM Workshop on Hot Topics in Networks, October 2014
59. Vissicchio, S., Vanbever, L., Rexford, J.: Sweet little lies: fake topologies for flexible routing. In: HotNets-XIII: Proceedings of the 13th ACM Workshop on Hot Topics in Networks, October 2014
60. Chandrasekaran, B., Benson, T.: Tolerating SDN application failures with LegoSDN. In: HotNets-XIII: Proceedings of the 13th ACM Workshop on Hot Topics in Networks, October 2014
61. Ludwig, A., Rost, M., Foucard, D., Schmid, S.: Good network updates for bad packets: waypoint enforcement beyond destination-based routing policies. In: HotNets-XIII: Proceedings of the 13th ACM Workshop on Hot Topics in Networks, October 2014
62. Donovan, S., Feamster, N.: Intentional network monitoring: finding the needle without capturing the haystack. In: HotNets-XIII: Proceedings of the 13th ACM Workshop on Hot Topics in Networks, October 2014

63. Schlinker, B., Zarifis, K., Cunha, I., Feamster, N., Katz-Bassett, E.: PEERING: an AS for Us. In: HotNets-XIII: Proceedings of the 13th ACM Workshop on Hot Topics in Networks, October 2014
64. Majumdar, R., Tetali, S.D., Wang, Z.: Kuai: a model checker for software-defined networks. In: FMCAD '14: Proceedings of the 14th Conference on Formal Methods in Computer-Aided Design, October 2014
65. Bonelli, N., Giordano, S., Procissi, G., Abeni, L.: A purely functional approach to packet processing. In: ANCS '14: Proceedings of the tenth ACM/IEEE Symposium on Architectures for Networking and Communications Systems, October 2014
66. Kaplan, M., Zheng, C., Monaco, M., Keller, E., Sicker, D.: WASP: a software-defined communication layer for hybrid wireless networks. In: ANCS '14: Proceedings of the tenth ACM/IEEE Symposium on Architectures for Networking and Communications Systems, October 2014
67. Kotani, D., Okabe, Y.: A packet-in message filtering mechanism for protection of control plane in openflow networks. In: ANCS '14: Proceedings of the tenth ACM/IEEE Symposium on Architectures for Networking and Communications Systems, October 2014
68. Arumaithurai, M., Chen, J., Monticelli, E., Fu, X., Ramakrishnan, K.K.: Exploiting ICN for flexible management of software-defined networks. In: INC '14: Proceedings of the 1st International Conference on Information-Centric Networking, September 2014
69. Casado, M., Foster, N., Guha, A.: Abstractions for software-defined networks. Commun. ACM **57**(10), 86–95 (2014)
70. Sapio, A., Liao, Y., Baldi, M., Ranjan, G., Risso, F., Tongaonkar, A., Torres, R., Nucci, A.: Per-user policy enforcement on mobile apps through network functions virtualization. In: MobiArch '14: Proceedings of the 9th ACM Workshop on Mobility in the Evolving Internet Architecture, September 2014
71. Kuo, Y.-S., Pannuto, P., Dutta, P.: System architecture directions for a software-defined lighting infrastructure. In: VLCS '14: Proceedings of the 1st ACM MobiCom Workshop on Visible Light Communication Systems, September 2014
72. Juhola, A., Ahola, T., Ahola, K.: Adaptive risk management with ontology linked evidential statistics and SDN. In: ECSAW '14: Proceedings of the 2014 European Conference on Software Architecture Workshops, August 2014
73. Casey, C.J., Sutton, A., Sprintson, A.: tinyNBI: distilling an API from essential OpenFlow abstractions. In: HotSDN '14: Proceedings of the Third Workshop on Hot Topics in Software Defined Networking, August 2014
74. Chandrasekaran, B., Benson, T.: Tolerating SDN application failures with LegoSDN. In: HotSDN '14: Proceedings of the Third Workshop on Hot Topics in Software Defined Networking, August 2014
75. Durairajan, R., Sommers, J., Barford, P.: OFf: bug spray for openflow. In: HotSDN '14: Proceedings of the Third Workshop on Hot Topics in Software Defined Networking, August 2014
76. Matsumoto, S., Hitz, S., Perrig, A.: Fleet: defending SDNs from malicious administrators. In: HotSDN '14: Proceedings of the Third Workshop on Hot Topics in Software Defined Networking, August 2014
77. Ghorbani, S., Godfrey, B.: Towards correct network virtualization. In: HotSDN '14: Proceedings of the Third Workshop on Hot Topics in Software Defined Networking, August 2014
78. Jagadeesan, N.A., Pal, R., Nadikuditi, K., Huang, Y., Shi, E., Yu, M.: A secure computation framework for SDNs. In: HotSDN '14: Proceedings of the Third Workshop on Hot Topics in Software Defined Networking, August 2014
79. Hand, R., Keller, E.: Closed flow: openflow-like control over proprietary devices. In: HotSDN '14: Proceedings of the Third Workshop on Hot Topics in Software Defined Networking, August 2014

80. Agarwal, K., Dixon, C., Rozner, E., Carter, J.: Shadow MACs: scalable label-switching for commodity Ethernet. In: HotSDN '14: Proceedings of the Third Workshop on Hot Topics in Software Defined Networking, August 2014
81. Heinonen, J., Partti, T., Kallio, M., Lappalainen, K., Flinck, H., Hillo, J.: Dynamic tunnel switching for SDN-based cellular core networks. In: AllThingsCellular '14: Proceedings of the 4th Workshop on All things Cellular: Operations, Applications, & Challenges, August 2014
82. Edwards, T.G., Belkin, W.: Using SDN to facilitate precisely timed actions on real-time data streams. In: HotSDN '14: Proceedings of the Third Workshop on Hot Topics in Software Defined Networking, August 2014
83. Fayaz, S.K., Sekar, V.: Testing stateful and dynamic data plans with Flow Test. In: HotSDN '14: Proceedings of the Third Workshop on Hot Topics in Software Defined Networking, August 2014
84. Jamjoom, H., Williams, D., Sharma, U.: Don't call them middleboxes, call them middlepipes. In: HotSDN '14: Proceedings of the Third Workshop on Hot Topics in Software Defined Networking, August 2014
85. Berde, P., Gerola, M., Hart, J., Higuchi, Y., Kobayashi, M., Koide, T., Lantz, B., O'Connor, B., Radoslavov, P., Snow, W., Parulkar, G.: ONOS: towards an open, distributed SDN OS. In: HotSDN '14: Proceedings of the Third Workshop on Hot Topics in Software Defined Networking, August 2014
86. Bailey, J., Pemberton, D., Linton, A., Pelsser, C., Bush, R.: Enforcing RPKI-based routing policy on the data plane at an internet exchange. In: HotSDN '14: Proceedings of the Third Workshop on Hot Topics in Software Defined Networking, August 2014
87. Baldin, I., Huang, S., Gopidi, R.: A resource delegation framework for software defined networks. In: HotSDN '14: Proceedings of the Third Workshop on Hot Topics in Software Defined Networking, August 2014
88. Klaedtke, F., Karame, G.O., Bifulco, R., Cui, H.: Access control for SDN controllers. In: HotSDN '14: Proceedings of the Third Workshop on Hot Topics in Software Defined Networking, August 2014
89. Mekky, H., Hao, F., Mukherjee, S., Zhang, Z.-L., Lakshman, T.V.: Application-aware data plane processing in SDN. In: HotSDN '14: Proceedings of the Third Workshop on Hot Topics in Software Defined Networking, August 2014
90. Nagaraj, K., Katti, S.: ProCel: smart traffic handling for a scalable software EPC. In: HotSDN '14: Proceedings of the Third Workshop on Hot Topics in Software Defined Networking, August 2014
91. Schlesinger, C., Greenberg, M., Walker, D.: Concurrent NetCore: from policies to pipelines. In: ICFP '14: Proceedings of the 19th ACM SIGPLAN International Conference on Functional Programming, August 2014
92. Parniewicz, D., Doriguzzi Corin, R., Ogrodowczyk, L., Rashidi Fard, M., Matias, J., Gerola, M., Fuentes, V., Toseef, U., Zaalouk, A., Belter, B., Jacob, E., Pentikousis, K.: Design and implementation of an OpenFlow hardware abstraction layer. In: DCC '14: Proceedings of the 2014 ACM SIGCOMM Workshop on Distributed Cloud Computing, August 2014
93. Shanmugam, P.K., Subramanyam, N.D., Breen, J., Roach, C., Van der Merwe, J.: DEIDtect: towards distributed elastic intrusion detection. In: DCC '14: Proceedings of the 2014 ACM SIGCOMM Workshop on Distributed Cloud Computing, August 2014
94. Cao, Z., Kodialam, M., Lakshma, T.V.: Traffic steering in software defined networks: planning and online routing. In: DCC '14: Proceedings of the 2014 ACM SIGCOMM Workshop on Distributed Cloud Computing, August 2014
95. Shirali-Shahreza, S., Ganjali, Y.: Traffic statistics collection with FleXam. In: SIGCOMM '14: Proceedings of the 2014 ACM Conference on SIGCOMM, August 2014
96. Alwabel, A., Yu, M., Zhang, Y., Mirkovic, J.: SENSS: observe and control your own traffic in the internet. In: SIGCOMM '14: Proceedings of the 2014 ACM Conference on SIGCOMM, August 2014

97. Rasley, J., Stephens, B., Dixon, C., Rozner, E., Felter, W., Agarwal, K., Carter, J., Fonseca, R.: Planck: millisecond-scale monitoring and control for commodity networks. In: SIGCOMM '14: Proceedings of the 2014 ACM Conference on SIGCOMM, August 2014
98. Sun, P., Mahajan, R., Rexford, J., Yuan, L., Zhang, M., Arefin, A.: A network-state management service. In: SIGCOMM '14: Proceedings of the 2014 ACM Conference on SIGCOMM, August 2014
99. Gupta, A., Vanbever, L., Shahbaz, M., Donovan, S.P., Schlinker, B., Feamster, N., Rexford, J., Shenker, S., Clark, R., Katz-Bassett, E.: SDX: a software defined internet exchange. In: SIGCOMM '14: Proceedings of the 2014 ACM Conference on SIGCOMM, August 2014
100. Alizadeh, M., Edsall, T., Dharmapurikar, S., Vaidyanathan, R., Chu, K., Fingerhut, A., Lam, V.T., Matus, F., Pan, R., Yadav, N., Varghese, G.: CONGA: distributed congestion-aware load balancing for datacenters. In: SIGCOMM '14: Proceedings of the 2014 ACM Conference on SIGCOMM, August 2014
101. Li, J., Berg, S., Zhang, M., Reiher, P., Wei, T.: Drawbridge: software-defined DDoS-resistant traffic engineering. In: SIGCOMM '14: Proceedings of the 2014 ACM Conference on SIGCOMM, August 2014
102. Gember-Jacobson, A., Viswanathan, R., Prakash, C., Grandl, R., Khalid, J., Das, S., Akella, A.: OpenNF: enabling innovation in network function control. In: SIGCOMM '14: Proceedings of the 2014 ACM Conference on SIGCOMM, August 2014
103. Miao, R., Yu, M., Jain, N.: NIMBUS: cloud-scale attack detection and mitigation. In: SIGCOMM '14: Proceedings of the 2014 ACM Conference on SIGCOMM, August 2014
104. Donovan, S., Feamster, N.: NetAssay: providing new monitoring primitives for network operators. In: SIGCOMM '14: Proceedings of the 2014 ACM Conference on SIGCOMM, August 2014
105. Ma, L., He, T., Leung, K.K., Swami, A., Towsley, D.: Inferring link metrics from end-to-end path measurements: identifiability and monitor placement. IEEE/ACM Trans. Networking **22** (4), 1351–1368 (2014)
106. Burgess, M.: Promise theory—what is it? Linux J. **2014**(244) (2014)
107. Wolf, T., Griffioen, J., Calvert, K.L., Dutta, R., Rouskas, G.N., Baldin, I., Nagurney, A.: ChoiceNet: toward an economy plane for the internet. SIGCOMMComput. Commun. Rev. **44**(3), 58–65 (2014)
108. Khan, K.R., Ahmed, Z., Ahmed, S., Syed, A., Khayam, S.A.: Rapid and scalable isp service delivery through a programmable middlebox. SIGCOMM Comput. Commun. Rev. **44**(3), 31–37 (2014)
109. DeBruhl, B., Kroer, C., Datta, A., Sandholm, T., Tague, P.: Power napping with loud neighbors: optimal energy-constrained jamming and anti-jamming. In: WiSec '14: Proceedings of the 2014 ACM Conference on Security and Privacy in Wireless & Mobile Networks, July 2014
110. Neeman, H., Akin, D., Alexander, J., Brunson, D., Calhoun, S.P., Deaton, J., Fotou, F.F., George, B., Gentis, D., Gray, Z., Huebsch, E., Louthan, G., Runion, M., Snow, J., Zimmerman, B.: The OneOklahoma friction free network: towards a multi-institutional science DMZ in an EPSCoR State. In: XSEDE '14: Proceedings of the 2014 Annual Conference on Extreme Science and Engineering Discovery Environment, July 2014
111. Tantar, A.-A., Tantar, E.: A survey on sustainability in ICT: a computing perspective. In: GECCO Comp '14: Proceedings of the 2014 Conference Companion on Genetic and Evolutionary Computation Companion, July 2014
112. Fontes, R.R., Oliveira, A.L.C., Sampaio, P.N.M., Pinheiro, T.R., Figueira, R.A.R.B.: Authoring of OpenFlow networks with visual network description (SDN version) (WIP). In: SummerSim '14: Proceedings of the 2014 Summer Simulation Multiconference, July 2014
113. Yu, Z., Li, M., Liu, Y., Li, X.: GatorCloud: a fine-grained and dynamic resource sharing architecture for multiple cloud services. In: BigSystem '14: Proceedings of the 2014 ACM International Workshop on Software-Defined Ecosystems, June 2014

114. Moody, W.C., Anderson, J., Wange, K.-C., Apon, A.: Reconfigurable network testbed for evaluation of datacenter topologies. In: DIDC '14: Proceedings of the Sixth International Workshop on Data Intensive Distributed Computing, June 2014
115. Farhadi, H., Du, P., Nakao, A.: User-defined actions for SDN. In: CFI '14: Proceedings of the Ninth International Conference on Future Internet Technologies, June 2014
116. Naik, M.: Large-scale configurable static analysis. In: SOAP '14: Proceedings of the 3rd ACM SIGPLAN International Workshop on the State of the Art in Java Program Analysis, June 2014
117. Lee, J., Uddin, M., Tourrilhes, J., Sen, S., Banerjee, S., Arndt, M., Kim, K.-H., Nadeem, T.: meSDN: mobile extension of SDN. In: MCS '14: Proceedings of the Fifth International Workshop on Mobile Cloud Computing & Services, June 2014
118. Crowley, P.: Author retrospective for characterizing processor architectures for programmable network interfaces. In: International Conference on Supercomputing 25th Anniversary Volume, June 2014
119. Ball, T., Bjørner, N., Gember, A., Itzhaky, S., Karbyshev, A., Sagiv, M., Schapira, M., Valadarsky, A.: VeriCon: towards verifying controller programs in software-defined networks. In: PLDI '14: Proceedings of the 35th ACM SIGPLAN Conference on Programming Language Design and Implementation, June 2014
120. Vadrevu, C.S.K., Wang, R., Tornatore, M., Martel, C.U., Mukherjee, B.: Degraded service provisioning in mixed-line-rate WDM backbone networks using multipath routing. IEEE/ACM Trans. Networking **22**(3), 840–849 (2014)
121. Vishnoi, A., Poddar, R., Mann, V., Bhattacharya, S.: Effective switch memory management in OpenFlow networks. In: DEBS '14: Proceedings of the 8th ACM International Conference on Distributed Event-Based Systems, May 2014
122. Jusko, J., Rehak, M., Pevny, T.: A memory efficient privacy preserving representation of connection graphs. In: ACySE '14: Proceedings of the 1st International Workshop on Agents and CyberSecurity, May 2014
123. Vissicchio, S., Vanbever, L., Bonaventure, O.: Opportunities and research challenges of hybrid software defined networks. SIGCOMM Comput. Commun. Rev. **44**(2), 70–75 (2014)
124. Bianchi, G., Bonola, M., Capone, A., Cascone, C.: OpenState: programming platform-independent stateful openflow applications inside the switch. SIGCOMM Comput. Commun. Rev. **44**(2), 44–51 (2014)
125. Han, Y., Lu, W., Xu, S.: Characterizing the power of moving target defense via cyber epidemic dynamics. In: HotSoS '14: Proceedings of the 2014 Symposium and Bootcamp on the Science of Security, April 2014
126. Feamster, N., Rexford, J., Zegura, E.: The road to SDN: an intellectual history of programmable networks. SIGCOMM Comput. Commun. Rev. **44**(2), 87–98 (2014)
127. Handigol, N., Heller, B., Jeyakumar, V., Mazières, D., McKeown, N.: I know what your packet did last hop: using packet histories to troubleshoot networks. In: NSDI'14: Proceedings of the 11th USENIX Conference on Networked Systems Design and Implementation, April 2014
128. Koponen, T., Amidon, K., Balland, P., Casado, M., Chanda, A., Fulton, B., Ganichev, I., Gross, J., Gude, N., Ingram, P., Jackson, E., Lambeth, A., Lenglet, R., Li, S.-H., Padmanabhan, A., Pettit, J., Pfaff, B., Ramanathan, R., Shenker, S., Shieh, A., Stribling, J., Thakkar, P., Wendlandt, D., Yip, A., Zhang, R.: Network virtualization in multi-tenant datacenters. In: NSDI'14: Proceedings of the 11th USENIX Conference on Networked Systems Design and Implementation, April 2014
129. Nelson, T., Ferguson, A.D., Scheer, M.J.G., Krishnamurthi, S.: Tierless programming and reasoning for software-defined networks. In: NSDI'14: Proceedings of the 11th USENIX Conference on Networked Systems Design and Implementation, April 2014
130. Hwang, J., Ramakrishnan, K.K., Wood, T.: NetVM: high performance and flexible networking using virtualization on commodity platforms. In: NSDI'14: Proceedings of the 11th USENIX Conference on Networked Systems Design and Implementation, April 2014

131. Khan, F., Hosein, N., Ghiasi, S., Chuah, C.-N., Sharma, P.: Streaming solutions for fine-grained network traffic measurements and analysis. IEEE/ACM Trans. Networking **22**(2), 377–390 (2014)
132. Zeng, H., Kazemian, P., Varghese, G., McKeown, N.: Automatic test packet generation. IEEE/ACM Trans. Networking **22**(2), 554–566 (2014)
133. Mushi, M., Dutta, R.: Data-driven study of network administration in the evolving landscape of software defined networking. In: HCBDR '14: Proceedings of the 2014 Workshop on Human Centered Big Data Research, April 2014
134. Gomes, R.L., Bittencourt, L.F., Madeira, E.R.M.: A similarity model for virtual networks negotiation. In: SAC '14: Proceedings of the 29th Annual ACM Symposium on Applied Computing, March 2014
135. Bumgardner, V.K.C., Marek, V.W.: Scalable hybrid stream and hadoop network analysis system. In: ICPE '14: Proceedings of the 5th ACM/SPEC International Conference on Performance Engineering, March 2014
136. Stecklina, J.: Shrinking the hypervisor one subsystem at a time: a userspace packet switch for virtual machines. In: VEE '14: Proceedings of the 10th ACM SIGPLAN/SIGOPS International Conference on Virtual Execution Environments, March 2014
137. Anderson, C.J., Foster, N., Guha, A., Jeannin, J.-B., Kozen, D., Schlesinger, C., Walker, D.: NetKAT: semantic foundations for networks. In: POPL '14: Proceedings of the 41st ACM SIGPLAN-SIGACT Symposium on Principles of Programming Languages, January 2014
138. Haw, R., Hong, C.S., Lee, S.: An efficient content delivery framework for SDN based LTE network. In: ICUIMC '14: Proceedings of the 8th International Conference on Ubiquitous Information Management and Communication, January 2014
139. Li, X., Freedman, M.J.: Scaling IP multicast on datacenter topologies. In: CoNEXT '13: Proceedings of the Ninth ACM Conference on Emerging Networking Experiments and Technologies, December 2013
140. Lee, S.B., Kang, M.S., Gligor, V.D.: CoDef: collaborative defense against large-scale link-flooding attacks. In: CoNEXT '13: Proceedings of the Ninth ACM Conference on Emerging Networking Experiments and Technologies, December 2013
141. Jin, X., Li, L.E., Vanbever, L., Rexford, J.: SoftCell: scalable and flexible cellular core network architecture. In: CoNEXT '13: Proceedings of the Ninth ACM Conference on Emerging Networking Experiments and Technologies, December 2013
142. Mysore, R.N., Porter, G., Vahdat, A.: FasTrak: enabling express lanes in multi-tenant data centers. In: CoNEXT '13: Proceedings of the Ninth ACM Conference on Emerging Networking Experiments and Technologies, December 2013
143. Sun, X., Xie, G.G.: Minimizing network complexity through integrated top-down design. In: CoNEXT '13: Proceedings of the Ninth ACM Conference on Emerging Networking Experiments and Technologies, December 2013
144. Lu, H., Arora, N., Zhang, H., Lumezanu, C., Rhee, J., Jiang, G.: HybNET: network manager for a hybrid network infrastructure. In: Middleware Industry '13: Proceedings of the Industrial Track of the 13th ACM/IFIP/USENIX International Middleware Conference, December 2013
145. Hand, R., Ton, M., Keller, E.: Active security. In: HotNets-XII: Proceedings of the Twelfth ACM Workshop on Hot Topics in Networks, November 2013
146. Monaco, M., Michel, O., Keller, E.: Applying operating system principles to SDN controller design. In: HotNets-XII: Proceedings of the Twelfth ACM Workshop on Hot Topics in Networks, November 2013
147. Sivaraman, A., Winstein, K., Subramanian, S., Balakrishnan, H.: No silver bullet: extending SDN to the data plane. In: HotNets-XII: Proceedings of the Twelfth ACM Workshop on Hot Topics in Networks, November 2013
148. Soulé, R., Basu, S., Kleinberg, R., Sirer, E.G., Foster, N.: Managing the network with Merlin. In: HotNets-XII: Proceedings of the Twelfth ACM Workshop on Hot Topics in Networks, November 2013

149. Mogul, J.C., AuYoung, A., Banerjee, S., Popa, L., Lee, J., Mudigonda, J., Sharma, P., Turner, Y.: Corybantic: towards the modular composition of SDN control programs. In: HotNets-XII: Proceedings of the Twelfth ACM Workshop on Hot Topics in Networks, November 2013
150. Levine, D., Katti, S., Oran D.: Proceedings of the Twelfth ACM Workshop on Hot Topics in Networks. In: HotNets-XII: Proceedings of the Twelfth ACM Workshop on Hot Topics in Networks, November 2013
151. Dart, E., Rotman, L., Tierney, B., Hester, M., Zurawski, J.: The Science DMZ: a network design pattern for data-intensive science. In: SC '13: Proceedings of the International Conference on High Performance Computing, Networking, Storage and Analysis, November 2013
152. Shin, S., Yegneswaran, V., Porras, P., Gu, G.: AVANT-GUARD: scalable and vigilant switch flow management in software-defined networks. In: CCS '13: Proceedings of the 2013 ACM SIGSAC Conference on Computer & Communications Security, November 2013
153. Chatzis, N., Smaragdakis, G., Feldmann, A., Willinger, W.: There is more to IXPs than meets the eye. SIGCOMM Comput. Commun. Rev. **43**(5), 19–28 (2013)
154. Crowcroft, J., Fidler, M., Nahrstedt, K., Steinmetz, R.: Is SDN the de-constraining constraint of the future internet? SIGCOMM Comput. Commun. Rev. **43**(5), 13–18 (2013)
155. TalebiFard, P., Nicanfar, H., Hu, X., Leung, V.C.M.: Semantic based networking of information in vehicular clouds based on dimensionality reduction. In: DIVANet '13: Proceedings of the Third ACM International Symposium on Design and Analysis of Intelligent Vehicular Networks and Applications, November 2013
156. Thereska, E., Ballani, H., O'Shea, G., Karagiannis, T., Rowstron, A., Talpey, T., Black, R., Zhu, T.: IOFlow: a software-defined storage architecture. In: SOSP '13: Proceedings of the Twenty-Fourth ACM Symposium on Operating Systems Principles, November 2013
157. Fitfield, T.: Introduction to OpenStack. Linux J. **2013**(235) (2013)
158. Ma, L., He, T., Leung, K.K., Swami, A., Towsley, D.: Identifiability of link metrics based on end-to-end path measurements. In: IMC '13: Proceedings of the 2013 Conference on Internet Measurement Conference, October 2013
159. Buddhikot, M.M.: Towards a virtual cellular network with variable grade spectrum: challenges and opportunities. In: MobiCom '13: Proceedings of the 19th Annual International Conference on Mobile Computing & Networking, September 2013
160. Fund, F., Korakis, T., Panwar, S.S.: Implementation of a protocol for cooperative packet recovery over hybrid networks. In: WiNTECH '13: Proceedings of the 8th ACM International Workshop on Wireless Network Testbeds, Experimental Evaluation & Characterization, September 2013
161. Kirkpatrick, K.: Software-defined networking. Commun. ACM **56**(9), 16–19 (2013)
162. Fayazbakhsh, S.K., Sekar, V., Yu, M., Mogul, J.C.: FlowTags: enforcing network-wide policies in the presence of dynamic middlebox actions. In: HotSDN '13: Proceedings of the Second ACM SIGCOMM Workshop on Hot Topics in Software Defined Networking, August 2013
163. Antonenko, V., Smelyanskiy, R.: Global network modelling based on mininet approach. In: HotSDN '13: Proceedings of the Second ACM SIGCOMM Workshop on Hot Topics in Software Defined Networking, August 2013
164. Shirali-Shahreza, S., Ganjali, Y.: FleXam: flexible sampling extension for monitoring and security applications in openflow. In: HotSDN '13: Proceedings of the Second ACM SIGCOMM Workshop on Hot Topics in Software Defined Networking, August 2013
165. Song, H.: Protocol-oblivious forwarding: unleash the power of SDN through a future-proof forwarding plane. In: HotSDN '13: Proceedings of the Second ACM SIGCOMM Workshop on Hot Topics in Software Defined Networking, August 2013
166. Martins, J., Ahmed, M., Raiciu, C., Huici, F.: Enabling fast, dynamic network processing with clickOS. In: HotSDN '13: Proceedings of the Second ACM SIGCOMM Workshop on Hot Topics in Software Defined Networking, August 2013

167. Wen, X., Chen, Y., Hu, C., Shi, C., Wang, Y.: Towards a secure controller platform for openflow applications. In: HotSDN '13: Proceedings of the Second ACM SIGCOMM Workshop on Hot Topics in Software Defined Networking, August 2013
168. Heller, B., Scott, C., McKeown, N., Shenker, S., Wundsam, A., Zeng, H., Whitlock, S., Jeyakumar, V., Handigol, N., McCauley, J., Zarifis, K., Kazemian, P.: Leveraging SDN layering to systematically troubleshoot networks. In: HotSDN '13: Proceedings of the Second ACM SIGCOMM Workshop on Hot Topics in Software Defined Networking, August 2013
169. Katta, N.P., Rexford, J., Walker, D.: Incremental consistent updates. In: HotSDN '13: Proceedings of the Second ACM SIGCOMM Workshop on Hot Topics in Software Defined Networking, August 2013
170. Benton, K., Camp, L.J., Small, C.: OpenFlow vulnerability assessment. In: HotSDN '13: Proceedings of the Second ACM SIGCOMM Workshop on Hot Topics in Software Defined Networking, August 2013
171. Barkai, S., Katz, R., Farinacci, D., Meyer, D.: Software defined flow-mapping for scaling virtualized network functions. In: HotSDN '13: Proceedings of the Second ACM SIGCOMM Workshop on Hot Topics in Software Defined Networking, August 2013
172. Hong, K., Lillethun, D., Ramachandran, U., Ottenwälder, B., Koldehofe, B.: Mobile fog: a programming model for large-scale applications on the internet of things. In: MCC '13: Proceedings of the Second ACM SIGCOMM Workshop on Mobile Cloud Computing, August 2013
173. Shin, S., Gu, G.: Attacking software-defined networks: a first feasibility study. In: HotSDN '13: Proceedings of the Second ACM SIGCOMM Workshop on Hot Topics in Software Defined Networking, August 2013
174. Reitblatt, M., Canini, M., Guha, A., Foster, N.: FatTire: declarative fault tolerance for software-defined networks. In: HotSDN '13: Proceedings of the Second ACM SIGCOMM Workshop on Hot Topics in Software Defined Networking, August 2013
175. Nelson, T., Guha, A., Dougherty, D.J., Fisler, K., Krishnamurthi, S.: A balance of power: expressive, analyzable controller programming. In: HotSDN '13: Proceedings of the Second ACM SIGCOMM Workshop on Hot Topics in Software Defined Networking, August 2013
176. Qazi, A., Tu, C.-C., Chiang, L., Miao, R., Sekar, V., Yu, M.: SIMPLE-fying middlebox policy enforcement using SDN Zafar. In: SIGCOMM '13: Proceedings of the ACM SIGCOMM 2013 Conference on SIGCOMM, August 2013
177. Crisan, D., Birke, R., Cressier, G., Minkenberg, C., Gusat, M.: Got loss? Get zOVN! In: SIGCOMM '13: Proceedings of the ACM SIGCOMM 2013 Conference on SIGCOMM, August 2013
178. Qazi, Z.A., Lee, J., Jin, T., Bellala, G., Arndt, M., Noubir, G.: Application-awareness in SDN. In: SIGCOMM '13: Proceedings of the ACM SIGCOMM 2013 Conference on SIGCOMM, August 2013
179. Patel, P., Bansal, D., Yuan, L., Murthy, A., Greenberg, A., Maltz, D.A., Kern, R., Kumar, H., Zikos, M., Wu, H., Kim, C., Karri, N.: Ananta: cloud scale load balancing. In: SIGCOMM '13: Proceedings of the ACM SIGCOMM 2013 Conference on SIGCOMM, August 2013
180. Ferguson, A.D., Guha, A., Liang, C., Fonseca, R., Krishnamurthi, S.: Participatory networking: an API for application control of SDNs. In: SIGCOMM '13: Proceedings of the ACM SIGCOMM 2013 Conference on SIGCOMM, August 2013
181. Jain, S., Kumar, A., Mandal, S., Ong, J., Poutievski, L., Singh, A., Venkata, S., Wanderer, J., Zhou, J., Zhu, M., Zolla, J., Hölzle, U., Stuart, S., Vahdat, A.: B4: experience with a globally-deployed software defined wan. In: SIGCOMM '13: Proceedings of the ACM SIGCOMM 2013 Conference on SIGCOMM, August 2013
182. Gember, A., Grandl, R., Khalid, J., Akella, A.: Design and implementation of a framework for software-defined middlebox networking. In: SIGCOMM '13: Proceedings of the ACM SIGCOMM 2013 Conference on SIGCOMM, August 2013
183. Ganegedara, T., Prasanna, V.: A comprehensive performance analysis of virtual routers on FPGA. Trans. Reconfigurable Technol. Syst. **6**(2), 1–2 (2013)

184. Koldehofe, B., Dürr, F., Tariq, M.A.: Tutorial: event-based systems meet software-defined networking. In: DEBS '13: Proceedings of the 7th ACM International Conference on Distributed Event-Based Systems, June 2013
185. Pescapé, A., Fernandes, S.: Proceedings of the first edition Workshop on High Performance and Programmable Networking. In: HPPN '13: Proceedings of the First Edition Workshop on High Performance and Programmable Networking, June 2013
186. Gill, H., Lin, D., Han, X., Nguyen, C., Gill, T., Loo, B.T.: Scalanytics: a declarative multi-core platform for scalable composable traffic analytics. In: HPDC '13: Proceedings of the 22nd International Symposium on High-performance Parallel and Distributed Computing, June 2013
187. Cui, Z., Bridges, P.G., Lange, J.R., Dinda, P.A.: Virtual TCP offload: optimizing ethernet overlay performance on advanced interconnects. In: HPDC '13: Proceedings of the 22nd International Symposium on High-performance Parallel and Distributed Computing, June 2013
188. Guha, A., Reitblatt, M., Foster, N.: Machine-verified network controllers. In: PLDI '13: Proceedings of the 34th ACM SIGPLAN Conference on Programming Language Design and Implementation, June 2013
189. Marotta, A., Carrozza, G., Avallone, S., Manetti, V.: An OpenFlow-based architecture for IaaS security. In: ATACCS '13: Proceedings of the 3rd International Conference on Application and Theory of Automation in Command and Control Systems, May 2013
190. Jin, D., Nicol, D.M.: Parallel simulation of software defined networks. In: SIGSIM-PADS '13: Proceedings of the 2013 ACM SIGSIM Conference on Principles of Advanced Discrete Simulation, May 2013
191. Zhang, Z., Wu, C., Cheung, D.W.L.: A survey on cloud interoperability, taxonomies, standards, and practice. SIGMETRICS Perform. Eval. Rev. **40**(4), 13–22 (2013)
192. Costa-Pérez, X., Festag, A., Kolbe, H.-J., Quittek, J., Schmid, S., Stiemerling, M., Swetina, J., van der Veen, H.: Latest trends in telecommunication standards. SIGCOMM Comput. Commun. Rev. **43**(2), 64–71 (2013)
193. Skowyra, R.W., Lapets, A., Bestavros, A., Kfoury, A.: Verifiably-safe software-defined networks for CPS. In: HiCoNS '13: Proceedings of the 2nd ACM International Conference on High Confidence Networked Systems, April 2013
194. Braun, T., Mauthe, A., Siris, V.: Service-centric networking extensions. In: SAC '13: Proceedings of the 28th Annual ACM Symposium on Applied Computing, March 2013
195. Madhavapeddy, A., Mortier, R., Rotsos, C., Scott, D., Singh, B., Gazagnaire, T., Smith, S., Hand, S., Crowcroft, J.: Unikernels: library operating systems for the cloud. In: ASPLOS '13: Proceedings of the Eighteenth International Conference on Architectural Support for Programming Languages and Operating Systems, April 2013
196. Malkhi, D., van Renesse, R.: Workshop report on LADIS 2012. SIGOPS Oper. Syst. Rev. **47** (1) (2013)
197. Rexford, J., Zave, P.: Report of the DIMACS working group on abstractions for network services, architecture, and implementation. SIGCOMM Comput. Commun. Rev. **43**(1), 56–59 (2013)
198. Brighten Godfrey, P.: Hotnets 2012 highlights. SIGCOMM Comput. Commun. Rev. **43**(1), 38–42 (2013)
199. Sen, S.: New products. Linux J. **2013**(225) (2013)
200. Sen, S.: Review of PODC 2012. SIGACT News **43**(4) (2012)
201. Sun, X., Rao, S.G., Xie, G.G.: Modeling complexity of enterprise routing design. In: CoNEXT '12: Proceedings of the 8th International Conference on Emerging Networking Experiments and Technologies, December 2012
202. Mendonca, M., Obraczka, K., Turletti, T.: The case for software-defined networking in heterogeneous networked environments. In: CoNEXT Student '12: Proceedings of the 2012 ACM Conference on CoNEXT Student Workshop, December 2012

203. Thai, P.W., de Oliveira, J.C.: Decoupling BGP policy from routing with programmable reactive policy control. In: CoNEXT Student '12: Proceedings of the 2012 ACM Conference on CoNEXT Student Workshop, December 2012
204. Stephens, B., Cox, A., Felter, W., Dixon, C., Carter, J.: PAST: scalable ethernet for data centers. In: CoNEXT '12: Proceedings of the 8th International Conference on Emerging Networking Experiments and Technologies, December 2012
205. Soliman, M., Nandy, B., Lambadaris, I., Ashwood-Smith, P.: Source routed forwarding with software defined control, considerations and implications. In: CoNEXT Student '12: Proceedings of the 2012 ACM Conference on CoNEXT Student Workshop, December 2012
206. Costa, P., Hu, W., Sekar, V.: Proceedings of the 2012 ACM Conference on CoNEXT Student Workshop. In: CoNEXT Student '12: Proceedings of the 2012 ACM Conference on CoNEXT Student Workshop, December 2012
207. Vanbever, L., Vissicchio, S., Pelsser, C., Francois, P., Bonaventure, O.: Lossless migrations of link-state IGPs. IEEE/ACM Trans. Networking **20**(6), 1842–1855 (2012)
208. Gember, A., Dragga, C., Akella, A.: ECOS: leveraging software-defined networks to support mobile application offloading. In: ANCS '12: Proceedings of the Eighth ACM/IEEE Symposium on Architectures for Networking and Communications Systems, October 2012
209. Kotronis, V., Dimitropoulos, X., Ager, B.: Outsourcing the routing control logic: better internet routing based on SDN principles. In: HotNets-XI: Proceedings of the 11th ACM Workshop on Hot Topics in Networks, October 2012
210. Reitblatt, M., Foster, N., Rexford, J., Schlesinger, C., Walker, D.: Abstractions for network update. SIGCOMM Comput. Commun. Rev. **42**(4), 323–334 (2012)
211. Ahmed, M., Huici, F., Jahanpanah, A.: Enabling dynamic network processing with clickOS. SIGCOMM Comput. Commun. Rev. **42**(4), 293–294 (2012)
212. Khurshid, A., Zhou, W., Matthew, C., Brighten Godfrey, P.: Veriflow: verifying network-wide invariants in real time. SIGCOMM Comput. Commun. Rev. **42**(4), 467–472 (2012)
213. Kim, N., Kim, J.W.: Prototype of a programmable computing/networking switch for multi-screen content consumption. In: CFI '12: Proceedings of the 7th International Conference on Future Internet Technologies, September 2012
214. Chang, D., Suh, J., Jung, H., Kwon, T.T., Choi, Y.: How to realize CDN interconnection (CDNI) over OpenFlow? In: CFI '12: Proceedings of the 7th International Conference on Future Internet Technologies, September 2012
215. Ko, B.J., Pappas, V., Raghavendra, R., Song, Y., Dilmaghani, R.B., Lee, K.-W., Verma, D.: An information-centric architecture for data center networks. In: ICN '12: Proceedings of the Second Edition of the ICN Workshop on Information-Centric Networking, August 2012
216. Gutz, S., Story, A., Schlesinger, C., Foster, N.: Splendid isolation: a slice abstraction for software-defined networks. In: HotSDN '12: Proceedings of the First Workshop on Hot Topics in Software Defined Networks, August 2012
217. Reitblatt, M., Foster, N., Rexford, J., Schlesinger, C., Walker, D.: Abstractions for network update. In: SIGCOMM '12: Proceedings of the ACM SIGCOMM 2012 Conference on Applications, Technologies, Architectures, and Protocols for Computer Communication, August 2012
218. Casado, M., Koponen, T., Shenker, S., Tootoonchian, A.: Fabric: a retrospective on evolving SDN. In: HotSDN '12: Proceedings of the First Workshop on Hot Topics in Software Defined Networks, August 2012
219. Jafarian, J.H., Al-Shaer, E., Duan, Q.: Openflow random host mutation: transparent moving target defense using software defined networking. In: HotSDN '12: Proceedings of the First Workshop on Hot Topics in Software Defined Networks, August 2012
220. Rothenberg, C.E., Nascimento, M.R., Salvador, M.R., Corrêa, C.N.A., de Lucena, S.C., Raszuk, R.: Revisiting routing control platforms with the eyes and muscles of software-defined networking. In: HotSDN '12: Proceedings of the First Workshop on Hot Topics in Software Defined Networks, August 2012

221. Mogul, J.C., Congdon, P.: Hey, you darned counters! get off my ASIC! In: HotSDN '12: Proceedings of the First Workshop on Hot Topics in Software Defined Networks, August 2012
222. Ahmed, M., Huici, F., Jahanpanah, A.: Enabling dynamic network processing with clickOS. In: SIGCOMM '12: Proceedings of the ACM SIGCOMM 2012 Conference on Applications, Technologies, Architectures, and Protocols for Computer Communication, August 2012
223. Ghorbani, S., Caesar, M.: Walk the line: consistent network updates with bandwidth guarantees. In: HotSDN '12: Proceedings of the First Workshop on Hot Topics in Software Defined Networks, August 2012
224. Porras, P., Shin, S., Yegneswaran, V., Fong, M., Tyson, M., Gu, G.: A security enforcement kernel for OpenFlow networks. In: HotSDN '12: Proceedings of the First Workshop on Hot Topics in Software Defined Networks, August 2012
225. Khurshid, A., Zhou, W., Caesar, M., Godfrey, P.B.: VeriFlow: verifying network-wide invariants in real time. In: HotSDN '12: Proceedings of the First Workshop on Hot Topics in Software Defined Networks, August 2012
226. Chetty, M., Feamster, N.: Refactoring network infrastructure to improve manageability: a case study of home networking. SIGCOMM Comput. Commun. Rev. **42**(3), 54–61 (2012)
227. Stabler, G., Rosen, A., Goasguen, S., Wang, K.-C.: Elastic IP and security groups implementation using OpenFlow. In: VTDC '12: Proceedings of the 6th International Workshop on Virtualization Technologies in Distributed Computing Date, June 2012
228. Monsanto, C., Foster, N., Harrison, R., Walker, D.: A compiler and run-time system for network programming languages. In: POPL '12: Proceedings of the 39th Annual ACM SIGPLAN-SIGACT Symposium on Principles of Programming Languages, January 2012
229. Ghodsi, A., Shenker, S., Koponen, T., Singla, A., Raghavan, B., Wilcox, J.: Intelligent design enables architectural evolution. In: HotNets-X: Proceedings of the 10th ACM Workshop on Hot Topics in Networks, November 2011
230. Reitblatt, M., Foster, N., Rexford, J., Walker, D.: Consistent updates for software-defined networks: change you can believe in! In: HotNets-X: Proceedings of the 10th ACM Workshop on Hot Topics in Networks, November 2011
231. Lin, P., Bi, J., Hu, H., Feng, T., Jiang, X.: A quick survey on selected approaches for preparing programmable networks. In: AINTEC '11: Proceedings of the 7th Asian Internet Engineering Conference, November 2011
232. Benson, T., Akella, A., Shaikh, A., Sahu, S.: CloudNaaS: a cloud networking platform for enterprise applications. In: SOCC '11: Proceedings of the 2nd ACM Symposium on Cloud Computing, October 2011
233. Mai, H., Khurshid, A., Agarwal, R., Caesar, M., Godfrey, P.B., King, S.T.: Debugging the data plane with anteater. In: SIGCOMM '11: Proceedings of the ACM SIGCOMM 2011 Conference, August 2011
234. Koponen, T., Shenker, S., Balakrishnan, H., Feamster, N., Ganichev, I., Ali, G., Godfrey, P. B., McKeown, N., Parulkar, G., Raghavan, B., Rexford, J., Arianfar, S., Kuptsov, D.: Architecting for innovation. SIGCOMM Comput. Commun. Rev. **41**(3) (2011)
235. Nascimento, M.R., Rothenberg, C.E., Salvador, M.R., Corrêa, C.N.A., de Lucena, S.C., Magalhães, M.F.: Virtual routers as a service: the RouteFlow approach leveraging software-defined networks. In: CFI '11: Proceedings of the 6th International Conference on Future Internet Technologies, June 2011
236. Koponen, T., Casado, M., Gude, N., Stribling, J., Poutievski, L., Zhu, M., Ramanathan, R., Iwata, Y., Inoue, H., Hama, T., Shenker, S.: Onix: a distributed control platform for large-scale production networks. In: OSDI'10: Proceedings of the 9th USENIX Conference on Operating Systems Design and Implementation, October 2010
237. Bremler-Barr, A., Hay, D., Hendler, D., Roth, R.M.: PEDS: a parallel error detection scheme for TCAM devices. IEEE/ACM Trans. Networking **18**(5), 1345–1358 (2010)
238. Rexford, J., Dovrolis, C.: Future Internet architecture: clean-slate versus evolutionary research. Commun. ACM **53**(9), 36–40 (2010)

239. Casado, M., Freedman, M.J., Pettit, J., Luo, J., McKeown, N., Shenker, S.: Ethane: taking control of the enterprise. In: ACM SIGCOMM'07, 2007
240. McKeown, N., et al.: OpenFlow: enabling innovation in campus networks. ACM SIGCOMM Comput. Commun. Rev. **38**(2), 69–74 (2008)
241. Song, S.: Improving network health monitoring accuracy based on data fusion for software defined networking. In: Park, J.J., Stojmenovic, I., Choi, M., Xhafa, F. (eds.) Future Information Technology. Springer, Berlin (2014). doi:10.1007/978-3-642-40861-8_65. http://link.springer.com/chapter/10.1007/978-3-642-40861-8_65
242. Tsugawa, M., Matsunaga, A., Fortes, J.A.B.: Cloud computing security: what changes with software-defined networking? In: Secure Cloud Computing, pp. 77–93. Springer, New York (2014). doi:10.1007/978-1-4614-9278-8_4. http://link.springer.com/chapter/10.1007/978-1-4614-9278-8_4
243. Mehdi, S.A., Khalid, J., Khayam, S.A.: Revisiting traffic anomaly detection using software defined networking. In: Recent Advances in Intrusion Detection, pp. 161–180. Springer, Berlin (2011). doi:10.1007/978-3-642-23644-0_9. http://link.springer.com/chapter/10.1007/978-3-642-23644-0_9
244. Monfared, A.T., Rong, C.: Multi-tenant network monitoring based on software defined networking. In: On the Move to Meaningful Internet Systems: OTM 2013 Conferences, pp. 327–341. Springer, Berlin (2013). doi:10.1007/978-3-642-41030-7_24. http://link.springer.com/chapter/10.1007/978-3-642-41030-7_24
245. Ruponen, S.: On software-defined networking for rural areas: controlling wireless networks with OpenFlow. In: e-Infrastructure and e-Services for Developing Countries. Springer, Berlin (2014). doi:10.1007/978-3-319-08368-1_5. http://link.springer.com/chapter/10.1007/978-3-319-08368-1_5
246. Lin, L., Lin, P.: Software-defined networking (SDN) for cloud applications. In: Cloud Computing. Springer, Berlin (2014). doi:10.1007/978-3-319-10530-7_9. http://link.springer.com/chapter/10.1007/978-3-319-10530-7_9
247. Singh, S., Khan, R.A., Alka, A.: Applicability of software defined networking in campus network. In: Proceedings of the 3rd International Conference on Frontiers of Intelligent Computing: Theory and Applications (FICTA) 2014, pp. 619–627. Springer, Berlin (2015). doi:10.1007/978-3-319-12012-6_68. http://link.springer.com/chapter/10.1007/978-3-319-12012-6_68
248. Karimzadeh, M., Sperotto, A., Pras, A.: Software defined networking to improve mobility management performance. In: Monitoring and Securing Virtualized Networks and Services, pp. 118–122. Springer, Berlin (2014). doi:10.1007/978-3-662-43862-6_14. http://link.springer.com/chapter/10.1007/978-3-662-43862-6_14
249. Venmani, D.P., Gourhant, Y., Reynaud, L., Chemouil, P., Zeghlache, D.: Substitution networks based on software defined networking. In: Ad Hoc Networks, pp. 242–259. Springer, Berlin (2013). doi:10.1007/978-3-642-36958-2_17. http://link.springer.com/chapter/10.1007/978-3-642-36958-2_17
250. Ryoo, I., Na, W., Kim, S.: Information exchange architecture based on software defined networking for cooperative intelligent transportation systems. Cluster Comput. **18**(2), 771–782 (2015). doi:10.1007/s10586-015-0442-z. http://link.springer.com/article/10.1007/s10586-015-0442-z
251. Vizváry, M., Vykopal, J.: Future of DDoS attacks mitigation in software defined networks. In: Monitoring and Securing Virtualized Networks and Services, pp. 123–127. Springer, Berlin (2014). doi:10.1007/978-3-662-43862-6_15. http://link.springer.com/chapter/10.1007/978-3-662-43862-6_15
252. Tantar, E., Palattella, M.R., Avanesov, T., Kantor, M., Engel, T.: Cognition: a tool for reinforcing security in software defined networks. In: EVOLVE—A Bridge Between Probability, Set Oriented Numerics, and Evolutionary Computation, pp. 61–78. Springer, Berlin (2014). doi:10.1007/978-3-319-07494-8_6. http://link.springer.com/chapter/10.1007/978-3-319-07494-8_6

253. Han, W., Hu, H., Ahn, G.-J.: LPM: layered policy management for software-defined networks. In: Data and Applications Security and Privacy XXVIII, pp. 356–363. Springer, Berlin (2014). doi:10.1007/978-3-662-43936-4_23. http://link.springer.com/chapter/10.1007/978-3-662-43936-4_23
254. Zakharov, V.A., Smelyansky, R.L., Chemeritsky, E.V.: A formal model and verification problems for software defined networks. Autom. Cont. Comp. Sci. **48**(7), 398–406 (2014). doi:10.3103/S0146411614070165. http://link.springer.com/article/10.3103/S0146411614070165
255. He, S., Jianwei, L., Mao, J., Jie, C.: Hierarchical solution for access control and authentication in software defined networks. In: Network and System Security, pp. 70–81. Springer, Berlin (2014). doi:10.1007/978-3-319-11698-3_6. http://link.springer.com/chapter/10.1007/978-3-319-11698-3_6
256. Kim, D., Byeon, O., Cho, K.: Federated software defined network operations for LHC experiments. J. Korean Phys. Soc. **63**(5), 975–981 (2013). doi:10.3938/jkps.63.975. http://link.springer.com/article/10.3938/jkps.63.975
257. Marconett, D., Yoo, S.J.B.: FlowBroker: a software-defined network controller architecture for multi-domain brokering and reputation. J. Netw. Syst. Manage. **23**(2), 328–359 (2015). doi:10.1007/s10922-014-9325-5. http://link.springer.com/article/10.1007/s10922-014-9325-5
258. Khasnabish, B., Choi, B.-Y., Feamster, N.: JONS: special issue on management of software-defined networks. J. Netw. Syst. Manage. **23**(2), 249–251 (2015). doi:10.1007/s10922-014-9339-z. http://link.springer.com/article/10.1007/s10922-014-9339-z
259. Kim, D., Gil, J.-M.: Reliable and fault-tolerant software-defined network operations scheme for remote 3D printing. J. Electron. Mater. **44**(3), 804–814 (2015). doi:10.1007/s11664-014-3548-9. http://link.springer.com/article/10.1007/s11664-014-3548-9
260. Rass, S., Rainer, B., Vavti, M., Göllner, J., Peer, A., Schauer, S.: Secure communication over software-defined networks. Mob. Netw. Appl. **20**(1), 105–110 (2015). doi:10.1007/s11036-015-0582-7. http://link.springer.com/article/10.1007/s11036-015-0582-7
261. Sun, P., Yu, M., Freedman, M.J., Rexford, J., Walker, D.: HONE: joint host-network traffic management in software-defined networks. J. Netw. Syst. Manage. **23**(2), 374–399 (2015). doi:10.1007/s10922-014-9321-9. http://link.springer.com/article/10.1007/s10922-014-9321-9
262. Heorhiadi, V., Fayaz, S.K., Reiter, M.K., Sekar, V.: SNIPS: a software-defined approach for scaling intrusion prevention systems via offloading. In: Information Systems Security, pp. 9–29. Springer, Berlin (2014). doi:10.1007/978-3-319-13841-1_2. http://link.springer.com/chapter/10.1007/978-3-319-13841-1_2
263. Bhaumik, P., Zhang, S., Chowdhury, P., Lee, S.-S., Lee, J.H., Mukherjee, B.: Software-defined optical networks (SDONs): a survey. Photonic Netw. Commun. **28**(1), 4–18 (2014). doi:10.1007/s11107-014-0451-5. http://link.springer.com/article/10.1007/s11107-014-0451-5
264. Yang, M., Li, Y., Jin, D., Zeng, L., Wu, X., Vasilakos, A.V.: Software-defined and virtualized future mobile and wireless networks: a survey. Mob. Netw. Appl. **20**(1), 4–18 (2015). doi:10.1007/s11036-014-0533-8. http://link.springer.com/article/10.1007/s11036-014-0533-8
265. Wang, Y., Bi, J., Zhang, K.: Design and implementation of a software-defined mobility architecture for IP networks. Mob. Netw. Appl. **20**(1), 40–52 (2015). doi:10.1007/s11036-015-0579-2. http://link.springer.com/article/10.1007/s11036-015-0579-2
266. Qadir, J., Ahmed, N., Ahad, N.: Building programmable wireless networks: an architectural survey. EURASIP J. Wireless Commun. Network. **1**, 172 (2014). doi:10.1186/1687-1499-2014-172. http://link.springer.com/article/10.1186/1687-1499-2014-172
267. Ben Yoo, S.J., Liu, L., Proietti, R., Scott, R.P.: Software defined elastic optical networking in temporal, spectral, and spatial domains. Photonic Netw. Commun. **28**(1), 19–33 (2014). doi:10.1007/s11107-014-0448-0. http://link.springer.com/article/10.1007/s11107-014-0448-0
268. Li, Y., Vasilakos, A.V.: Editorial: software-defined and virtualized future wireless networks. Mob. Netw. Appl. **20**(1), 1–3 (2015). doi:10.1007/s11036-015-0569-4. http://link.springer.com/article/10.1007/s11036-015-0569-4
269. Rückert, J., Blendin, J., Hausheer, D.: Software-defined multicast for over-the-top and overlay-based live streaming in ISP networks. J. Netw. Syst. Manage. **23**(2), 280–308

(2015). doi:10.1007/s10922-014-9322-8. http://link.springer.com/article/10.1007/s10922-014-9322-8
270. Bae, H.-B., Park, M.-W., Kim, S.-H., Chung, T.-M.: Zombie PC detection and treatment model on software-defined network. In: Computer Science and Its Applications, pp. 837–843. Springer, Berlin (2015). doi:10.1007/978-3-662-45402-2_119. http://link.springer.com/chapter/10.1007/978-3-662-45402-2_119
271. Bruni, R., Montanari, U., Sammartino, M.: Reconfigurable and software-defined networks of connectors and components. In: Software Engineering for Collective Autonomic Systems, pp. 73–106. Springer, Berlin (2015). doi:10.1007/978-3-319-16310-9_2. http://link.springer.com/chapter/10.1007/978-3-319-16310-9_2
272. Xiao, X.F., Kui, X.: The characterizes of communication contacts between vehicles and intersections for software-defined vehicular networks. Mob. Netw. Appl. **20**(1), 98–104 (2015). doi:10.1007/s11036-014-0535-6. http://link.springer.com/article/10.1007/s11036-014-0535-6
273. Yu, C., Lumezanu, C., Sharma, A., Xu, Q., Jiang, G., Madhyastha, H.V.: Software-defined latency monitoring in data center networks. In: Passive and Active Measurement, pp. 360–372. Springer, Berlin (2015). doi:10.1007/978-3-319-15509-8_27. http://link.springer.com/chapter/10.1007/978-3-319-15509-8_27
274. Nikitinskiy, M.A., Alekseev, I.V.: Analyzing the possibility of applying asymmetric transport protocols in terms of software defined networks. Autom. Cont. Comput. Sci. **49**(2), 94–102 (2015). doi:10.3103/S0146411615020042. http://link.springer.com/article/10.3103/S0146411615020042
275. Geske, J., Stanchev, P.: Next-generation internet projects. In: Distributed Computer and Communication Networks, pp. 1–10. Springer, Berlin (2014). doi:10.1007/978-3-319-05209-0_1. http://link.springer.com/chapter/10.1007/978-3-319-05209-0_1
276. Jung, J.K., Hong, J.H., Chung, T.M.: A study of reducing resource waste for mobile grid with software defined network. In: Computational Science and Its Applications—ICCSA 2014, pp. 755–765. Springer, Berlin (2014). doi:10.1007/978-3-319-09147-1_55. http://link.springer.com/chapter/10.1007/978-3-319-09147-1_55
277. Simmons, J.M.: Dynamic optical networking. In: Optical Network Design and Planning, pp. 349–399. Springer, Berlin (2014). doi:10.1007/978-3-319-05227-4_8. http://link.springer.com/chapter/10.1007/978-3-319-05227-4_8
278. De Turck, F., Kiriha, Y., Hong, J.W.-K.: Management of the future internet: status and challenges. J. Netw. Syst. Manage. **20**(4), 616–624 (2012). doi:10.1007/s10922-012-9245-1. http://link.springer.com/article/10.1007/s10922-012-9245-1
279. Chávez-Santiago, R., Szydełko, M., Kliks, A., Foukalas, F., Haddad, Y., Nolan, K.E., Kelly, M.Y., Masonta, M.T., Balasingham, I.: 5G: the convergence of wireless communications. Wireless Pers. Commun. (2015). doi:10.1007/s11277-015-2467-2. http://link.springer.com/article/10.1007/s11277-015-2467-2
280. Carrozza, G., Manetti, V., Marotta, A., Canonico, R., Avallone, S.: Exploiting SDN approach to tackle cloud computing security issues in the ATC scenario. In: Dependable Computing, pp. 54–60. Springer, Berlin (2013). doi:10.1007/978-3-642-38789-0_5. http://link.springer.com/chapter/10.1007/978-3-642-38789-0_5
281. Kozen, D.: NetKAT—a formal system for the verification of networks. In: Programming Languages and Systems, pp. 1–18. Springer, Berlin (2014). doi:10.1007/978-3-319-12736-1_1. http://link.springer.com/chapter/10.1007/978-3-319-12736-1_1
282. Sim, J.-H., Kim, S.-H., Park, M.-W., Chung, T.-M.: Eliminating duplicated paths to reduce computational cost of rule generation by using SDN. In: Computational Science and Its Applications—ICCSA 2014, pp. 603–613. Springer, Berlin (2014). doi:10.1007/978-3-319-09153-2_45. http://link.springer.com/chapter/10.1007/978-3-319-09153-2_45
283. Anderson, R., Hall, C.: Collaborating with the enemy on network management (transcript of discussion). In: Security Protocols XXII, pp. 163–171. Springer, Berlin (2014). doi:10.1007/978-3-319-12400-1_16. http://link.springer.com/chapter/10.1007/978-3-319-12400-1_16

284. Yeluri, R., Castro-Leon, E.: Network security in the cloud. In: Building the Infrastructure for Cloud Security, pp. 123–140. Springer, Berlin (2014). doi:10.1007/978-1-4302-6146-9_6. http://link.springer.com/chapter/10.1007/978-1-4302-6146-9_6
285. Clemm, A.: Network-embedded management. In: Network-Embedded Management and Applications, pp. 59–78. Springer, Berlin (2013). doi:10.1007/978-1-4419-6769-5_3. http://link.springer.com/chapter/10.1007/978-1-4419-6769-5_3
286. Hurel, G., Badonnel, R., Lahmadi, A., Festor, O.: Outsourcing mobile security in the cloud. In: Monitoring and Securing Virtualized Networks and Services, pp. 69–73. Springer, Berlin (2014). doi:10.1007/978-3-662-43862-6_9. http://link.springer.com/chapter/10.1007/978-3-662-43862-6_9
287. Ho, Q.-D., Gao, Y., Rajalingham, G., Le-Ngoc, T.: SGCN: further aspects and issues. In: Wireless Communications Networks for the Smart Grid, pp. 99–108. Springer, Berlin (2014). doi:10.1007/978-3-319-10347-1_6. http://link.springer.com/chapter/10.1007/978-3-319-10347-1_6
288. Zaalouk, A., Pentikousis, K.: Network configuration in OpenFlow networks. In: Mobile Networks and Management, pp. 91–104. Springer, Berlin (2015). doi:10.1007/978-3-319-16292-8_7. http://link.springer.com/chapter/10.1007/978-3-319-16292-8_7
289. Eckert, M., Knoll, T.M.: QoE management framework for internet services in SDN enabled mobile networks. In: Advances in Communication Networking, pp. 112–123. Springer, Berlin (2013). doi:10.1007/978-3-642-40552-5_11. http://link.springer.com/chapter/10.1007/978-3-642-40552-5_11
290. Wang, J., Wang, Y., Hu, H., Sun, Q., He, S., Zeng, L.: Towards a security-enhanced firewall application for OpenFlow networks. In: Cyberspace Safety and Security, pp. 92–103. Springer, Berlin (2013). doi:10.1007/978-3-319-03584-0_8. http://link.springer.com/chapter/10.1007/978-3-319-03584-0_8
291. Minerva, R., Manzalini, A., Moiso, C., Crespi, N.: Virtualizing network. In: Evolution of Telecommunication Services, pp. 227–256. Springer, Berlin (2013). doi:10.1007/978-3-642-41569-2_12. http://link.springer.com/chapter/10.1007/978-3-642-41569-2_12
292. Choi, T., Lee, B., Kang, S., Song, S., Park, H., Yoon, S., Yang, S.: IRIS-CoMan: scalable and reliable control and management architecture for SDN-enabled large-scale networks. J. Netw. Syst. Manage. **23**(2), 252–279 (2015). doi:10.1007/s10922-015-9341-0. http://link.springer.com/article/10.1007/s10922-015-9341-0
293. Wen, H., Tiwary, P.K., Le-Ngoc, T.: Network virtualization technologies and techniques. In: Wireless Virtualization, pp. 25–40. Springer, Berlin (2013). doi:10.1007/978-3-319-01291-9_4. http://link.springer.com/chapter/10.1007/978-3-319-01291-9_4
294. Matsubara, D., Egawa, T., Nishinaga, N., Shin, M.-K., Kafle, V.P., Galis, A.: Open the way to future networks—a viewpoint framework from ITU-T. In: The Future Internet, pp. 27–38. Springer, Berlin (2013). doi:10.1007/978-3-642-38082-2_3. http://link.springer.com/chapter/10.1007/978-3-642-38082-2_3
295. Lin, P., Bi, J., Wang, Y.: East-west bridge for SDN network peering. In: Frontiers in Internet Technologies, pp. 170–181. Springer, Berlin (2013). doi:10.1007/978-3-642-53959-6_16. http://link.springer.com/chapter/10.1007/978-3-642-53959-6_16
296. Banjar, A., Pupatwibul, P., Braun, R.: Comparison of TCP/IP routing versus OpenFlow table and implementation of intelligent computational model to provide autonomous behavior. In: Computational Intelligence and Efficiency in Engineering Systems, pp. 121–142. Springer, Berlin (2015). doi:10.1007/978-3-319-15720-7_9. http://link.springer.com/chapter/10.1007/978-3-319-15720-7_9
297. Alberti, A.M.: A conceptual-driven survey on future internet requirements, technologies, and challenges. J. Braz. Comput. Soc. **19**(3), 291–311 (2013). doi:10.1007/s13173-013-0101-2. http://link.springer.com/article/10.1007/s13173-013-0101-2
298. Derakhshan, F., Grob-Lipski, H., Roessler, H., Schefczik, P., Soellner, M.: Enabling cloud connectivity using SDN and NFV technologies. In: Mobile Networks and Management,

pp. 245–258. Springer, Berlin (2013). doi:10.1007/978-3-319-04277-0_19. http://link.springer.com/chapter/10.1007/978-3-319-04277-0_19
299. Owezarski, P., Lobo, J., Medhi, D.: Network and service management for cloud computing and data centers: a report on CNSM 2012. J. Netw. Syst. Manage. **21**(4), 707–712 (2013). doi:10.1007/s10922-013-9281-5. http://link.springer.com/article/10.1007/s10922-013-9281-5
300. Stewart, G.: Computational verification of network programs in Coq. In: Certified Programs and Proofs, pp. 33–49. Springer, Berlin (2013). doi:10.1007/978-3-319-03545-1_3. http://link.springer.com/chapter/10.1007/978-3-319-03545-1_3
301. Dobrijevic, O., Kassler, A.J., Skorin-Kapov, L., Matijasevic, M.: Q-POINT: QoE-driven path optimization model for multimedia services. In: Wired/Wireless Internet Communications, pp. 134–147. Springer, Berlin (2014). doi:10.1007/978-3-319-13174-0_11. http://link.springer.com/chapter/10.1007/978-3-319-13174-0_11
302. Galis, A., Rubio-Loyola, J., Clayman, S., Mamatas, L., Kukliński, S., Serrat, J., Zahariadis, T.: Software enabled future internet—challenges in orchestrating the future internet. In: Mobile Networks and Management, pp. 228–244. Springer, Berlin (2013). doi:10.1007/978-3-319-04277-0_18. http://link.springer.com/chapter/10.1007/978-3-319-04277-0_18
303. Sperotto, A., Doyen, G., Latré, S., Charalambides, M., Famaey, J., Velan, P., Čeleda, P.: Report on the 8th International Conference on Autonomous Infrastructure, Management, and Security (AIMS 2014). J. Netw. Syst. Manage. (2015). doi:10.1007/s10922-015-9346-8. http://link.springer.com/article/10.1007/s10922-015-9346-8
304. Gavrilovska, L., Rakovic, V., Atanasovski, V.: Visions towards 5G: technical requirements and potential enablers. Wireless Pers. Commun. (2015). doi:10.1007/s11277-015-2632-7. http://link.springer.com/article/10.1007/s11277-015-2632-7
305. Song, Y.J., Shin, S., Choi, Y.: Network iron curtain: hide enterprise networks with OpenFlow. In: Information Security Applications, pp. 218–230. Springer, Berlin (2014). doi:10.1007/978-3-319-05149-9_14. http://link.springer.com/chapter/10.1007/978-3-319-05149-9_14
306. Hall, C., Yu, D., Zhang, Z.-l., Stout, J., Odlyzko, A., Moore, A.W., Camp, J., Benton, K., Anderson, R.: Collaborating with the enemy on network management. In: Security Protocols XXII, pp. 154–162. Springer, Berlin (2014). doi:10.1007/978-3-319-12400-1_15. http://link.springer.com/chapter/10.1007/978-3-319-12400-1_15
307. Watashiba, Y., Date, S., Abe, H., Kido, Y., Ichikawa, K., Yamanaka, H., Kawai, E., Shimojo, S., Takemura, H.: Efficacy analysis of a SDN-enhanced resource management system through NAS parallel benchmarks. Rev. Socionetw. Strateg. **8**(2), 69–84 (2014). doi:10.1007/s12626-014-0045-9. http://link.springer.com/article/10.1007/s12626-014-0045-9
308. Chau, N.-T., Nguyen, M.-D., Jung, S., Jung, S.: SecaaS framework and architecture: a design of dynamic packet control. In: Information Security Applications, pp. 190–201. Springer, Berlin (2015). doi:10.1007/978-3-319-15087-1_15. http://link.springer.com/chapter/10.1007/978-3-319-15087-1_15
309. Armando, A., Castiglione, A., Costa, G., Fiore, U., Merlo, A., Verderame, L., You, I.: Trustworthy opportunistic access to the internet of services. In: Information and Communication Technology, pp. 469–478. Springer, Berlin (2013). doi:10.1007/978-3-642-36818-9_52. http://link.springer.com/chapter/10.1007/978-3-642-36818-9_52
310. Tsugawa, M., Matsunaga, A., Fortes, J.A.B.: Cloud networking to support data intensive applications. In: Cloud Computing for Data-Intensive Applications, pp. 61–81. Springer, Berlin (2014). doi:10.1007/978-1-4939-1905-5_3. http://link.springer.com/chapter/10.1007/978-1-4939-1905-5_3
311. Lo, C.-C., Chin, H.-H., Horng, M.-F., Kuo, Y.-H., Hsu, J.-P.: A flexible network management framework based on OpenFlow technology. In: Modern Advances in Applied Intelligence, pp. 298–307. Springer, Berlin (2014). doi:10.1007/978-3-319-07467-2_32. http://link.springer.com/chapter/10.1007/978-3-319-07467-2_32
312. Rak, J., Pickavet, M., Trivedi, K.S., Lopez, J.A., Koster, A.M.C.A., Sterbenz, J.P.G., Çetinkaya, E.K., Gomes, T., Gunkel, M., Walkowiak, K., Staessens, D.: Future research

directions in design of reliable communication systems. Telecommun. Syst. (2015). doi:10.1007/s11235-015-9987-7. http://link.springer.com/article/10.1007/s11235-015-9987-7

313. Maini, E., Manzalini, A.: Management and orchestration of virtualized network functions. In: Monitoring and Securing Virtualized Networks and Services, pp. 52–56. Springer, Berlin (2014). doi:10.1007/978-3-662-43862-6_6. http://link.springer.com/chapter/10.1007/978-3-662-43862-6_6
314. Tomovic, S., Pejanovic-Djurisic, M., Radusinovic, I.: SDN based mobile networks: concepts and benefits. Wireless Pers. Commun. **78**(3), 1629–1644 (2014). doi:10.1007/s11277-014-1909-6. http://link.springer.com/article/10.1007/s11277-014-1909-6
315. Chen, M., Mao, S., Zhang, Y., Leung, V.C.M.: Open issues and outlook. In: Big Data, pp. 81–89. Springer, Berlin (2014). doi:10.1007/978-3-319-06245-7_7. http://link.springer.com/chapter/10.1007/978-3-319-06245-7_7
316. de Oliveira Silva, F., Dias, A., Ferreira, C.C., De Souza Santos, E., Pereira, F.S.F., de Andrade, I.C., de Souza Pereira, J.H., Camargos, L.J., Theodoro, L.C., Gonçalves, M.A., Pasquini, R., José, A., Neto, V., Rosa, P.F., Kofuji, S.T.: Semantically enriched services to understand the need of entities. In: The Future Internet, pp. 142–153. Springer, Berlin (2012). doi:10.1007/978-3-642-30241-1_13. http://link.springer.com/chapter/10.1007/978-3-642-30241-1_13
317. Benamrane, F., Mamoun, M.B., Benaini, R.: Short: a case study of the performance of an OpenFlow controller. In: Networked Systems, pp. 330–334. Springer, Berlin (2014). doi:10.1007/978-3-319-09581-3_25. http://link.springer.com/chapter/10.1007/978-3-319-09581-3_25
318. Antikainen, M., Aura, T., Särelä, M.: Spook in your network: attacking an SDN with a compromised OpenFlow switch. In: Secure IT Systems, pp. 229–244. Springer, Berlin (2014). doi:10.1007/978-3-319-11599-3_14. http://link.springer.com/chapter/10.1007/978-3-319-11599-3_14
319. Aldaya, I., Cafini, R., Cerroni, W., Raffaelli, C., Savi, M.: Optical switch emulation in programmable software router testbed. Photonic Netw. Commun. **25**(1), 10–23 (2013). doi:10.1007/s11107-012-0386-7. http://link.springer.com/article/10.1007/s11107-012-0386-7
320. Kim, D., Gil, J.-M., Wang, G., Kim, S.-H.: Integrated SDN and non-SDN network management approaches for future internet environment. In: Multimedia and Ubiquitous Engineering, pp. 529–536. Springer, Berlin (2013). doi:10.1007/978-94-007-6738-6_64. http://link.springer.com/chapter/10.1007/978-94-007-6738-6_64
321. Fernandez, E.B., Monge, R., Hashizume, K.: Building a security reference architecture for cloud systems. Requir. Eng. (2015). doi:10.1007/s00766-014-0218-7. http://link.springer.com/article/10.1007/s00766-014-0218-7
322. Dongting, Y.: Authentication for resilience: the case of SDN (transcript of discussion). In: Security Protocols XXI, pp. 45–53. Springer, Berlin (2013). doi:10.1007/978-3-642-41717-7_7. http://link.springer.com/chapter/10.1007/978-3-642-41717-7_7
323. Gardiner, J., Nagaraja, S.: On the reliability of network measurement techniques used for malware traffic analysis. In: Security Protocols XXII, pp. 321–333. Springer, Berlin (2014). doi:10.1007/978-3-319-12400-1_31. http://link.springer.com/chapter/10.1007/978-3-319-12400-1_31
324. Zhang, S., Malik, S.: SAT based verification of network data planes. In: Automated Technology for Verification and Analysis, pp. 496–505. Springer, Berlin (2013). doi:10.1007/978-3-319-02444-8_43. http://link.springer.com/chapter/10.1007/978-3-319-02444-8_43
325. Kukliński, S., Wytrębowicz, J., Dinh, K.T., Tantar, E.: Application of cognitive techniques to network management and control. In: EVOLVE—A Bridge Between Probability, Set Oriented Numerics, and Evolutionary Computation V, pp. 79–93. Springer, Berlin (2014). doi:10.1007/978-3-319-07494-8_7. http://link.springer.com/chapter/10.1007/978-3-319-07494-8_7
326. Jafarian, J.H., Al-Shaer, E., Duan, Q.: Formal approach for route agility against persistent attackers. In: Computer Security—ESORICS 2013, pp. 237–254. Springer, Berlin (2013).

doi:10.1007/978-3-642-40203-6_14. http://link.springer.com/chapter/10.1007/978-3-642-40203-6_14
327. Park, B., Lee, W., Kim, T.Y., Kim, H.C.: A virtualization and management architecture of micro-datacenter. In: Information Science and Applications, pp. 181–187. Springer, Berlin (2015). doi:10.1007/978-3-662-46578-3_22. http://link.springer.com/chapter/10.1007/978-3-662-46578-3_22
328. Bertier, M., Desprez, F., Fedak, G., Lebre, A., Orgerie, A.-C., Pastor, J., Quesnel, F., Rouzaud-Cornabas, J., Tedeschi, C.: Beyond the clouds: how should next generation utility computing infrastructures be designed? In: Cloud Computing, pp. 325–345. Springer, Berlin (2014). doi:10.1007/978-3-319-10530-7_14. http://link.springer.com/chapter/10.1007/978-3-319-10530-7_14
329. Alberti, A.M.: Searching for synergies among future internet ingredients. In: Convergence and Hybrid Information Technology, pp. 61–68. Springer, Berlin (2012). doi:10.1007/978-3-642-32692-9_9. http://link.springer.com/chapter/10.1007/978-3-642-32692-9_9
330. Seppänen, K., Kilpi, J., Suihko, T.: Integrating WMN based mobile backhaul with SDN control. Mob. Netw. Appl. **20**(1), 32–39 (2015). doi:10.1007/s11036-015-0574-7. http://link.springer.com/article/10.1007/s11036-015-0574-7
331. Lv, G., Sun, Z., Chen, Y., Li, T.: Towards internet innovation: software defined data plane. In: Frontiers in Internet Technologies, pp. 236–247. Springer, Berlin (2013). doi:10.1007/978-3-642-53959-6_21. http://link.springer.com/chapter/10.1007/978-3-642-53959-6_21
332. Bozakov, Z., Sander, V.: OpenFlow: a perspective for building versatile networks. In: Network-Embedded Management and Applications, pp. 217–245. Springer, Berlin (2013). doi:10.1007/978-1-4419-6769-5_11. http://link.springer.com/chapter/10.1007/978-1-4419-6769-5_11
333. Kuźniar, M., Perešíni, P., Kostić, D.: What you need to know about SDN flow tables. In: Passive and Active Measurement, pp. 347–359. Springer, Berlin (2015). doi:10.1007/978-3-319-15509-8_26. http://link.springer.com/chapter/10.1007/978-3-319-15509-8_26
334. Yu, D., Moore, A.W., Hall, C., Anderson, R.: Authentication for resilience: the case of SDN. In: Security Protocols XXI, pp. 39–44. Springer, Berlin (2013). doi:10.1007/978-3-642-41717-7_6. http://link.springer.com/chapter/10.1007/978-3-642-41717-7_6
335. Fortino, G., Di Fatta, G., Pathan, M., Vasilakos, A.V.: Cloud-assisted body area networks: state-of-the-art and future challenges. Wireless Netw. **20**(7), 1925–1938 (2014). doi:10.1007/s11276-014-0714-1. http://link.springer.com/article/10.1007/s11276-014-0714-1
336. Minerva, R., Moiso, C., Manzalini, A., Crespi, N.: Virtualizing platforms. In: Evolution of Telecommunication Services, pp. 203–226. Springer, Berlin (2013). doi:10.1007/978-3-642-41569-2_11. http://link.springer.com/chapter/10.1007/978-3-642-41569-2_11
337. De Turck, F., Gaspary, L.P., Betser, J., Zuckerman, D.N., Moore, T.: Managing the next wave of information and communications technologies: a report on NOMS 2012. J. Netw. Syst. Manage. **21**(3), 510–516 (2013). doi:10.1007/s10922-013-9274-4. http://link.springer.com/article/10.1007/s10922-013-9274-4
338. Bjørner, N., Jayaraman, K.: Checking Cloud Contracts in Microsoft Azure. In: Xxxx, X. (ed.) Distributed Computing and Internet Technology, pp. 21–32. Springer, Berlin (2015). doi:10.1007/978-3-319-14977-6_2. http://link.springer.com/chapter/10.1007/978-3-319-14977-6_2
339. Borokhovich, M., Schmid, S.: How (not) to shoot in your foot with SDN local fast failover. In: Principles of Distributed Systems, pp. 68–82. Springer, Berlin (2013). doi:10.1007/978-3-319-03850-6_6. http://link.springer.com/chapter/10.1007/978-3-319-03850-6_6
340. Fernandes, D.A.B., Soares, L.F.B., Gomes, J.V., Freire, M.M., Inácio, P.R.M.: Security issues in cloud environments: a survey. Int. J. Inform. Secur. **13**(2), 113–170 (2014). doi:10.1007/s10207-013-0208-7. http://link.springer.com/article/10.1007/s10207-013-0208-7
341. Logota, E., Saghezchi, F.B., Marques, H., Rodriguez, J.: Cooperative strategies for end-to-end energy saving and QoS control. In: Novel 3D Media Technologies, pp. 135–161. Springer, Berlin (2015). doi:10.1007/978-1-4939-2026-6_8. http://link.springer.com/chapter/10.1007/978-1-4939-2026-6_8

342. Schultz, G.: Security aspects and principles. In: Architecture and Design for the Future Internet, pp. 115–131. Springer, Berlin (2011). doi:10.1007/978-90-481-9346-2_6. http://link.springer.com/chapter/10.1007/978-90-481-9346-2_6
343. Pawar, P.S., Sajjad, A., Dimitrakos, T., Chadwick, D.W.: Security-as-a-service in multi-cloud and federated cloud environments. In: Trust Management IX, pp. 251–261. Springer, Berlin (2015). doi:10.1007/978-3-319-18491-3_21. http://link.springer.com/chapter/10.1007/978-3-319-18491-3_21
344. Bays, L.R., Oliveira, R.R., Barcellos, M.P., Gaspary, L.P., Madeira, E.R.M.: Virtual network security: threats, countermeasures, and challenges. J. Internet Serv. Appl. **6**, 1 (2015). doi:10.1186/s13174-014-0015-z. http://link.springer.com/article/10.1186/s13174-014-0015-z
345. Blefari Melazzi, N., Detti, A., Pomposini, M.: Scalability measurements in an information-centric network. In: Measurement Methodology and Tools, pp. 81–106. Springer, Berlin (2013). doi:10.1007/978-3-642-41296-7_6. http://link.springer.com/chapter/10.1007/978-3-642-41296-7_6
346. Lee, B., Murray, N., Qiao, Y.: Active accounting and charging for programmable wireless networks. Mob. Netw. Appl. **20**(1), 111–120 (2015). doi:10.1007/s11036-015-0585-4. http://link.springer.com/article/10.1007/s11036-015-0585-4
347. Melazzi, N.B., Andrade, T., Walker, R., Chiariglione, L., Venieris, I.S., Hussmann, H.: Conclusions and future research topics. In: Enhancing the Internet with the CONVERGENCE System, pp. 263–266. Springer, Berlin (2014). doi:10.1007/978-1-4471-5373-3_10. http://link.springer.com/chapter/10.1007/978-1-4471-5373-3_10
348. Ahamed, S.V., Lawrence, V.B.: The BCR view of intelligent networks (INs). In: Intelligent Broadband Multimedia Networks, pp. 229–249. Springer, Berlin (1997). doi:10.1007/978-1-4615-6341-9_9. http://link.springer.com/chapter/10.1007/978-1-4615-6341-9_9
349. Khasnabish, B., Huang, D., Bai, X., Bellavista, P., Martinez, G., Antonopoulos, N.: Cloud computing, networking, and services. J. Netw. Syst. Manage. **20**(4), 463–467 (2012). doi:10.1007/s10922-012-9254-0. http://link.springer.com/article/10.1007/s10922-012-9254-0
350. Casola, V., De Benedictis, A., Albanese, M.: A multi-layer moving target defense approach for protecting resource-constrained distributed devices. In: Integration of Reusable Systems, pp. 299–324. Springer, Berlin (2014). doi:10.1007/978-3-319-04717-1_14. http://link.springer.com/chapter/10.1007/978-3-319-04717-1_14
351. Ma, Z., Zhang, Z.Q., Ding, Z.G., Fan, P.Z., Li, H.C.: Key techniques for 5G wireless communications: network architecture, physical layer, and MAC layer perspectives. Sci. China Inform. Sci. **58**(4), 1–20 (2015). doi:10.1007/s11432-015-5293-y. http://link.springer.com/article/10.1007/s11432-015-5293-y
352. da Costa, F.: Small data, big data, and human interaction. In: Rethinking the Internet of Things, pp. 77–94. Springer, Berlin (2013). doi:10.1007/978-1-4302-5741-7_5. http://link.springer.com/chapter/10.1007/978-1-4302-5741-7_5
353. Zhou, S., Qu, Y.R., Prasanna, V.K.: Multi-core implementation of decomposition-based packet classification algorithms. In: Parallel Computing Technologies, pp. 105–119. Springer, Berlin (2013). doi:10.1007/978-3-642-39958-9_9. http://link.springer.com/chapter/10.1007/978-3-642-39958-9_9
354. Bhaumik, P., Thota, S., Zhangli, K., Chen, J., ElBakoury, H., Fang, L., Mukherjee, B.: On downstream transmissions in EPON Protocol over Coax (EPoC): an analysis of Coax framing approaches and other relevant considerations. Photonic Netw. Commun. **28**(2), 178–189 (2014). doi:10.1007/s11107-014-0468-9. http://link.springer.com/article/10.1007/s11107-014-0468-9
355. Anantha Narayanan, V., Rajeswari, A., Sureshkumar, V.: Service-adaptive fuzzy multi criteria based intelligent vertical handover decision algorithm for heterogeneous wireless networks. In: Artificial Intelligence and Evolutionary Algorithms in Engineering Systems, pp. 297–304. Springer, Berlin (2015). doi:10.1007/978-81-322-2126-5_33. http://link.springer.com/chapter/10.1007/978-81-322-2126-5_33

356. Wen, H., Tiwary, P.K., Le-Ngoc, T.: Wireless virtualization. In: Wireless Virtualization, pp. 41–81. Springer, Berlin (2013). doi:10.1007/978-3-319-01291-9_5. http://link.springer.com/chapter/10.1007/978-3-319-01291-9_5
357. Dai, Q.-L., Shou, G.-C., Hu, Y.-H., Guo, Z.-G.: Performance improvement for applying network virtualization in fiber-wireless (FiWi) access networks. J. Zhejiang Univ. Sci. C **15** (11), 1058–1070 (2014). doi:10.1631/jzus.C1400044. http://link.springer.com/article/10.1631/jzus.C1400044
358. Hiller, W., Budich, R.: Future perspectives. In: Earth System Modelling, vol. 6, pp. 79–81. Springer, Berlin (2013). doi:10.1007/978-3-642-37244-5_8. http://link.springer.com/chapter/10.1007/978-3-642-37244-5_8
359. Sobeslav, V., Horalek, J., Pavlik, J.: Utilization of cloud computing in education with focus on open-source technologies. In: Proceedings of the 4th International Conference on Computer Engineering and Networks, pp. 813–819. Springer, Berlin (2015). doi:10.1007/978-3-319-11104-9_94. http://link.springer.com/chapter/10.1007/978-3-319-11104-9_94
360. Evancich, N., Lu, Z., Li, J., Cheng, Y., Tuttle, J., Xie, P.: Network-wide awareness. In: Cyber Defense and Situational Awareness, pp. 63–91. Springer, Berlin (2014). doi:10.1007/978-3-319-11391-3_5. http://link.springer.com/chapter/10.1007/978-3-319-11391-3_5
361. Stephanakis, I.M., Chochliouros, I.P.: Multimedia content distribution over next-generation heterogeneous networks featuring a service architecture of sliced resources. In: Artificial Intelligence Applications and Innovations, pp. 300–310. Springer, Berlin (2012). doi:10.1007/978-3-642-33412-2_31. http://link.springer.com/chapter/10.1007/978-3-642-33412-2_31
362. Karame, G.O.: Towards trustworthy network measurements. In: Trust and Trustworthy Computing, pp. 83–91. Springer, Berlin (2013). doi:10.1007/978-3-642-38908-5_6. http://link.springer.com/chapter/10.1007/978-3-642-38908-5_6
363. Ho, Q.-D., Gao, Y., Rajalingham, G., Le-Ngoc, T.: Introduction. In: Wireless Communications Networks for the Smart Grid, pp. 1–14. Springer, Berlin (2014). doi:10.1007/978-3-319-10347-1_1. http://link.springer.com/chapter/10.1007/978-3-319-10347-1_1
364. Tronco, T.R., Tome, T., Rothenberg, C.E., Alberti, A.M.: New generation internet architectures: recent and ongoing projects. In: New Network Architectures, pp. 121–140. Springer, Berlin (2010). doi:10.1007/978-3-642-13247-6_6. http://link.springer.com/chapter/10.1007/978-3-642-13247-6_6
365. Al-Shaer, E., Duan, Q., Jafarian, J.H.: Random host mutation for moving target defense. In: Security and Privacy in Communication Networks, pp. 310–327. Springer, Berlin (2013). doi:10.1007/978-3-642-36883-7_19. http://link.springer.com/chapter/10.1007/978-3-642-36883-7_19
366. Huang, C., Zhu, S., Erbacher, R.: Toward software diversity in heterogeneous networked systems. In: Data and Applications Security and Privacy XXVIII, pp. 114–129. Springer, Berlin (2014). doi:10.1007/978-3-662-43936-4_8. http://link.springer.com/chapter/10.1007/978-3-662-43936-4_8
367. Clement, M.R., Volpano, D.: Programmable diagnostic network measurement with localization and traffic observation. In: Automated Security Management, pp. 153–167. Springer, Berlin (2013). doi:10.1007/978-3-319-01433-3_9. http://link.springer.com/chapter/10.1007/978-3-319-01433-3_9
368. Natal, A.R., Jakab, L., Portolés, M., Ermagan, V., Natarajan, P., Maino, F., Meyer, D., Aparicio, A.C.: LISP-MN: mobile networking through LISP. Wireless Pers. Commun. **70** (1), 253–266 (2013). doi:10.1007/s11277-012-0692-5. http://link.springer.com/article/10.1007/s11277-012-0692-5
369. Qaisar, S.B., Ali, S., Felemban, E.A.: Wireless sensor networks in next generation communication infrastructure: vision and challenges. In: Computational Science and Its Applications—ICCSA 2014, pp. 790–803. Springer, Berlin (2014). doi:10.1007/978-3-319-09147-1_58. http://link.springer.com/chapter/10.1007/978-3-319-09147-1_58

370. Qu, Y.R., Zhou, S., Prasanna, V.K.: Packet classification on multi-core platforms. In: Handbook on Data Centers, pp. 425–447. Springer, Berlin (2015). doi:10.1007/978-1-4939-2092-1_13. http://link.springer.com/chapter/10.1007/978-1-4939-2092-1_13
371. Horsmanheimo, S., Maskey, N., Tuomimäki, L.: Interdependency between mobile and electricity distribution networks: outlook and prospects. In: Smart Device to Smart Device Communication, pp. 281–308. Springer, Berlin (2014). doi:10.1007/978-3-319-04963-2_10. http://link.springer.com/chapter/10.1007/978-3-319-04963-2_10
372. Pignolet, Y.A., Schmid, S., Tredan, G.: Adversarial topology discovery in network virtualization environments: a threat for ISPs? Distrib. Comput. **28**(2), 91–109 (2015). doi:10.1007/s00446-014-0217-4. http://link.springer.com/article/10.1007/s00446-014-0217-4
373. Kushida, K.E., Murray, J., Zysman, J.: Cloud computing: from scarcity to abundance. J. Indus. Compet. Trade **15**(1), 5–19 (2015). doi:10.1007/s10842-014-0188-y. http://link.springer.com/article/10.1007/s10842-014-0188-y
374. Lalanda, P., McCann, J.A., Diaconescu, A.: Future of autonomic computing and conclusions. In: Autonomic Computing, pp. 263–278. Springer, Berlin (2013). doi:10.1007/978-1-4471-5007-7_10. http://link.springer.com/chapter/10.1007/978-1-4471-5007-7_10
375. Jiang, W., Prasanna, V.K.: Network virtualization in data centers: a data plane perspective. In: Handbook on Data Centers, pp. 327–349. Springer, Berlin (2015). doi:10.1007/978-1-4939-2092-1_10. http://link.springer.com/chapter/10.1007/978-1-4939-2092-1_10
376. Cohen, R., Wang, T.: Intel embedded hardware platform. In: Android Application Development for the Intel® Platform, pp. 19–46. Springer, Berlin (2014). doi:10.1007/978-1-4842-0100-8_2. http://link.springer.com/chapter/10.1007/978-1-4842-0100-8_2
377. Bonomi, F., Milito, R., Natarajan, P., Zhu, J.: Fog computing: a platform for internet of things and analytics. In: Big Data and Internet of Things: A Roadmap for Smart Environments, pp. 169–186. Springer, Berlin (2014). doi:10.1007/978-3-319-05029-4_7. http://link.springer.com/chapter/10.1007/978-3-319-05029-4_7
378. Awad, M., Khanna, R.: Bioinspired computing: swarm intelligence. In: Efficient Learning Machines, pp. 105–125. Springer, Berlin (2015). doi:10.1007/978-1-4302-5990-9_6. http://link.springer.com/chapter/10.1007/978-1-4302-5990-9_6
379. Petcu, D.: Consuming resources and services from multiple clouds. J. Grid Computing. **12**(2), 321–345 (2014). doi:10.1007/s10723-013-9290-3. http://link.springer.com/article/10.1007/s10723-013-9290-3
380. Sonchack, J., Aviv, A.J.: LESS is more: host-agent based simulator for large-scale evaluation of security systems. In: Computer Security—ESORICS 2014, pp. 365–382. Springer, Berlin (2014). doi:10.1007/978-3-319-11212-1_21. http://link.springer.com/chapter/10.1007/978-3-319-11212-1_21
381. Dongre, D., Gourav, S., Kurhekar, M.P., Deshpande, U.A., Keskar, R.B., Radke, M.A.: Scalable cloud deployment on commodity hardware using OpenStack. In: Advanced Computing, Networking and Informatics, vol. 2, pp. 415–424. Springer, Berlin (2014). doi:10.1007/978-3-319-07350-7_46. http://link.springer.com/chapter/10.1007/978-3-319-07350-7_46
382. Campbell, R.H., Montanari, M., Farivar, R.: A middleware for assured clouds. J. Internet Serv. Appl. **3**(1), 87–94 (2012). doi:10.1007/s13174-011-0044-9. http://link.springer.com/article/10.1007/s13174-011-0044-9
383. Lin, G., Liu, E.: High performance network architectures for data intensive computing. In: Handbook of Data Intensive Computing, pp. 3–23. Springer, Berlin (2011). doi:10.1007/978-1-4614-1415-5_1. http://link.springer.com/chapter/10.1007/978-1-4614-1415-5_1
384. Janczukowicz, E., Tuffin, S., Braud, A., Bouabdallah, A., Fromentoux, G., Bonnin, J.-M.: Approaches for offering QoS and specialized traffic treatment for WebRTC. In: Advances in Communication Networking, pp. 59–69. Springer, Berlin (2014). doi:10.1007/978-3-319-13488-8_6. http://link.springer.com/chapter/10.1007/978-3-319-13488-8_6
385. Hahn, W., Sanneck, H.: Centralized GW control and IP address management for 3GPP networks. In: Mobile Networks and Management, pp. 13–27. Springer, Berlin (2013).

doi:10.1007/978-3-642-37935-2_2. http://link.springer.com/chapter/10.1007/978-3-642-37935-2_2

386. Datenschutz und Datensicherheit—DuD. **39**(3), 202–207 (2015). doi:10.1007/s11623-015-0394-8. http://link.springer.com/article/10.1007/s11623-015-0394-8
387. Zhou, S., Qu, Y.R., Prasanna, V.K.: Multi-core implementation of decomposition-based packet classification algorithms. J. Supercomput. **69**(1), 34–42 (2014). doi:10.1007/s11227-014-1205-y. http://link.springer.com/article/10.1007/s11227-014-1205-y
388. Waschke, M.: Network and internet standards. In: Cloud Standards, pp. 199–240. Springer, Berlin (2012). doi:10.1007/978-1-4302-4111-9_9. http://link.springer.com/chapter/10.1007/978-1-4302-4111-9_9
389. Hwang, H., Ata, S., Murata, M.: Resource name-based routing in the network layer. J. Netw. Syst. Manage. **22**(1), 1–22 (2014). doi:10.1007/s10922-012-9257-x. http://link.springer.com/article/10.1007/s10922-012-9257-x
390. Simmons, J.M.: Introduction to optical networks. In: Optical Network Design and Planning, pp. 1–23. Springer, Berlin (2014). doi:10.1007/978-3-319-05227-4_1. http://link.springer.com/chapter/10.1007/978-3-319-05227-4_1
391. Craus, M., Butincu, C.: The potential of cloud computing for analysis and finding solutions in disasters. In: Improving Disaster Resilience and Mitigation—IT Means and Tools, pp. 239–252. Springer, Berlin (2014). doi:10.1007/978-94-017-9136-6_15. http://link.springer.com/chapter/10.1007/978-94-017-9136-6_15
392. Ge, J., Wang, S., Wu, Y., Tang, H., Yuepeng, E.: Performance improvement for source mobility in named data networking based on global—local FIB updates. Peer-to-Peer Network. Appl. (2015). doi:10.1007/s12083-015-0353-z. http://link.springer.com/article/10.1007/s12083-015-0353-z
393. Lee, J., Park, M.-W., Chung, T.-M.: Path information based packet verification for authentication of SDN network manager. In: Computer Science and Its Applications, pp. 861–866. Springer, Berlin (2015). doi:10.1007/978-3-662-45402-2_122. http://link.springer.com/chapter/10.1007/978-3-662-45402-2_122
394. Martinez-Julia, P., Skarmeta, A.F., Galis, A.: Towards a secure network virtualization architecture for the future internet. In: The Future Internet, pp. 141–152. Springer, Berlin (2013). doi:10.1007/978-3-642-38082-2_12. http://link.springer.com/chapter/10.1007/978-3-642-38082-2_12
395. Cerroni, W., Raffaelli, C.: Analytical model of quality of service scheduling for optical aggregation in data centers. Photonic Netw. Commun. **28**(3), 264–275 (2014). doi:10.1007/s11107-014-0449-z. http://link.springer.com/article/10.1007/s11107-014-0449-z
396. Ravi, A., Peddoju, S.K.: Handoff strategy for improving energy efficiency and cloud service availability for mobile devices. Wireless Pers. Commun. **81**(1), 101–132 (2015). doi:10.1007/s11277-014-2119-y. http://link.springer.com/article/10.1007/s11277-014-2119-y
397. Ös, M.D., Bressan, G.: A community cloud for a real-time financial application—requirements, architecture and mechanisms. In: Algorithms and Architectures for Parallel Processing, pp. 364–377. Springer, Berlin (2014). doi:10.1007/978-3-319-11197-1_28. http://link.springer.com/chapter/10.1007/978-3-319-11197-1_28
398. Ahamed, S.V., Lawrence, V.B.: Recent American intelligent networks. In: Intelligent Broadband Multimedia Networks, pp. 152–183. Springer, Berlin (1997). doi:10.1007/978-1-4615-6341-9_6. http://link.springer.com/chapter/10.1007/978-1-4615-6341-9_6
399. Aßmann, J., Kiontke, A., Roller, S.: Requirements for modern network infrastructures. In: Sustained Simulation Performance 2014, pp. 141–150. Springer, Berlin (2015). doi:10.1007/978-3-319-10626-7_12. http://link.springer.com/chapter/10.1007/978-3-319-10626-7_12
400. Heck, A.: S. In: StarBriefs Plus, pp. 855–957. Springer, Berlin (2004). doi:10.1007/978-0-306-48603-6_19. http://link.springer.com/chapter/10.1007/978-0-306-48603-6_19
401. Rost, M., Schmid, S.: VirtuCast: multicast and aggregation with in-network processing. In: Principles of Distributed Systems, pp. 221–235. Springer, Berlin (2013). doi:10.1007/978-3-319-03850-6_16. http://link.springer.com/chapter/10.1007/978-3-319-03850-6_16

402. Terplan, K.: Integrated network management. In: Network Management and Control, pp. 31–57. Springer, Berlin (1990). doi:10.1007/978-1-4613-1471-4_4. http://link.springer.com/chapter/10.1007/978-1-4613-1471-4_4
403. Anderson, T., Birman, K., Broberg, R., Caesar, M., Comer, D., Cotton, C., Freedman, M.J., Haeberlen, A., Ives, Z.G., Krishnamurthy, A., Lehr, W., Loo, B.T., Mazières, D., Nicolosi, A., Smith, J.M., Stoica, I.: The NEBULA future internet architecture. In: The Future Internet, pp. 16–26. Springer, Berlin (2013). doi:10.1007/978-3-642-38082-2_2. http://link.springer.com/chapter/10.1007/978-3-642-38082-2_2
404. Szymanski, T.H.: Impact of future trends on exascale grid and cloud computing. In: Supercomputing, pp. 215–231. Springer, Berlin (2014). doi:10.1007/978-3-319-07518-1_14. http://link.springer.com/chapter/10.1007/978-3-319-07518-1_14
405. Liu, Y., Wu, J.P., Zhang, Z., Ke, X.: Research achievements on the new generation Internet architecture and protocols. Sci. China Inform. Sci. **56**(11), 1–25 (2013). doi:10.1007/s11432-013-5021-4. http://link.springer.com/article/10.1007/s11432-013-5021-4
406. Budka, K.C., Deshpande, J.G., Thottan, M.: Future of smart grid communication networks. In: Communication Networks for Smart Grids, pp. 325–330. Springer, Berlin (2014). doi:10.1007/978-1-4471-6302-2_12. http://link.springer.com/chapter/10.1007/978-1-4471-6302-2_12
407. Baby Nirmala, M.: Cloud based big data analytics: WAN optimization techniques and solutions. In: Computational Intelligence for Big Data Analysis, pp. 237–254. Springer, Berlin (2015). doi:10.1007/978-3-319-16598-1_11. http://link.springer.com/chapter/10.1007/978-3-319-16598-1_11
408. Xu, Z.-W.: Cloud-sea computing systems: towards thousand-fold improvement in performance per watt for the coming zettabyte era. J. Comput. Sci. Technol. **29**(2), 177–181 (2014). doi:10.1007/s11390-014-1420-2. http://link.springer.com/article/10.1007/s11390-014-1420-2
409. Ali, S.: Virtualization with KVM. In: Practical Linux Infrastructure, pp. 53–80. Springer, Berlin (2015). doi:10.1007/978-1-4842-0511-2_3. http://link.springer.com/chapter/10.1007/978-1-4842-0511-2_3
410. Xu, K., Zhu, M., Hu, G.W., Liang, Z., Zhong, Y.F., Liu, Y., Wu, J.P., Wang, N.: Towards evolvable Internet architecture-design constraints and models analysis. Sci. China Inform. Sci. **57**(11), 1–24 (2014). doi:10.1007/s11432-014-5134-4. http://link.springer.com/article/10.1007/s11432-014-5134-4
411. Heck, A.: S. In: StarBriefs 2001, pp. 619–693. Springer, Berlin (2001). doi:10.1007/978-94-011-4351-6_19. http://link.springer.com/chapter/10.1007/978-94-011-4351-6_19
412. Geller, H.: Telecommunications policy issues: the new money delivery modes. In: Electronic Funds Transfers and Payments: The Public Policy Issues, pp. 63–78. Springer, Berlin (1987). doi:10.1007/978-94-015-7738-0_4. http://link.springer.com/chapter/10.1007/978-94-015-7738-0_4
413. Ahmedin, A., Pandit, K., Ghosal, D., Ghosh, A.: Exploiting scalable video coding for content aware downlink video delivery over LTE. In: Distributed Computing and Networking, pp. 423–437. Springer, Berlin (2014). doi:10.1007/978-3-642-45249-9_28. http://link.springer.com/chapter/10.1007/978-3-642-45249-9_28
414. Su, G., Hidell, M., Abrahamsson, H., Ahlgren, B., Li, D., Sjödin, P., Tanyingyong, V., Ke, X.: Resource management in radio access and IP-based core networks for IMT advanced and beyond. Sci. China Inform. Sci. **56**(2), 1–16 (2013). doi:10.1007/s11432-012-4777-2. http://link.springer.com/article/10.1007/s11432-012-4777-2
415. Paolucci, F., Castro, A., Fresi, F., Imran, M., Giorgetti, A., Bhownik, B.B., Berrettini, G., Meloni, G., Cugini, F., Velasco, L., Potì, L., Castoldi, P.: Active PCE demonstration performing elastic operations and hitless defragmentation in flexible grid optical networks. Photonic Netw. Commun. **29**(1), 57–66 (2015). doi:10.1007/s11107-014-0464-0. http://link.springer.com/article/10.1007/s11107-014-0464-0
416. Kliem, A., Renner, T.: Towards on-demand resource provisioning for IoT environments. In: Intelligent Information and Database Systems, pp. 484–493. Springer, Berlin (2015). doi:10.

1007/978-3-319-15705-4_47. http://link.springer.com/chapter/10.1007/978-3-319-15705-4_47
417. Bodei, C., Brodo, L., Bruni, R.: Open multiparty interaction. In: Recent Trends in Algebraic Development Techniques, pp. 1–23. Springer, Berlin (2013). doi:10.1007/978-3-642-37635-1_1. http://link.springer.com/chapter/10.1007/978-3-642-37635-1_1
418. Zhu, J.: Computing styles. In: China Cloud Rising, pp. 41–56. Springer, Berlin (2014). doi:10.1007/978-3-642-53745-5_6. http://link.springer.com/chapter/10.1007/978-3-642-53745-5_6
419. Kannan, K., Banerjee, S.: Compact TCAM: flow entry compaction in TCAM for power aware SDN. In: Distributed Computing and Networking, pp. 439–444. Springer, Berlin (2013). doi:10.1007/978-3-642-35668-1_32. http://link.springer.com/chapter/10.1007/978-3-642-35668-1_32
420. Rothenberg, C.E.: Re-architected cloud data center networks and their impact on the future internet. In: New Network Architectures, pp. 179–187. Springer, Berlin (2010). doi:10.1007/978-3-642-13247-6_10. http://link.springer.com/chapter/10.1007/978-3-642-13247-6_10
421. Ahamed, S.V., Lawrence, V.B.: Networks and the information society. In: Intelligent Broadband Multimedia Networks, pp. 60–93. Springer, Berlin (1997). doi:10.1007/978-1-4615-6341-9_3. http://link.springer.com/chapter/10.1007/978-1-4615-6341-9_3
422. Chatterjee, P., Ghosh, U., Sengupta, I., Ghosh, S.K.: A trust enhanced secure clustering framework for wireless ad hoc networks. Wireless Netw. **20**(7), 1669–1684 (2014). doi:10.1007/s11276-014-0701-6. http://link.springer.com/article/10.1007/s11276-014-0701-6
423. Detti, A., Salsano, S., Melazzi, N.B.: The network level (CONET). In: Enhancing the Internet with the CONVERGENCE System, pp. 49–72. Springer, Berlin (2014). doi:10.1007/978-1-4471-5373-3_3. http://link.springer.com/chapter/10.1007/978-1-4471-5373-3_3
424. Monserrat, J.F., Mange, G., Braun, V., Tullberg, H., Zimmermann, G., Bulakci, Ö.: METIS research advances towards the 5G mobile and wireless system definition. EURASIP J. Wireless Commun. Network. (2015). doi:10.1186/s13638-015-0302-9. http://link.springer.com/article/10.1186/s13638-015-0302-9
425. Fragkiadakis, A., Askoxylakis, I., Chatziadam, P.: Denial-of-service attacks in wireless networks using off-the-shelf hardware. In: Distributed, Ambient, and Pervasive Interactions, pp. 427–438. Springer, Berlin (2014). doi:10.1007/978-3-319-07788-8_40. http://link.springer.com/chapter/10.1007/978-3-319-07788-8_40
426. Joel Jr., A.E.: Communication switching architectures for business, industry, and government. In: Fundamentals of Digital Switching, pp. 429–446. Springer, Berlin (1990). doi:10.1007/978-1-4684-9880-6_11. http://link.springer.com/chapter/10.1007/978-1-4684-9880-6_11
427. Vlietstra, J.: S. In: Dictionary of Acronyms and Technical Abbreviations, pp. 538–604. Springer, Berlin (1997). doi:10.1007/978-1-4471-3396-4_19. http://link.springer.com/chapter/10.1007/978-1-4471-3396-4_19
428. Klein, D., Zinner, T., Borchert, K., Lange, S., Singeorzan, V., Schmid, M.: Evaluation of video quality monitoring based on pre-computed frame distortions. In: Advances in Communication Networking, pp. 100–111. Springer, Berlin (2013). doi:10.1007/978-3-642-40552-5_10. http://link.springer.com/chapter/10.1007/978-3-642-40552-5_10
429. Vlietstra, J.: S. In: Dictionary of Acronyms and Technical Abbreviations, pp. 545–610. Springer, Berlin (2001). doi:10.1007/978-1-4471-0263-2_19. http://link.springer.com/chapter/10.1007/978-1-4471-0263-2_19
430. Hong, J.W.-K., Yuan-Kuang, T., Hong, C.S., Tseng, S.-S., Kiriha, Y., Chao, H.-C., Zhanikeev, M., Song, W.-C.: Managing clouds, smart networks and services: a report on APNOMS 2011. J. Netw. Syst. Manage. **20**(1), 134–142 (2012). doi:10.1007/s10922-011-9221-1. http://link.springer.com/article/10.1007/s10922-011-9221-1
431. Fang, Y.-C., Gao, Y., Stap, C.: Future enterprise computing looking into 2020. In: Frontier and Innovation in Future Computing and Communications, pp. 127–134. Springer, Berlin

(2014). doi:10.1007/978-94-017-8798-7_16. http://link.springer.com/chapter/10.1007/978-94-017-8798-7_16
432. Kannan, K., Banerjee, S.: FlowMaster: early eviction of dead flow on SDN switches. In: Distributed Computing and Networking, pp. 484–498. Springer, Berlin (2014). doi:10.1007/978-3-642-45249-9_32. http://link.springer.com/chapter/10.1007/978-3-642-45249-9_32
433. Tong, X.C.: Perspectives and future trends. In: Advanced Materials for Integrated Optical Waveguides, pp. 509–543. Springer, Berlin (2014). doi:10.1007/978-3-319-01550-7_12. http://link.springer.com/chapter/10.1007/978-3-319-01550-7_12
434. Zhang, S., Malik, S., McGeer, R.: Verification of computer switching networks: an overview. In: Automated Technology for Verification and Analysis, pp. 1–16. Springer, Berlin (2012). doi:10.1007/978-3-642-33386-6_1. http://link.springer.com/chapter/10.1007/978-3-642-33386-6_1
435. daCosta, F.: It's different out here. In: Rethinking the Internet of Things, pp. 1–21. Springer, Berlin (2013). doi:10.1007/978-1-4302-5741-7_1. http://link.springer.com/chapter/10.1007/978-1-4302-5741-7_1
436. Elmallah, E.S., Acharya, H.B., Gouda, M.G.: Incremental verification of computing policies. In: Stabilization, Safety, and Security of Distributed Systems, pp. 226–236. Springer, Berlin (2014). doi:10.1007/978-3-319-11764-5_16. http://link.springer.com/chapter/10.1007/978-3-319-11764-5_16
437. Huang, W.: The development of cloud computing in pacific rim. In: Web Services and Formal Methods, pp. 3–12. Springer, Berlin (2014). doi:10.1007/978-3-319-08260-8_1. http://link.springer.com/chapter/10.1007/978-3-319-08260-8_1
438. Astorga, J., Jacob, E., Toledo, N., Aguado, M., Higuero, M.: A lossy channel aware parameterisation of a novel security protocol for wireless IP-enabled sensors. Wireless Netw. **21**(4), 1289–1308 (2015). doi:10.1007/s11276-014-0854-3. http://link.springer.com/article/10.1007/s11276-014-0854-3
439. Antonenko, V.A., Smelyanskiy, R.L.: Simulation of malicious activity in wide area networks. Program. Comput. Softw. **39**(1), 25–33 (2013). doi:10.1134/S0361768813010027. http://link.springer.com/article/10.1134/S0361768813010027
440. Corradi, A., Fanelli, M., Foschini, L.: Management infrastructures for power-efficient cloud computing architectures. In: Cloud Computing, pp. 133–152. Springer, Berlin (2013). doi:10.1007/978-1-4471-5107-4_7. http://link.springer.com/chapter/10.1007/978-1-4471-5107-4_7
441. Kumaran, S.: A perspective of the cellular network of the future: cloud-RAN. In: Afro-European Conference for Industrial Advancement, pp. 27–41. Springer, Berlin (2015). doi:10.1007/978-3-319-13572-4_3. http://link.springer.com/chapter/10.1007/978-3-319-13572-4_3
442. Liang, J., Lin, Z., Ma, Y.: OF-NEDL: an OpenFlow networking experiment description language based on XML. In: Web Information Systems and Mining, pp. 686–697. Springer, Berlin (2012). doi:10.1007/978-3-642-33469-6_85. http://link.springer.com/chapter/10.1007/978-3-642-33469-6_85
443. Ardagna, C.A., Damiani, E.: Network and storage latency attacks to online trading protocols in the cloud. In: On the Move to Meaningful Internet Systems: OTM 2014 Workshops, pp. 192–201. Springer, Berlin (2014). doi:10.1007/978-3-662-45550-0_20. http://link.springer.com/chapter/10.1007/978-3-662-45550-0_20
444. Nikolik, D.: Wide area networks. In: A Manager's Primer on e-Networking, pp. 163–180. Springer, Berlin (2003). doi:10.1007/978-94-007-0862-4_11. http://link.springer.com/chapter/10.1007/978-94-007-0862-4_11
445. Industriegipfel Feldafing—System Leadership 2030"—ein Resümee erster Strategiegespräche zu Industrie 4.0. Informatik-Spektrum. **37**(1), 54–72 (2014). doi:10.1007/s00287-013-0763-3. http://link.springer.com/article/10.1007/s00287-013-0763-3
446. Yu, C., Lumezanu, C., Zhang, Y., Singh, V., Jiang, G., Madhyastha, H.V.: FlowSense: monitoring network utilization with zero measurement cost. In: Passive and Active

Measurement, pp. 31–41. Springer, Berlin (2013). doi:10.1007/978-3-642-36516-4_4. http://link.springer.com/chapter/10.1007/978-3-642-36516-4_4

447. Shu, L., Zhang, Y., Chen, X., Wang, S.: Editorial for special issue on industrial networks and intelligent systems. Mob. Netw. Appl. **20**(2), 121–123 (2015). doi:10.1007/s11036-015-0594-3. http://link.springer.com/article/10.1007/s11036-015-0594-3
448. Gifre, L., Paolucci, F., Aguado, A., Casellas, R., Castro, A., Cugini, F., Castoldi, P., Velasco, L., López, V.: Experimental assessment of in-operation spectrum defragmentation. Photonic Netw. Commun. **27**(3), 128–140 (2014). doi:10.1007/s11107-014-0433-7. http://link.springer.com/article/10.1007/s11107-014-0433-7
449. Ahamed, S.V., Lawrence, V.B.: Optical lightwave systems in existing networks. In: Design and Engineering of Intelligent Communication Systems, pp. 559–593. Springer, Berlin (1997). doi:10.1007/978-1-4615-6291-7_17. http://link.springer.com/chapter/10.1007/978-1-4615-6291-7_17
450. Stiller, B., Hausheer, D., Hoßfeld, T.: Towards a socially-aware management of new overlay application traffic combined with energy efficiency in the internet (SmartenIT). In: The Future Internet, pp. 3–15. Springer, Berlin (2013). doi:10.1007/978-3-642-38082-2_1. http://link.springer.com/chapter/10.1007/978-3-642-38082-2_1
451. Ahamed, S.V., Lawrence, V.B.: Current digital networks. In: Design and Engineering of Intelligent Communication Systems, pp. 85–122. Springer, Berlin (1997). doi:10.1007/978-1-4615-6291-7_4. http://link.springer.com/chapter/10.1007/978-1-4615-6291-7_4
452. Togawa, S., Kanenishi, K.: Private cloud collaboration framework for e-learning environment for disaster recovery using smartphone alert notification. In: Human Interface and the Management of Information. Information and Knowledge in Applications and Services, pp. 118–126. Springer, Berlin (2014). doi:10.1007/978-3-319-07863-2_13. http://link.springer.com/chapter/10.1007/978-3-319-07863-2_13
453. Ahamed, S.V., Lawrence, V.B.: Advanced intelligent networks (AINs). In: Intelligent Broadband Multimedia Networks, pp. 318–336. Springer, Berlin (1997). doi:10.1007/978-1-4615-6341-9_13. http://link.springer.com/chapter/10.1007/978-1-4615-6341-9_13
454. Togawa, S., Kanenishi, K.: Private cloud cooperation framework for reducing the earthquake damage on e-learning environment. In: Human-Computer Interaction. Applications and Services, pp. 503–510. Springer, Berlin (2013). doi:10.1007/978-3-642-39262-7_57. http://link.springer.com/chapter/10.1007/978-3-642-39262-7_57
455. Wagh, A., Hou, Y., Qiao, C., Zhang, L., Xu, L., Sadek, A., Hulme, K., Wu, C., Xu, H.-L., Huang, L.-S.: Emerging applications for cyber transportation systems. J. Comput. Sci. Technol. **29**(4), 562–575 (2014). doi:10.1007/s11390-014-1450-9. http://link.springer.com/article/10.1007/s11390-014-1450-9
456. Santos, M.A.S., Porras, D.E.T., Silveira, R.M., Margi, C.B.: Multipath source routing strategies for video transmission in ad hoc wireless networks. Wireless Netw. **21**(3), 859–869 (2015). doi:10.1007/s11276-014-0823-x. http://link.springer.com/article/10.1007/s11276-014-0823-x
457. Chowdhury, S.R., Roy, A.R., Shaikh, M., Daudjee, K.: A taxonomy of decentralized online social networks. Peer-to-Peer Network. Appl. **8**(3), 367–383 (2015). doi:10.1007/s12083-014-0258-2. http://link.springer.com/article/10.1007/s12083-014-0258-2
458. Yoon, S., Lee, S.S., Kim, S.-H.: Robust mutual trust architecture for safety critical service in heterogeneous mobile network environment. Telecommun. Syst. (2015). doi:10.1007/s11235-015-0029-2. http://link.springer.com/article/10.1007/s11235-015-0029-2
459. Said Seddiki, M., Frikha, M., Song, Y.-Q.: A non-cooperative game-theoretic framework for resource allocation in network virtualization. Telecommun. Syst. (2015). doi:10.1007/s11235-015-9995-7. http://link.springer.com/article/10.1007/s11235-015-9995-7
460. Zinner, T., Hoßfeld, T., Fiedler, M., Liers, F., Volkert, T., Khondoker, R., Schatz, R.: Requirement driven prospects for realizing user-centric network orchestration. Multimedia Tools Appl. **74**(2), 413–437 (2015). doi:10.1007/s11042-014-2072-5. http://link.springer.com/article/10.1007/s11042-014-2072-5

461. Zhao, D., Wu, C., Hu, X., Liu, H.: LSC2: an extended link state protocol with centralized control. Peer-to-Peer Network. Appl. (2013). doi:10.1007/s12083-013-0226-2. http://link.springer.com/article/10.1007/s12083-013-0226-2
462. Kitamura, M., Kimiyama, H., Ogura, T., Fujii, T.: A basic study on high bandwidth streaming in realtime over multipath using LDPC-IRA codes. In: Internet and Distributed Computing Systems, pp. 217–226. Springer, Berlin (2014). doi:10.1007/978-3-319-11692-1_19. http://link.springer.com/chapter/10.1007/978-3-319-11692-1_19
463. Bayzelon, G., Yang, S., Xu, M., Li, Q.: Multi-dimensional forwarding tables. In: Frontiers in Internet Technologies, pp. 68–79. Springer, Berlin (2015). doi:10.1007/978-3-662-46826-5_6. http://link.springer.com/chapter/10.1007/978-3-662-46826-5_6
464. Aguiar, E., Riker, A., Cerqueira, E., Antônio Abelém, M.M., Braun, T., Curado, M., Zeadally, S.: A real-time video quality estimator for emerging wireless multimedia systems. Wireless Netw. **20**(7), 1759–1776 (2014). doi:10.1007/s11276-014-0709-y. http://link.springer.com/article/10.1007/s11276-014-0709-y
465. Sousa, B., Marques, C., Palma, D., Gonçalves, J., Simões, P., Bohnert, T., Cordeiro, L.: Towards a high performance DNSaaS deployment. In: Mobile Networks and Management, pp. 77–90. Springer, Berlin (2015). doi:10.1007/978-3-319-16292-8_6. http://link.springer.com/chapter/10.1007/978-3-319-16292-8_6
466. Trestian, R., Ormond, O., Muntean, G.-M.: Performance evaluation of MADM-based methods for network selection in a multimedia wireless environment. Wireless Netw. (2014). doi:10.1007/s11276-014-0882-z. http://link.springer.com/article/10.1007/s11276-014-0882-z
467. Haileselassie Hagos, D.: The performance of network-controlled mobile data offloading from LTE to WiFi networks. Telecommun. Syst. (2015). doi:10.1007/s11235-015-0061-2. http://link.springer.com/article/10.1007/s11235-015-0061-2
468. Ju, H., Hong, C.S., Takano, M., Yoo, J.-H., Chang, K.-Y., Yoshihara, K., Jeng, J.-Y.: Management in the big data & IoT era: a report on APNOMS 2012. J. Netw. Syst. Manage. **21**(3), 517–524 (2013). doi:10.1007/s10922-013-9270-8. http://link.springer.com/article/10.1007/s10922-013-9270-8
469. Al-Maqri, M.A., Mohamed, O., Mohd Ali, B., Mohd Hanapi, Z.: Adaptive multi-polling scheduler for QoS support of video transmission in IEEE 802.11e WLANs. Telecommun. Syst. (2015). doi:10.1007/s11235-015-0020-y. http://link.springer.com/article/10.1007/s11235-015-0020-y
470. Sedef Savas, S., Ferhat, D., Farhan Habib, M., Tornatore, M., Mukherjee, B.: Disaster-aware service provisioning with manycasting in cloud networks. Photonic Netw. Commun. **28**(2), 123–134 (2014). doi:10.1007/s11107-014-0457-z. http://link.springer.com/article/10.1007/s11107-014-0457-z
471. Li, L., Xu, X., Wang, J., Wang, W.: DS2: a DHT-based substrate for distributed services. Peer-to-Peer Network. Appl. **6**(4), 380–396 (2013). doi:10.1007/s12083-013-0228-0. http://link.springer.com/article/10.1007/s12083-013-0228-0
472. Jiang, J.-R., Wu, J.-W., Fan, C.-W., Jie-Yi, W.: Immersive voice communication for massively multiplayer online games. Peer-to-Peer Network. Appl. (2014). doi:10.1007/s12083-014-0312-0. http://link.springer.com/article/10.1007/s12083-014-0312-0
473. Tarasiuk, H., Hanczewski, S., Kaliszan, A., Szuman, R., Ogrodowczyk, Ł., Olszewski, I., Giertych, M., Wiśniewski, P.: The IPv6 QoS system implementation in virtual infrastructure. Telecommun. Syst. (2015). doi:10.1007/s11235-015-9996-6. http://link.springer.com/article/10.1007/s11235-015-9996-6
474. Abdeljaouad, I., Karmouch, A.: A game-theoretic approach for dynamic and adaptive managers selection in service specific overlay networks. J. Netw. Syst. Manage. (2014). doi:10.1007/s10922-014-9316-6. http://link.springer.com/article/10.1007/s10922-014-9316-6
475. Liu, Y., Chen, W.: Multicast storage and forwarding method for distributed router. In: Frontiers in Internet Technologies, pp. 106–117. Springer, Berlin (2015). doi:10.1007/978-3-662-46826-5_9. http://link.springer.com/chapter/10.1007/978-3-662-46826-5_9

476. Javaid, N., Ahmad, A., Khan, Y., Khan, Z.A., Alghamdi, T.A.: A relay based routing protocol for wireless in-body sensor networks. Wireless Pers. Commun. **80**(3), 1063–1078 (2015). doi:10.1007/s11277-014-2071-x. http://link.springer.com/article/10.1007/s11277-014-2071-x
477. Li, J., Zeng, W., Arora, A.: The configuration space of duty-cycled CSMA-based wireless MACs. Wireless Netw. **20**(8), 2561–2579 (2014). doi:10.1007/s11276-014-0750-x. http://link.springer.com/article/10.1007/s11276-014-0750-x
478. Alsmadi, I.: the integration of access control levels based on SDN. Int. J. High Perform. Comput. Network. (2016). doi:10.1504/IJHPCN.2016.077820
479. Aleroud, A., Alsmadi, I.: Identifying DoS attacks on software defined networks: a relation context approach. NOMS (2016)
480. Alsmadi, I., Munakami, M., Xu, D.: Model-based testing of SDN firewalls: a case study. In: Proceedings of The Second International Conference on Trustworthy Systems and Their Applications, (TSA'15), Taiwan, July 2015
481. Alsmadi, I., Xu, D.: Security of software defined networks: a survey. Comput. Secur. **53**, 79–108 (2015)

Chapter 12
SDN-Based Real-Time IDS/IPS Alerting System

Izzat M. Alsmadi and Ahmed AlEroud

12.1 Introduction

Intrusion detection systems (IDSs) perform thorough network traffic analysis to make intelligent detection of possible network attacks. Their methods to detect network attacks or harmful traffic can vary from simple methods that can do that based on direct flow-related information (e.g., port number, IP, MAC address). They can also perform complex intelligent methods to conduct signature or pattern analysis for some types of complex attacks. Intrusion protection systems (IPSs) complement IDSs through taking countermeasures or responses to stop detected network attacks. In most cases, systems perform IDS/IPS as dual modes especially as IPS requires IDS tasks in all cases as part of their protection or countermeasure process.

In comparison with firewalls, firewalls include access control lists (ACLs) that network administrators added to control traffic flows. Those firewall rules are static, written manually by network administrators, and require continuous monitoring or updates, or most of those rules can be obsolete given the current complex and dynamic networks. Further, firewall rules are written in direct traffic terminologies including those attributes included on flow headers (e.g., port number, IP, MAC address). That is typically considered as the most primitive task that IDS/IPS perform.

I.M. Alsmadi
Department of Computing and Cyber Security, University of Texas A&M, One University Way, San Antonio 78224, TX, USA
e-mail: ialsmadi@tamusa.edu

A. AlEroud (✉)
Yarmouk University, Irbid 21163, Jordan
e-mail: Ahmed.aleroud@yu.edu.jo

I.M. Alsmadi et al. (eds.), *Information Fusion for Cyber-Security Analytics*, Studies in Computational Intelligence 691, DOI 10.1007/978-3-319-44257-0_12

One of the features that SDN has which may support SDN tasks is the controller global view and knowledge of the network. One or more IDS appliances connected to the controller can receive overall global traffic and information related to the network. This may facilitate some tasks that were much harder and complex to accomplish in IDS/IPS given the traditional networking architecture. SDN architecture can contribute also to solving earlier problems or challenges related to packets or traffic collection. This was related to the ability to collect all network traffic and monitor the network without impacting its performance.

A report by Kerner [1] discussed Instate state experience with implementing an SDN-based IDS. The author showed the utilization of SDN for the university IPS. The major advantage was related to load balancing and the ability of the network to handle and distribute traffic based on security controls.

Rather than including ACLs, IDS/IPS includes rule engines where those rules study real-time traffic and judge whether that traffic can be classified as harmful or not based on rules, signatures, and patterns included in IDS/IPS engines. Firewalls act on L2–L3 level information, while IDS/IPS may work at all or most OSI layers' information.

12.2 Challenges/Research Problem

The task of IDS/IPS to distinguish harmful traffic from normal traffic is very complex and includes several challenges. The first challenge is related to the compromise between security and performance. Having several complex rules to investigate traffic thoroughly before allowing it to pass through the IDS/IPS can pose significant overhead and performance concerns. For example, in IDS there are two modes related to where to add the IDS/IPS in the network: On-path where every packet is tested through the IDS before it is authorized to the system. While this mode is better in terms of security, however, from a performance perspective, a significant overhead can be used by this on-path IDS/IPS. On the other hand, an of-path IDS/IPS may not have a significant performance overhead, while it may miss some serious real-time threats.

The second challenge is related to false-positive and false-negative occurrences. False positive occurs when IDS/IPS by mistake decides a normal traffic as harmful. This is also called false alarms. False negative occurs when IDS/IPS falsely misses harmful traffic and judges it as normal traffic. A good IDS/IPS is the one that can have a high accuracy in both cases (i.e., permitting all normal to pass and denying all harmful traffic to pass). Most IDS/IPS algorithms improve one of those two parts on the account of the other one.

The third challenge is related to timing issues. This is not about performance issue that is mentioned earlier; rather real-time detection of security threats or attacks is very important to detect and prevent attacks as soon as they occur without any delay. Some detection algorithms are very complex and may require long processing times. In addition, some threat patterns are detected over a long sequence of flows or packets. While SDN or programmable networks may present solutions to all three challenges, in this project, we will focus on the third challenge and how could programmable networks improve real-time detection and response of network attacks or threats.

12.3 Related Work

Existing research tried to integrate some popular IDSs such as Snort with SDN (e.g., Ballard et al. [2] and Xing et al. [3]). In general, with SDN (and in particular OpenFlow), a small subset of packets are actually handled by the controller. For example, a common way to set things up is for the controller to receive the first (or first few) packets of a given flow and once having received those, install rules in the switches that will handle the rest of the packets in that flow. In reality, the rules are set up with an "idle time-out" where if a rule is unused (i.e., matched no packet) for a certain amount of time, it is automatically removed from the switch. A hard time-out is also defined to remove the rule after the time-out value without the need for any condition. This is done because typically sending each packet to the controller is impractical. The challenge is then since Snort expects to see each and every packet in a flow, we will not be able to put Snort inside the controller effectively without vastly impacting the performance and the structure of OpenFlow network. An alternative design would be to create a service in the controller to manage a set of machines running Snort and to install rules that redirect traffic to the machines running Snort. There are also some proposals to include basic security mechanisms including IDS in switches. This however requires intelligence and control in those switches which contradicts with basic SDN architecture.

A Language for Arbitrary Route Management for Security (ALARMS) is proposed in Ballard et al. [2] to steer traffic for monitoring and security purposes.

Is it scalable to have the controller to set up connections and push traffic policies to switches and service nodes for layers above L2–L3 (e.g., L7 service)? Snort while being a good open-source IPS/IDS (with a rule-based language combining signature-, protocol-, and anomaly-based inspection) is still reliant on regular signature updates. Snort has no way to detect higher-level exploits such as web exploits (e.g., malicious JavaScript). Snort may not also help with attacks such as advanced persistent threats (APTs).

In traditional networking, switched port analyzer (SPAN) ports are used to reroute traffic for monitoring or security applications. In SDN, data can be extracted from the controller through northbound APIs. Filters can be applied to extract traffic based on certain criteria and command controller to rewrite traffic based on those criteria. Several filters can be also applied for complex queries or traffic.

The work of IDS/IPS to detect and protect against network security attacks can be considered large and complex. Several projects under SDN proposed more focused or small-scale security controls or services. For example, Defense4All is an example of SDN-based security service that is focused on detecting DoS or DDoS attacks. The project is implemented using the open-source controller OpenDaylight. The project includes also methods for real-time mitigation of discovered attacks.

There are some proposals on a small scale (e.g., Skowyra et al. [4]; Chung, Cui, et al. [5]; Chung, Khatkar, et al. [6]) to implement SDN-based IDS where attacks are detected and mitigated dynamically without human intervention. However, in order to extend this to a large-scale, IDS/IPS that can handle a large spectrum of different attacks is still a long pending research area. Chung, Cui, et al. [5] and Chung, Khatkar, et al. [6] presented a system on NIDS/NIPS in the virtual networks using OpenFlow-based programmable APIs. A graph-based analytical attack model is proposed to prevent attacks on VMs. The attacker's main goal is to find vulnerabilities in VMs and attack them based on those vulnerabilities. The developed system (NICE) periodically scans VMs and decides based on the severity of detected vulnerability to put the VM in an inspection state or not. In the inspection state, deep packet inspection (DPI) is conducted and decisions can be made based on the results. The significant contribution of the paper is the utilization of SDN in NIDS/NIPS. There are several distinguished differences in comparison with traditional approaches. First, traffic monitoring is accomplished easier by virtual switches with least intervention or interruption with operational network services. Further, the flexibility software solutions' offer can help switching dynamically a particular VM from one state to another in cases of suspicious attacks (e.g., quarantine state where the VM will be disabled from the rest of the network).

Xing et al. [3] investigated integrating Snort with OpenFlow networks. SnortFlow is capable of reconfiguring the cloud system on the fly to detect and counter intrusions. This work came as an extension or enhancement for the previously described one (i.e., NICE in Chung, Khatkar, et al. [6]). Several components between the two systems are the same. An earlier system (i.e., NICE) was missing a module that acts as Snort for sniffing and monitoring. Authors used Snort for coordinated attacks' detection which was not possible in early system. SnortFlow module includes three components: a daemon to collect alerts data from Snort agent, alert interpreter to parse alerts and decide which traffic to target, and finally rules' generator that will inject rules in OpenFlow switches. Changes caused by the new rules are saved to allow possible rollback or restoration. Countermeasures to take are classified based on cost and intrusiveness. Careful consideration should be made on the proper countermeasure to take so that it will not interrupt normal operations.

Previously surveyed papers focused on information related to (L2–L3) layers without looking at the actual packets' contents. Shirali-Shahreza and Ganjali [7, 8] proposed an extension to current OpenFlow protocol to allow controller to have access to packets' contents. Current information exchanged between the controller and switches is related to routing information only. The goal is to use such information for security applications including NIDS/NIPS or anomaly detection tools. In some cases, samples rather than the complete traffic are sent to the controller. Different sampling methods (e.g., deterministic or stochastic sampling) can be requested by the controller based on the nature of the security application or middle box. Full packets are only requested under certain conditions. The evolution of OpenFlow protocol should take into consideration all quality factors and not only security. This is since such addition of information to be exchanged with the controller is expected to cause a significant overhead for an already exhausted controller.

12.4 Approach/Leveraging SDN Infrastructure

SDN enables network administrators to control and program the network and its flows. Programs can be developed on top of the SDN network to program and control traffic to make very intelligent real-time queries for network traffic as it goes through the network. SDN controller already includes responsive methods to write flow rules in switches' flow tables as a response to real-time traffic. We think that this same task can be extended for SDN-based IDS/IPS. The controller takes two actions based on incoming traffic. The first action is the actual decision on what to do with the subject traffic (e.g., drop, forward, flood), and the second task is writing a flow rule in the switch flow table as a response to the subject flow or traffic. SDN-based IDS/IPS should perform similar tasks on a broader context. The single or simple flow in the case of control flow tables should be extended to include complex flow in the SDN-SDN inference engine.

Here are some design goals that should be fulfilled in the SDN-based IDS/IPS system:

1. Openness and extensibility: While there are some open-source IDS such as Snort, however, in most cases, those are not easy to extend. This is particularly true as many constructs that are related to the development environment are dependent or tied to the network. SDN open vendor programmable architecture opens the possibility to develop programming languages to write programs and APIs to interact with the network.
2. Flexibility: The developed IDS/IPS system should be flexible to use, configure, and interact with. Users, network administrators, and programmers should be able to access network functionalities that were used to be locked in in traditional networks (e.g., customizing or adding network flow and access protocols). The system should be also flexible to allow adding new rules, datasets of known vulnerabilities, malwares, etc.
3. Dynamic: This is one of the most ambitious goals of developing SDN-based IDS/IPS. The security system should be able to interact with the network with the least human intervention. The system should be able to automatically adjust configuration based on network changes. The user interaction with the system (e.g., to add rules, datasets, etc.) should be handled with well-defined interfaces that can be easily used by both users and other systems. Dynamic can have several other advanced interpretations. For example, the system can be dynamic

to adjust its own inference rules in response to network changes or traffic. For example, a new type of threat can be possibly detected by system traffic monitoring sub-module. In response to this detection, some rules will be adjusted to accommodate this new threat (e.g., add a new rule to inspect or block traffic that math certain characteristics).

The figure below shows the high-level component diagram of the proposed SDN-based IDS/IPS security control (Fig. 12.1).

An SDN IDS/IPS module as part of SDN controller should include two sub-modules to perform the following tasks:

1. Traffic monitor: This module should be able to communicate in two ways with OpenFlow switches to make customized traffic queries and receive customized traffic that is requested from the network or the switches. This is largely a customized traffic monitoring module. Hence, this task can be outsourced to applications such as sFlow that can interact with SDN controllers for traffic monitoring purposes.

   There are different modes on how to pass traffic to the IDS/IPS system. In one approach, thc controllcr can be programmed to pass the traffic to the IDS/IPS in addition to forwarding it to its destination. In another architecture, the special ports in OpenFlow switches [dedicated for management and monitoring] can be used. "Establishing a packet monitoring system is one of the use cases for Software-Defined Networking (SDN)" [9]. This is as in particular, SDN allows simple and dynamic programming of the packet monitoring system. Switched port analyzer (SPAN) is used for monitoring in traditional networking components. However, it has several problems and limitations.

   For IDS/IPS, SDN-based customized real-time monitoring can be the most significant contribution that SDN can bring to IDS/IPS. This can be categorized by three major aspects: real time, customized, and beyond L2–L4 information. We will describe each one of them shortly:

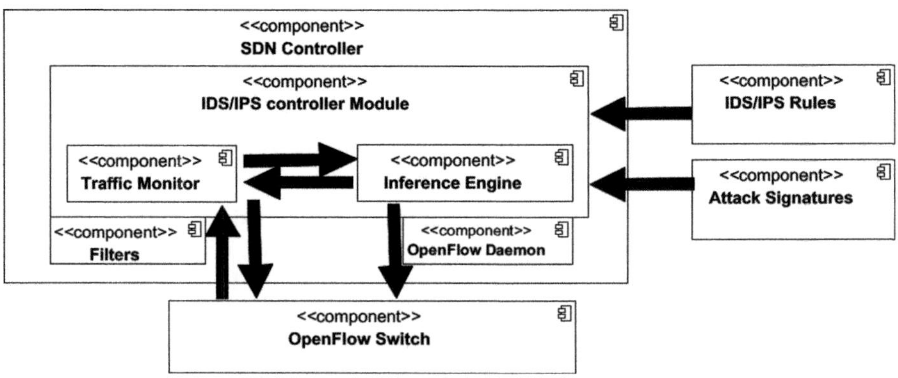

Fig. 12.1 High-level component diagram of the proposed SDN-based IDS/IPS security control

- Real time: As described earlier, both on-path and of-path IDS/IPS traditional solutions have strengths and limitations. The SDN controller can provide monitoring system on-path or real-time traffic information without significantly causing overhead in comparison with traditional approach. Many IDS/IPS systems require to mirror network traffic to be investigated.
- Customized traffic monitoring or packet processing: SDN is largely labeled as "control programmable networks" where the entire network can be programmed in SDN. As part of this programmability, network administrators and programmers can define monitoring programs to *steer* and interact with traffic in real time rather than just pulling the whole traffic to the monitoring system.
- Beyond L2–L4 information or intelligence: Current network requirements for security controls, policy management, etc. showed clear needs to extend traffic intelligence up to layer 7. This can in particular help develop smarter control and security services and allow high-level user abstractions communicate with low-level network-dependent information. As a result, network and traffic monitoring should evolve from being an embedded to a network function that can be managed and programmed by users through the SDN controller. One further advantage of changing the monitoring process from an embedded to a network function in addition to those listed earlier is that the same monitoring function can be used by different management or security services differently or in different contexts. Due to the fact that control in SDN is centralized, the controller can have a global view of the network and its traffic in comparison with localized views in traditional networks.

  Deep packet inspection (DPI) is a term used to refer to the process of examining packets as they pass an inspection point. Security-related goals can be specified behind this inspection process such as IDS, spam detection, worms, viruses, etc. Rather than based on rules in the firewall case, in the DPI the tool or the system decided whether to allow the packet to pass or not based on its deep inspection outcome. Its inference engine may include rules, datasets (of known malwares), algorithms, patterns, etc. to help in the inspection process.

2. Inference engine: This is the core module of the SDN IDS/IPS system. Rules can vary in complexity between simple flow-related ones to more complex pattern recognition algorithms. This module can be further divided into several sub-modules in which each sub-module is dedicated to one way of detecting security attacks (e.g., signature based, dictionary based, rule based). For example, a dictionary-based module may include a large dataset of known threats in terms of their known ports, IP addresses, etc.

   OpenFlow daemon: This module is responsible for inserting and removing the OpenFlow entries that will direct traffic from the outward-facing ports to the necessary module hosts and processing modules and then back out.

   Communication between detection and mitigation (i.e., traffic monitoring and inference engine modules) is a two-way communication. Attack analysis may require those two modules to contact back and forth for more than one cycle. The algorithm to detect the type of the attack and mitigate it by proposing a solution should try to optimize time and accuracy. This can be classified as an NP-hard problem or multiple-goal optimization process given those two competing goals along with a matrix of possible variables that will be used and analyzed to come up with a final decision. For mitigation, the system may come up with several possible mitigations and should then weight which one of them to use or apply. Machine learning algorithms can be used dynamically for classification and prediction. The performance of each algorithm can be evaluated based on ROC metrics to judge which one of them should be applied. The number of algorithms to evaluate and the detail of the variables in each algorithm to assess can depend on the time specially as decision on such traffic should be made in real time. This can be a continuous learning process where if, for example, some algorithms with certain setup are evaluated previously and showed good performance metrics, they can be given priority or precedence in future evaluations. For experimentation, there are some open datasets that can be used to evaluate IDS/IPS systems and algorithms (e.g., DARPA [10], KDD CUP [11], NSL, KDD [12].

Inputs to the inference engine are:

- IDS/IPS rules: This can be added and edited by users or security experts to control the IDS/IPS system. Those can be written by natural language terms or using semi-structured formats such as SDN policy languages (e.g., Frenetic and Pyretic).
- Attack signatures: Network experts can find open datasets for known security attacks and their signatures or how they typically access networks. This can be used as one of the inputs for the inference engine. We should in general then expect attacks for the inference engine to be either of known types or signatures or unknown types and signatures. For known security attacks, inference engine will use knowledge from attack signature module, and for unknown types of attacks, inference engine will use knowledge from IDS/IPS rule module.

  There are several problems with traditional IDS/IPS inference engines. For example, there is a need for fine-grained decisions beyond the (block/permit) binary decisions. This is since the majority of IDS/IPS decisions may fall in the gray area between those two binary decisions. This is why false-positive and false-negative alarms are considered the most significant problems related to those decisions. For example, QoS decision can be added to allow further investigation of suspect packets, holding them aside while allowing the next traffic to be investigated. The fact that SDN allows control programmability helps in defining fine-grained decisions between block and permit.

3. IDS/IPS rules. This is the IDS/IPS component that is accessed and updated by network administrators. Network administrators should be able to add or modify certain rules to override built-in or dynamic algorithms. In some other cases,

those added rules can be used as low-level or default rules that will be applied if no match of dynamic rules occurs.

The controller or any of its inner modules acts as an information flow processor (IFP) to collect information on behalf of the security controls. The concept of information flow processing presented pre-SDN to process IDS/IPS continuous real-time requests. However, implementing such IFP based on traditional architectures requires extensive network resources. In one definition IFP is defined as "A tool capable of timely processing large amount of information as it flows from the peripheral to the center of the system" [13].

12.5 Impacts

A real-time, zero day IDS/IPS alerting system can help networks preventing many security attacks that use new unknown vulnerabilities to attack those systems. IDS/IPS systems are more intelligent and complex if compared with firewalls. Their tasks include thorough traffic investigation and may require collecting information from OSI layers beyond the typical L2–L3 in firewall case.

In the current very complex and dynamic networks, building IDS/IPS security control and alerting systems with real-time response is very important and urgent. Traditional networks are largely static in terms of the interaction between security rules in firewalls, IDS/IPS or security controls in general, and real-time network traffic. SDN programmable networks have the ability to enable network administrators to interact with the network and its traffic and program algorithms to enable the network to dynamic response to network traffic or security threats. SDN IDS/IPS area is largely new in both academia/research and industry, and the future can show how much such new architectures can help in improving network dynamics to automate security control activities.

Enhanced IDS/IPS and DPI systems can provide critical information about the health and security of the network. Those new features are expected to change significantly the work of several players in systems and networking including network manufacturers, network security tools/software vendors, network administrators, programmers, and testers. The term or the role "network programmer" is expected to extend significantly with the evolution of programmable networks. Other new areas or roles may arise related to network information specialists, analysts, etc.

References

1. Kerner, S.M.: OpenFlow Can Provide Security. http://www.enterprisenetworkingplanet.com/datacenter/openflow-can-provide-security-too.html (2012)
2. Ballard, J.R., Rae, I., Akella, A.: Extensible and scalable network monitoring using OpenSAFE. Paper presented at the INM/WREN (2008)
3. Xing, T., Huang, D., Xu, L., Chung, C.-J., Khatkar, P.: Snortflow: a openflow-based intrusion prevention system in cloud environment. Paper presented at the research and educational experiment workshop (GREE), 2013 second GENI, pp. 89–92 (2013)
4. Skowyra, R., Bahargam, S., Bestavros, A.: Software-defined ids for securing embedded mobile devices. Paper presented at the high performance extreme computing conference (HPEC), 2013 IEEE, pp. 1–7 (2013)
5. Chung, C.-J., Cui, J., Khatkar, P., Huang, D.: Non-intrusive process-based monitoring system to mitigate and prevent VM vulnerability explorations. Paper presented at the Collaborative Computing: Networking, Applications and Worksharing (Collaboratecom), 2013 9th International Conference Conference on, pp. 21–30 (2013)
6. Chung, C.-J., Khatkar, P., Xing, T., Lee, J., Huang, D.: NICE: Network intrusion detection and countermeasure selection in virtual network systems. IEEE Trans. Depend. Secure Comput. **10** (4), 198–211 (2013)
7. Shirali-Shahreza, S., Ganjali, Y.: Efficient implementation of security applications in openflow controller with flexam. Paper presented at the High-Performance Interconnects (HOTI), 2013 I.E. 21st Annual Symposium on, pp. 49–54 (2013)
8. Shirali-Shahreza, S., Ganjali, Y.: Flexam: flexible sampling extension for monitoring and security applications in openflow. Paper presented at the proceedings of the second ACM SIGCOMM workshop on Hot topics in software defined networking, pp. 167–168 (2013)
9. Hogg, S.: Using SDN to create a packet monitoring system: packet-level monitoring use case with Cisco XNC and monitor manager. http://www.networkworld.com/article/2226003/cisco-subnet/using-sdn-to-create-a-packet-monitoring-system.html (2013)
10. MIT: DARPA intrusion detection evaluation. http://www.ll.mit.edu/mission/communications/CST/darpa.html (2012)
11. Stolfo, S.J., Fan, W., Lee, W., Prodromidis, A., Chan, P.K.: Costbased modeling for fraud and intrusion detection: results from the jam project. Discex 2, 1130 (2000)
12. Tavallaee, M., Bagheri, E., Lu, W., Ghorbani, A. A Detailed Analysis of the KDD CUP 99 Data Set. In proceedings of the 2009 IEEE Symposium on Computational Intelligence in Security and Defense Applications (CISDA 2009). (2009)
13. Cugola, G., Margara, A.: Processing flows of information: from data stream to complex event processing. ACM. Comput. Surv. (CSUR) **44**(3), 15 (2012)

Chapter 13
Digital Forensics: Implementation and Analysis for Google Android Framework

Harleen Kaur and Khairaj Ram Choudhary

Abstract Google Android forensics is a piece of digital forensics that offers many opportunities and challenges. Taking at the volume of newer Android devices and the abuse of these devices by criminals introduced challenges how to adequately extract and examine the information for forensic purposes. An exceptionally itemized comprehension of both the stage and forensic tools is required for procurement and examination of versatile information. There are various instruments on making smartphone crime scene investigation simpler. This research venture assesses the main instruments accessible in the market that back Android gadgets against its capacity to remove and analyze the information on numerous parameters. Acquiring data from a Google Android mobile is presently a critical issue in many criminal investigations. Android mobile can contain a lot of data, which can be useful in an investigation. These include typical Android device information such as SMS, browser history and searches, call log data, and device accounts. This proposed research focuses on the implementation, analysis, and performance of Google Android mobile forensic software and results of the finding. We have presented our research finding after execution of forensic procedures. However, during the proposed study, the artifacts obtained are applicable to use as evidences in the court of law against any criminal incident.

13.1 Introduction

Google Android forensics is a piece of digital forensics that offers many opportunities and challenges. Taking at the volume of newer Android devices and the abuse of these devices by criminals introduced challenges how to adequately extract and examine the information for forensic purposes. An exceptionally itemized comprehension of both the stage and forensic tools is required for procurement and examination of versatile information. There are various instruments on making smartphone crime scene investigation simpler. This research venture assesses the

H. Kaur (✉) • K.R. Choudhary
Faculty of Engineering and Technology, Department of Computer Science and Engineering, Hamdard University, New Delhi, India
e-mail: harleen_k1@rediffmail.com; khairaj.chowdhary@gmail.com

I.M. Alsmadi et al. (eds.), *Information Fusion for Cyber-Security Analytics*, Studies in Computational Intelligence 691, DOI 10.1007/978-3-319-44257-0_13

main instruments accessible in the market that back Android gadgets against its capacity to remove and analyze the information on numerous parameters. In such cases, innovation helps us to investigate the matter or the true facts to bring digital evidence acceptable in the court of law. Mobile forensics also includes the same strategies for the normal forensic examination. Mobile forensics is the process of collection identification, acquisition, preservation, and analyzing and presenting the evidence, which is legally acceptable.

13.2 Google Android Operating System Framework

The Google Android architecture framework (Fig.13.1) describes the communication of basic Android device parts in four particular layers: applications, application structure, libraries, and the Linux kernel. Every layer depends on the following layer to work appropriately.

The fundamental layer is the Linux kernel. The entire Android OS is based on top of the Linux kernel with some further building changes. It basically implies that Android at its center is Linux. Be that as it may, you can't run any Linux bundles on Android. It is a very surprising OS. It is this Linux piece that cooperates with the equipment, and it contains all the crucial equipment drivers. Drivers are projects that control and speak with the equipment. For instance, consider the Bluetooth capacity. Thusly, the portion must incorporate a Bluetooth driver to speak with the Bluetooth equipment. The Linux bit additionally goes about as a reflection layer between the equipment and other programming layers.

The libraries are written in C/C++. The center force of Android stage is used through the libraries like surface manager for making different drawing surfaces onto the screen. The OpenGL/ES and SGL make the core of the 3D and 2D graphics, and the media system constitutes different media packs like MPEG, AAC, mp3, mp4, SQLite for the information storage, and WebKit for the open-source browser engine [1].

Android RunTime comprises Dalvik virtual machine and core Java libraries. The Dalvik virtual machine is a sort of JVM utilized as a part of Android device to run applications and is advanced for low processing power and low memory environment. It is run dex files that are bytecode changed over from jar file. Dalvik machines are customized version of Java suited in the little stage environment. Dalvik uses highly optimized CPU, data shared to different applications in such a way that there may be multiple instances of Dalvik virtual machines at a given time. The center libraries are all the Java libraries for all the classes.

This is the toolkit for all application utilizations may it be the application written by Google or the application written by the developers. Every one of the applications utilizes the same framework and same APIs making it easy to write applications. By looking every one of the segments, the activity manager manages the application life cycle. Window administrator deals with the Windows UI. Telephony manager utilizes APIs that used to build the telephone applications;

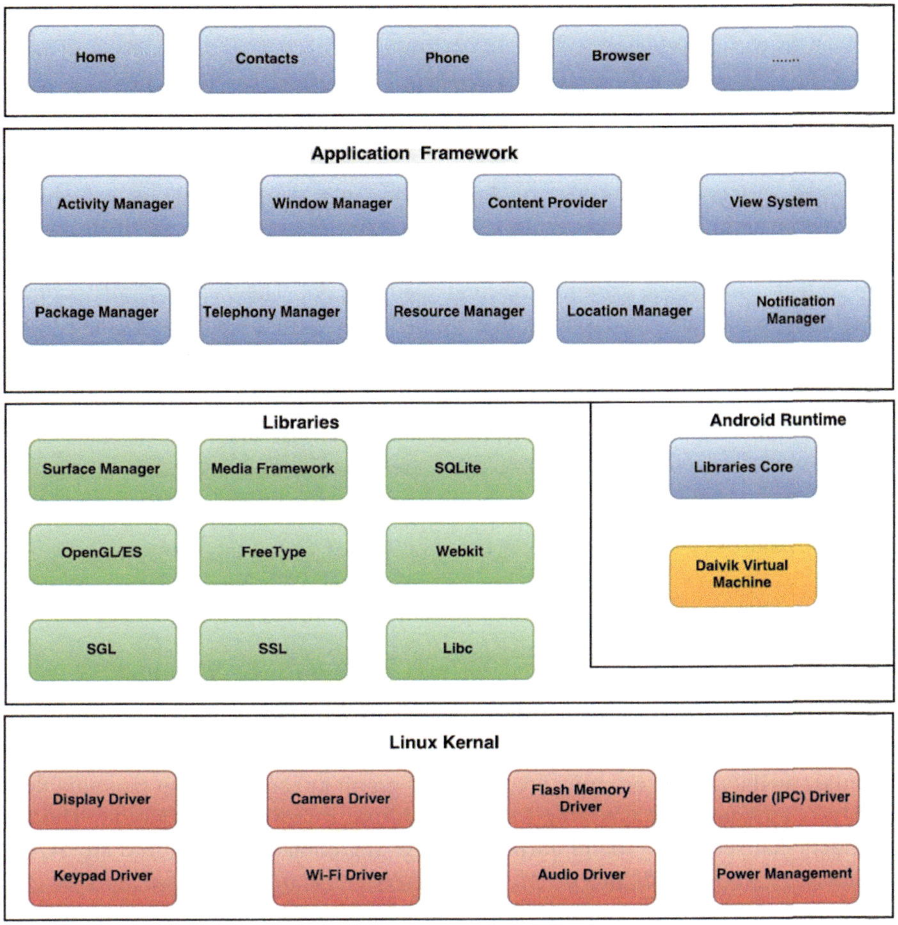

Fig. 13.1 Google Android architecture framework

the content provider is one of a kind that shares the data over every one of the applications like contacts that are shared to every one of the applications. Location manager and notification manager permit developers to develop new, energizing, and innovative applications [2, 3].

These programs deal with the basic elements of phone like resource management, voice call management, contacts, calendar, maps, SMS, Email client, and browser. These applications are written in Java programming language. These applications are multitasking. All applications run its own processes. Android gives a feature-rich APIs to building imaginative applications. APIs like location chief and XMPP services are given to developers to fabricate applications and notification manager of APIs to construct the notification-related applications.

13.3 Database Schema of Google Android Device

13.3.1 Proposed Forensic Analysis of Database Schema of Browser on Google Android Device

Browser artifacts such as history, bookmarks, and searches can be valuable to analysis searching for recovering evidences during forensic investigation. The key artifacts that should be found during investigating browser on Google Android are SQLite databases *browser2.db* and *webview.db*. The *browser2.db* contains details of bookmarks and search history [4]. The webview.db stores data user's ID and password. Figure 13.2 shows browser database schema containing *browser2.db*. Both databases can be found in locations:

Browser2.db→ *data/data/com.android.browser/database/browser2.db*
Webview.db→ *data/data/com.android.browser/database/webview.db*

Browser stores a copy of browser2.db and webview.db on the memory (flash/SD card) of Android device [5–7].

13.3.2 Proposed Forensic Analysis of Database Schema of Contacts on Google Android Device

Contact artifacts such as calls, contacts, and group attachments can be valuable to analysis looking for recovering evidences during forensic investigation. The key artifacts that should be found during investigating contacts on Android are SQLite databases *contacts2.db* and *profile.db*. The *contacts2.db* contains details of calls and contact data. The *webview.db* stores data user's accounts and activity

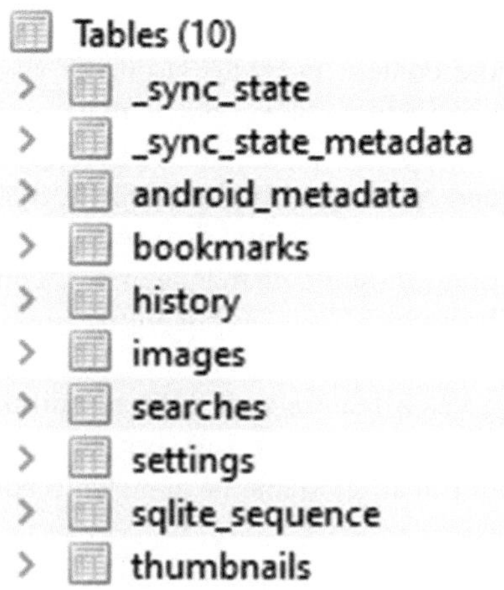

Fig. 13.2 Database structure in SQLite browser

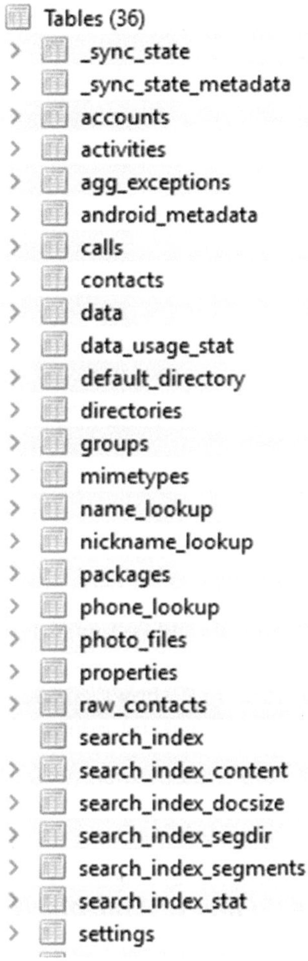

Fig. 13.3 Database structure in SQLite browser of contacts

[8, 9]. Figure 13.3 shows the contact database schema containing *contact2.db*. Both databases can be found in locations:

contacts2.db→ *data/data/com.android.providers.contact/database/contacts2.db*
profile.db→ *data/data/com.android.providers.contact/database/profile.db*

13.3.3 Proposed Forensic Analysis of Database Schema of Calendar on Google Android Device

Calendar artifacts such as reminders and event attachments can be valuable to analysis looking for recovering evidences during forensic investigation. The key

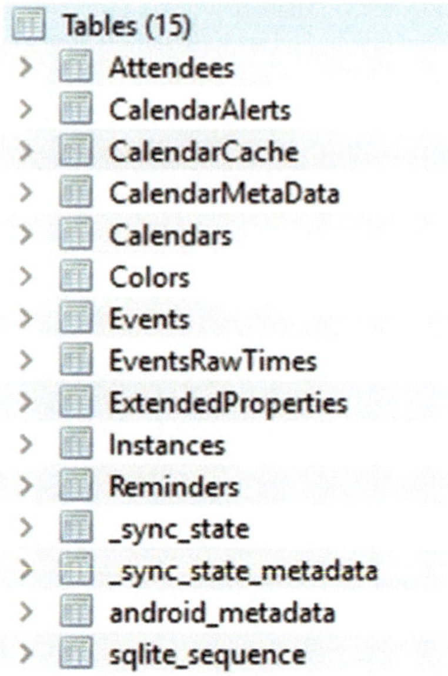

Fig. 13.4 Database structure in SQLite browser of calendar

artifacts that should be found during investigating calendar on Android are SQLite databases *calendar.db*. The *calendar.db* contains details of events and reminder data. Figure 13.4 shows contact database schema containing *calendar.db*. Databases can be found in location (Table 13.1):

calendae.db→ data/data/com.android.providers.calendar/database/calendar.db

13.3.4 Google Android Data Storage

Google Android device stores more information than any standard phone. Android stores data in five strategies. Forensic expert analysts search for the data in five formats. The five techniques in storing data are shared preference, internal storage, external storage, SQLite, and network [10].

Shared preferences allow to store key-value pair of primitive data type in XML format. These files can only be visible if a root access is gained (Table 13.2).

13.4 Testing Implementation of Forensic Procedures

The testing implementation has been conducted in this stage. Many different extraction tools are tested on the Android device. The forensic artifacts within the Android device were initially counted, and every tools was compared based on

Table 13.1 Application and artifact database location

Application	Folder location	Folder name
Browser history	*data/data/com.android.browser/database/browser2.db*	/database
Browser searches	*android.provider.Browser.SEARCHES*	/database
Calendar events	*data/data/com.android.providers.calendar/database/calendar.db*	/database
Call logs (incoming/outgoing/missed)	*android.provider.CallLog.Calls.CONTENT*	/database
Contacts	*data/data/com.android.providers.contact/database/contacts2.db*	/database
Device accounts	N/A	N/A
Device ID (e.g., IMEI, MEID, ESN)	N/A	N/A
MMS (incoming/outgoing)	*content://mms-sms/conversations*	/database
SMS (incoming/outgoing)	*content://sms*	/database
WhatsApp	*/data/data/com.whatsapp*	/database

Table 13.2 Google Android application data storage options

		Google Android application data storage options			
	Shared preference	Internal storage	External storage	SQLite	Network
File type	Key-value pairs of primitive data stored in lightweight XML format	Files of different formats. Developer based, no restriction	Files of different formats. No restriction	SQLite format (.db). Compact single cross-platform file	Config- and network-based files mainly. No restriction
Data type	Boolean, float, int, long, strings	Complicated data structures allowed	Complicated data structures allowed	SQLite-supported data types	Complicated data structures allowed
Location	*/data/data/com.android.phone/shared_prefs*	*/data/data* subdirectory	*/mnt/sdcard* or emulated SD card on */mnt/emmc*	Internal storage */data/data/<packageName>/databases*	Depends on network settings, info from log files in *data/data/files*
Access level	Owner can access	Developer controlled unless owner has root access	Owner can access MS FAT32 file system, no security mechanism	Encrypted unless owner has root access	Network level
Forensic use	Rich source of forensic data	Rich source of forensic data if root access	Rich source of forensic data	Rich source of forensic data	Forensic data from Java.net and Android.net

Table 13.3 List of software required for forensic procedures

Application	Available at	Paid/free
Titanium Backup Android Application	Google Play Store	Free
Framaroot	www.framaroot.net	Free
Root Checker	Google Play Store	Free
WhatsApp Viewer	http://andreas-mausch.github.io/whatsapp-viewer/	Free
Cerbero Profiler 2.4	http://cerbero.io/profiler/	Free (trial version)
SQLite Browser	sqlitebrowser.org/	Free
MPE+	http://accessdata.com/	Free (trial version)
MOBILedit	http://www.MOBiledit.com/forensic	Free (trial version)
Andriller	http://www.andriller.com/	Free (trial version)

what artifact could discover. A manual forensic extraction too was assessed along with these tools. For investigation, we have taken a sample device Sony Live with Walkman WT19i smartphone running Android 4.0.4 Ice Cream Sandwich OS as acquired device. The required software and tools for forensic procedures are arranged in Table 13.3.

13.4.1 Proposed Methodology on Google Android Device

Step 1: Android device is rooted using Android app Framaroot [11]. Table 13.4 indicates point by point techniques for rooting the Android device using Framaroot apps.

Further, we can verify that Android device is successfully rooted or not by installing Root Checker App and running it.

Step 2: Backup of messages using Titanium Backup apps. After rooting Android device, backup of message is done using Titanium Backup apps. Manual setting of Titanium Backup is done to corresponding each of application.

Step 3: Identify and find zipped folder within Titanium Backup folder on memory card location of Android device. Copy the identified zipped folder from Titanium Backup folder to the computer after connecting Android device to the computer.

Step 4: Backup the extracted folder from the Titanium Backup application on the computer [13].

After copying the zipped folder, we unzipped the required folder. Figure 13.5 shows the required folder from *com.android.browser* (browser backup using Titanium App), and Fig. 13.6 shows the required folder from *com.android.providers.contacts*. In Fig. 13.5 shows *browser2.db* and *webview.db* files and in Fig. 13.6 shows *contacts2.db* and *profile.db* database files for forensic analysis. Same procedure is applicable for other application databases.

Table 13.4 Rooting Android device using Framaroot application

S. no.	Methods	Snapshots
1	Enable installation of third party apps on Android device. To enable, open Settings > Security > Device Administration > Unknown Sources (check to enable)	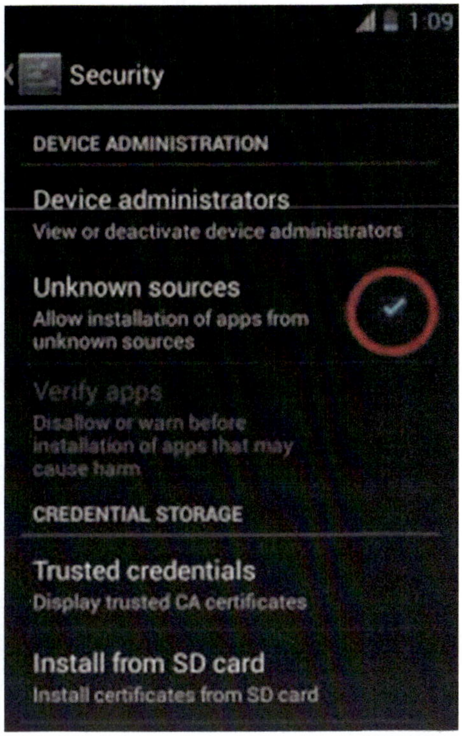

(continued)

Table 13.4 (continued)

S. no.	Methods	Snapshots
2	Download and install Framaroot App on Android device [12] After installation, Framaroot icon is displayed in the App Menu	

(continued)

Table 13.4 (continued)

S. no.	Methods	Snapshots
3	Run Framaroot by tapping on the Framaroot App icon	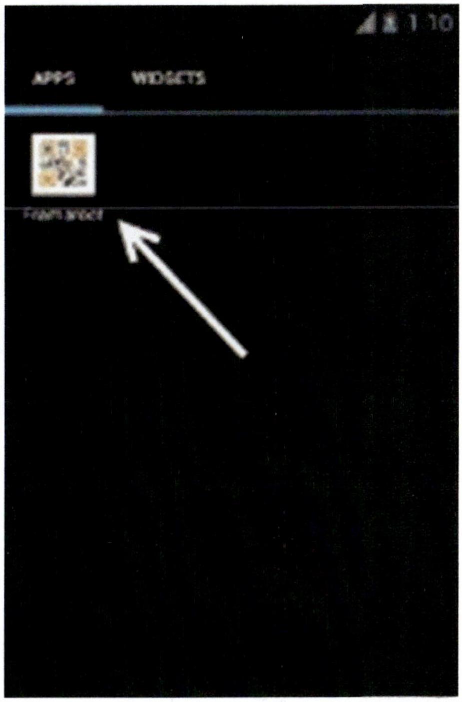

(continued)

Table 13.4 (continued)

S. no.	Methods	Snapshots
4	When Framaroot App is launched, the image shown beside is visible on the screen of mobile device	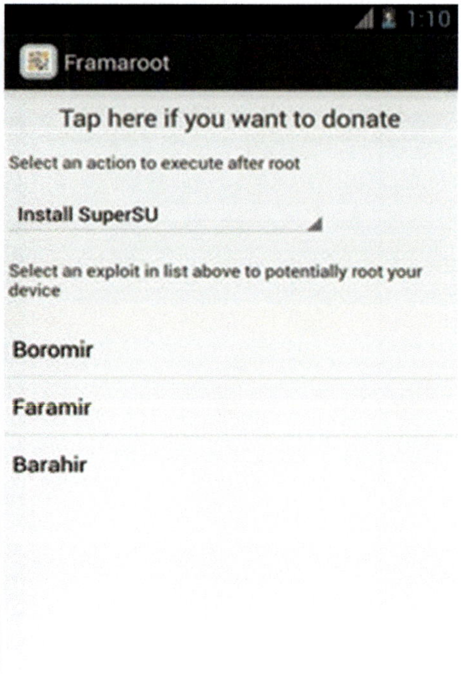

(continued)

Table 13.4 (continued)

S. no.	Methods	Snapshots
5	In the earlier snapshot, two options (Boromir and Faramir) are available Select any option to start the rooting process (e.g., we select Boromir). A success message will be displayed and prompt to restart the mobile device	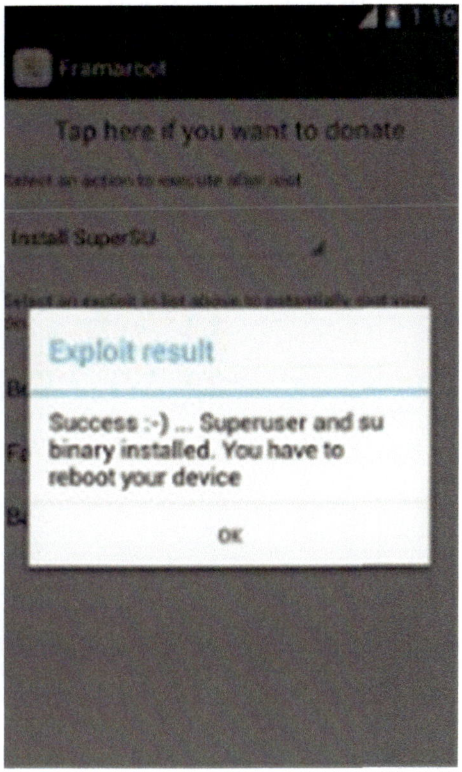

(continued)

Table 13.4 (continued)

S. no.	Methods	Snapshots
6	After restarting, the Android device is rooted, and an additional application SuperSU is available in the App Menu confirming successful rooting of the mobile device	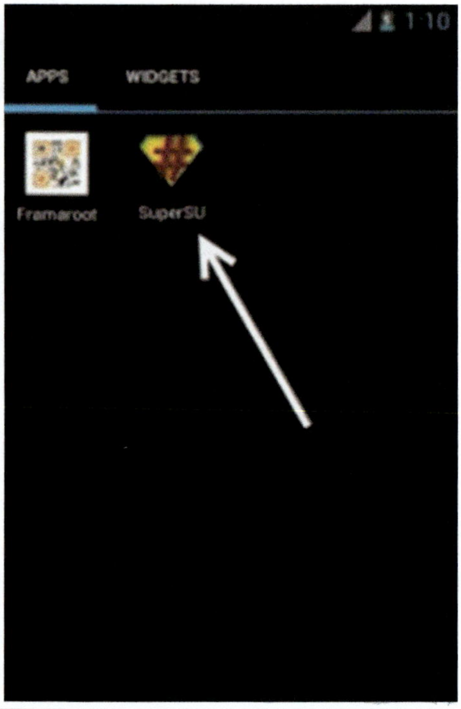

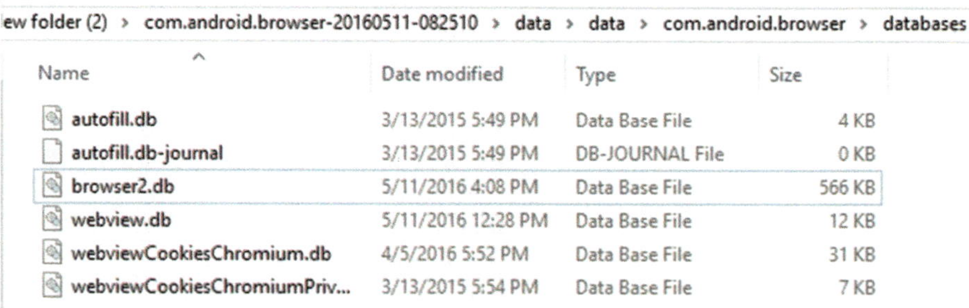

Fig. 13.5 Unzipped *com.android.browser* folder directory

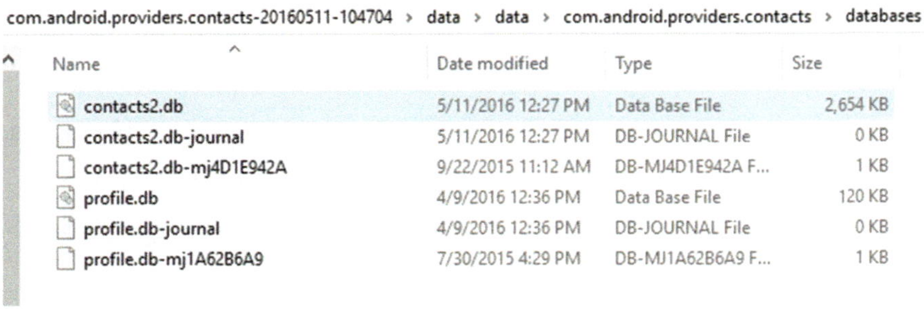

Fig. 13.6 Unzipped *com.android.providers.contents* folder directory

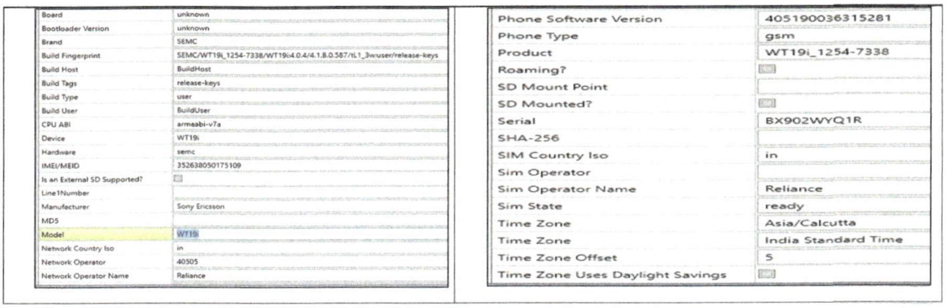

Fig. 13.7 Device info and artifacts using MPE+

13.4.1.1 Google Android Forensic Extraction of Phone ID Using MPE+

A phone ID is a unique identifier that differentiates the Android device from different telephones on an enterprise. The application uses an alarm that calls the *getDeviceld()* method from the Android API's *Telephony Manag*er class to directly get to a phone's device ID [14].

1. International mobile equipment identity (IMEI)
2. Electronic serial number (ESN)
3. Mobile equipment identifier (MEID)

An IMEI is returned for Global System for Mobile Communications (GSM) mobile phones, while an ESN or MEID is returned for Code Division Multiple Access (CDMA) mobile phones.

The *android.permission.READ_PHONE_STATE* permission must be declared in the *AndroidManifest* for an application to extract the phone ID. Figure 13.7 shows snapshot of extracting device information.

Fig. 13.8 Extracting SIM data artifacts using MOBILedit

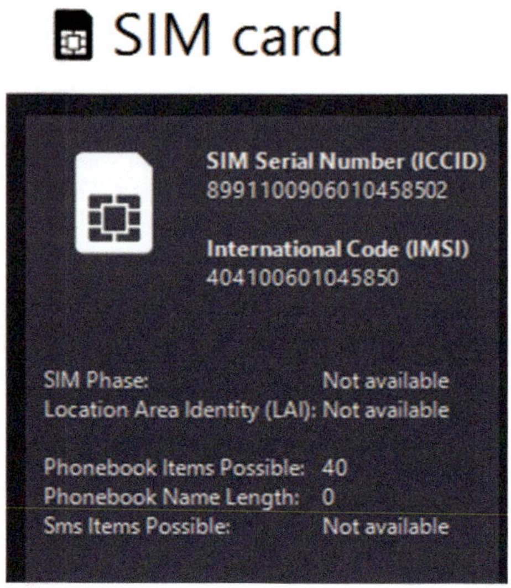

13.4.1.2 Google Android Forensic Analysis of Extracting SIM Data

Device details, including hardware information, SIM number, IMEI, and/or IMSI, were recovered using MOBILedit. Figure 13.8 shows snapshot of extracting SIM data [15].

13.4.1.3 Android Forensic Analysis of Extracting Wi-Fi Profiles

Android mobile phone stores Wi-Fi profiles so they can automatically reconnect when range of the Wi-Fi router or access point. Forensic investigator can recover the SSID, network name, security mode, password, username, WEP key, and MAC adds of the router. Forensic examiner can use this information to help find out where a suspect may have been, since this information is stored indefinitely until the profile is erased or the device is wiped. Figure 13.9 shows snapshot of extracting Wi-Fi artifacts [11].

13.4.1.4 Google Android Forensic Analysis of Extracting Bluetooth Profile

Android phones keep a list of any saved Bluetooth phones that were connected to the Android phones. Using MPE+ we will recover the MAC address, device name, phone class, last seen date, and time stamp [16]. These profiles can be helpful if an examiner is searching for evidence found on other connecting phones, such as another PC or car. Figure 13.10 shows snapshot of extracting Bluetooth artifacts.

Select	Status	Pre-shared Key	Priority	SSID
☐	enabled	☑	18	Blue Fluffy Monsters
☐	enabled	☐	8	Bb3b-bWFub2prYXRodXJpYTEyNw
☐	enabled	☐	9	ADYYWGlhb21p
☐	enabled	☐	12	"CShare_Rakesh_sony"
☐	enabled	☐	13	ADYYc2Ftc3VuZw
☐	enabled	☑	14	Kejriwal free wifi
☐	enabled	☐	15	ADYU29ueQ
☐	enabled	☐	16	Jamia Hamdard
☐	enabled	☐	23	BVSy-S2FyYm9ubiBUaXRhbml1bSBTNSB
☐	enabled	☑	24	shiv
☐	enabled	☐	25	komal
☐	enabled	☑	26	Realince.enQ.9310463000
☐	enabled	☑	32	D-Link
☐	enabled	☐	28	Bdra-aXJpcyBYMQ
☐	enabled	☑	29	IMPRESSION
☐	enabled	☑	33	CA deepak
☐	enabled	☑	64	NSB
☐	enabled	☑	34	AndroidAP
☐	enabled	☑	66	MKD

Fig. 13.9 Extracting Wi-Fi artifacts using MPE+

Select	Address	Name	Major Class	Major (number)	Device	Device (number)	Bond State
☐	4C:7F:62:09:15:DE	riyaz	phone	512	phone_smart	524	bonded
☐	E0:2A:82:02:48:EC	AUQIB	computer	256	computer_laptop	268	bonded
☐	3C:25:34:15:10:52	Micromax A74	phone	512	phone_smart	524	bonded
☐	B0:35:8D:92:97:47	Vish...	phone	512	phone_cellular	516	bonded
☐	18:3F:47:3F:8A:41	HM1100	audio_video	1024	audio_video_wearable_headset	1028	bonded
☐	20:D6:07:8E:94:22	Maxter slave	phone	512	phone_cellular	516	bonded
☐	14:F6:5A:7D:D7:E6	Redmi	phone	512	phone_smart	524	bonded

Fig. 13.10 Extracting Bluetooth artifacts using MPE+

13.4.1.5 Google Android Forensics of Extracting SMS

Android stores your text messages in a SQLite database in the *data/data/com.android.providers.telephony/databases/mms-sms.db*, and you can stack this file into a SQLite database viewer to see the SMS. The Android applications, including SMS and contacts, use SQLite databases to store information. The apps present

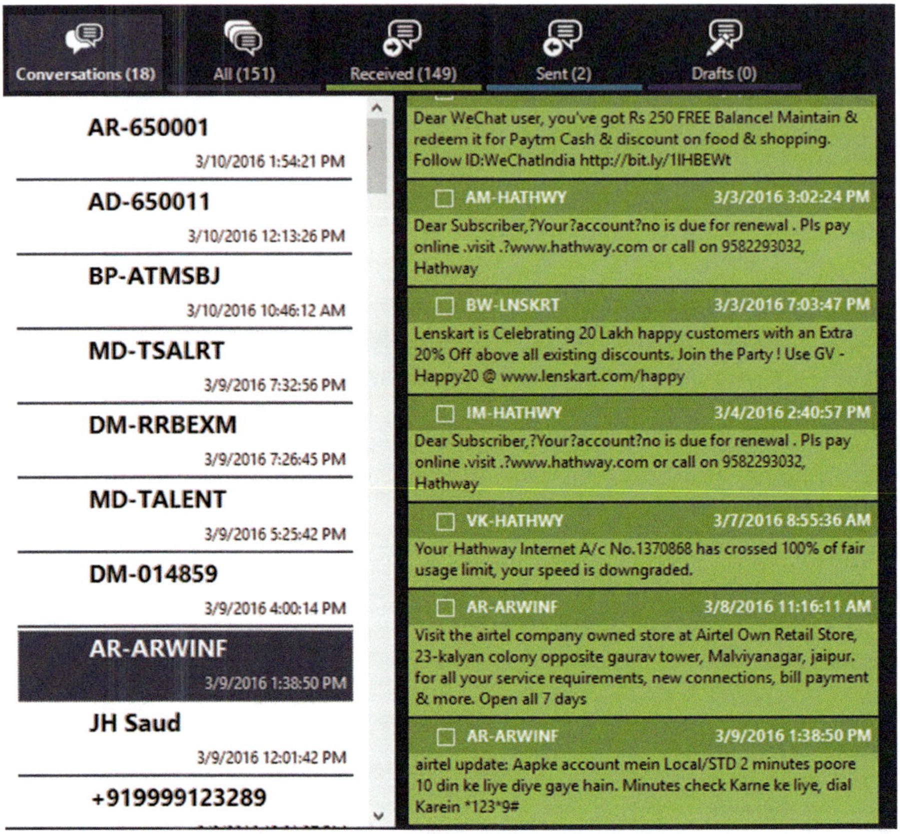

Fig. 13.11 Extracting SMS using MOBILedit

information in their own particular manners, such as SMS conversation; however, you can view the raw data stored as it is stored in a SQLite database viewer. To collect the SMS data in MOBILedit requires that the *android.permission.RECEIVE_SMS*, *android.permission.READ_SMS*, and *android.permission.READ_CONTACTS* permission be declared in the AndroidManifest. Figure 13.11 shows snapshot of extracting SMS artifacts [17].

13.4.1.6 Google Android Forensics of Extracting Calendar Events

Different applications store calendar information, which is then accessed by various applications. The Calendar's URI is accessed, and afterward the information is extricated, to be specific on the begin date of the event. This information is then stored in the database. A user can without much of a stretch know all the data on the calendar database. The alarm is configured to filter the *content://com.android.calendar/event_entities* content provided by URI.

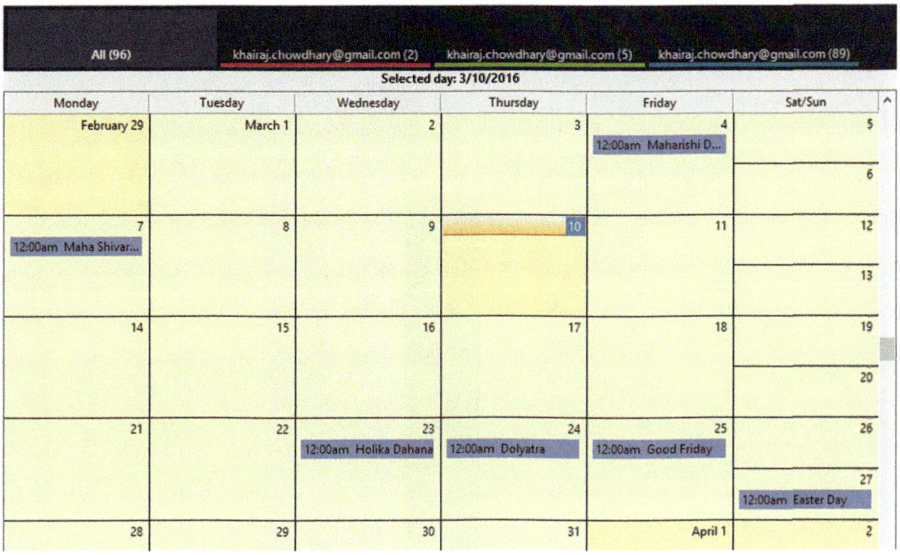

Fig. 13.12 Extracting calendar events using MOBILedit

The access to perform these activities on the calendar information set requires the *android.permission.READ_CALENDAR* permission be the Android Manifest. Figure 13.12 shows snapshot of extracting calendar events.

13.4.1.7 Google Android Forensics of Extracting Call Logs

The call log data of the Android device are stored in call log. It contains details about all the calls. This class is accessed to discover details about the call log, and these details are put away in the database [18]. These details include the inbuilt ID of every call; the cached number; the number stored in the mobile phone; the cached name, i.e., the date and time when the call was made; and the cached number sort related such as home, work, etc. Figure 13.13 shows snapshot of extracting call logs [19].

13.4.1.8 Google Android Forensics of Extracting Installed Applications

Using MOBILedit, list the installed apps on Android device. Examiners can recover package name and show name, platform, classification, and both the internal and display version for the apps. Using MOBILedit, we can recover the installed apps, and it gives you access to all app data, for example, Viber, Evernote, Skype, WhatsApp, and so forth. We retrieve deleted data from these apps. Figure 13.14 shows snapshot of extracting installed apps.

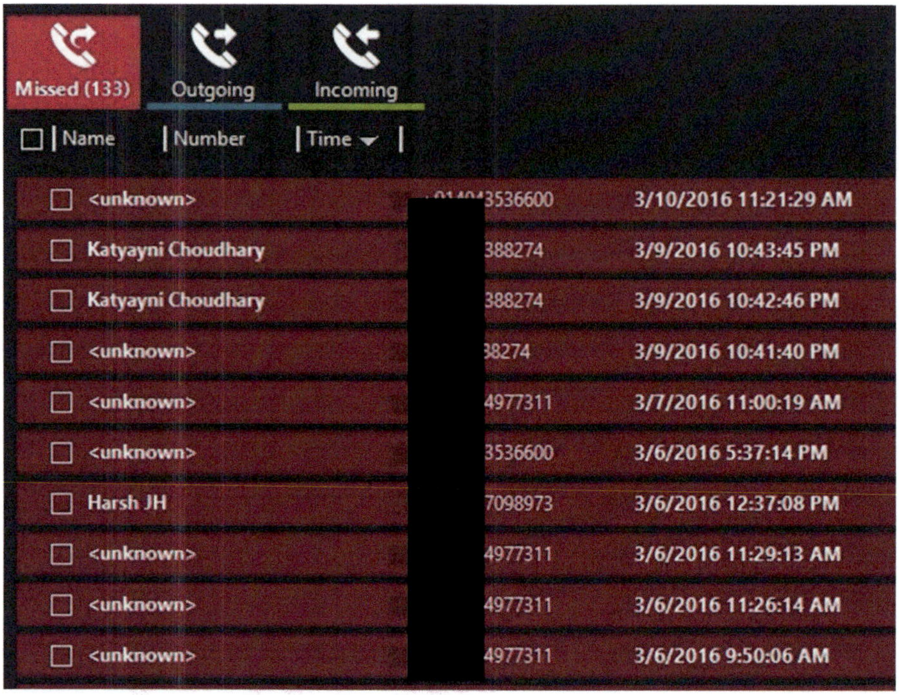

Fig. 13.13 Extracting call logs using MOBILedit

13.4.1.9 Google Android Forensic Analysis of WhatsApp Conversation and WhatsApp Calls

We concentrate on WhatsApp apps for forensic investigation to determine artifacts from the Android device [20]. After taking backup of WhatsApp application information, software applications like WhatsXtract, SQLite browser, WhatsApp Viewer, and so on are required to determine artifacts. These software applications present artifacts to the examiner in human readable and present format. WhatsApp Viewer is a small tool to show chats from WhatsApp files such as msgstore.db.crypt5, msgstore.db.crypt7, and msgstore.db.crypt8. Among all available tools, WhatsApp Viewer is the most convenient and simple to use because of the following features [21]:

- View WhatsApp chat on computer.
- Mobile phone backup.
- Conveniently read old conversation or "load older messages."
- Search all messages.
- No need to install Python, SQLite, or extra libraries.

For investigation of WhatsApp artifacts, we have used WhatsApp Viewer. Figure 13.15 shows snapshot of WhatsApp messages in WhatsApp Viewer. Analyst

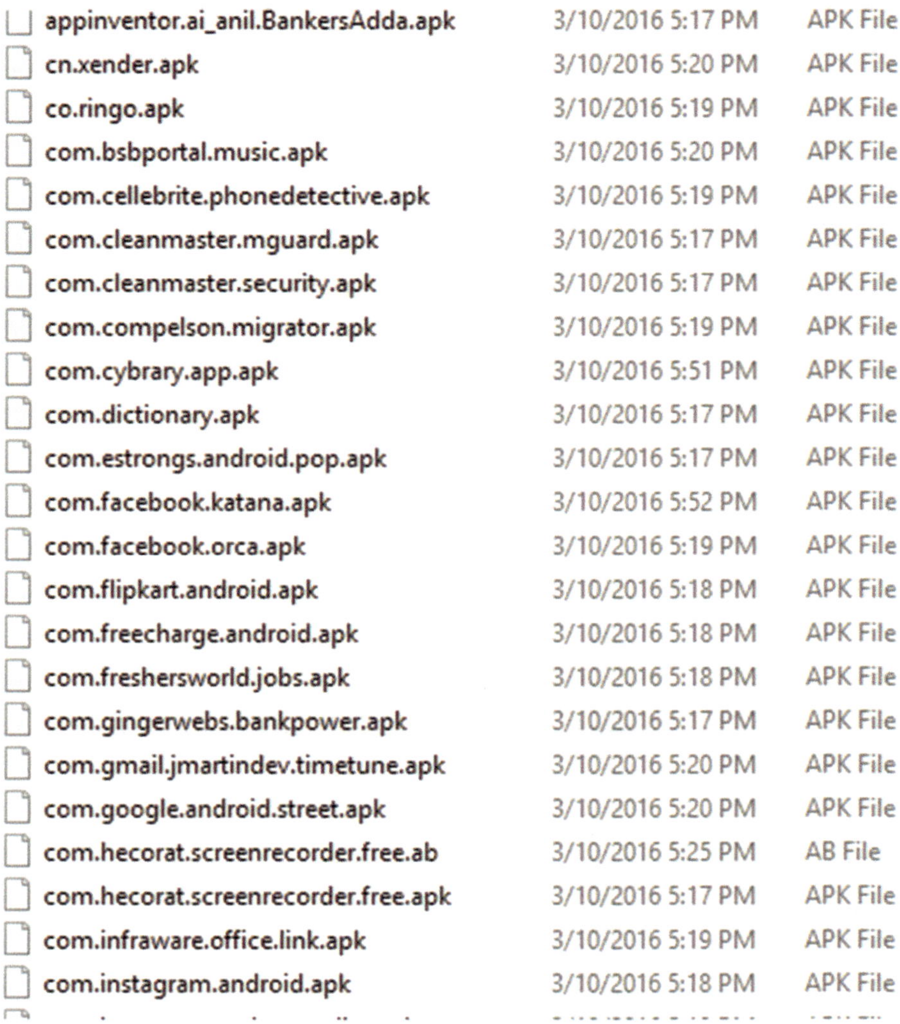

Fig. 13.14 Extracting installed apps

can browse through accessible contacts and read messages exchanged between the user of phone and contacts. For the analysis of WhatsApp calls artifacts, we have used Andriller [22]. Figure 13.16 shows snapshot of WhatsApp calls in Andriller forensic tool [23].

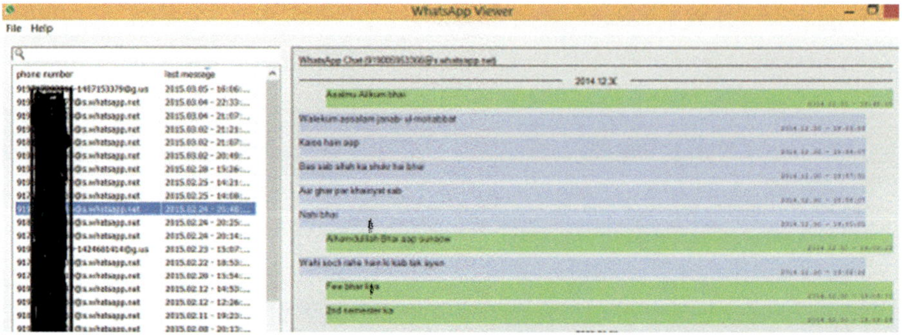

Fig. 13.15 WhatsApp artifacts using WhatsApp Viewer

[WhatsApp Calls]

Total items: 65

#	Type	Number	Time	Duration
6099	Dialled	3969370	2015-09-01 07:22:05 UTC+00:00	0:00:00
6098	Missed	3969370	2015-09-01 07:20:33 UTC+00:00	0:00:00
5670	Missed	0231461	2015-08-12 09:35:26 UTC+00:00	0:00:00
5456	Dialled	71618082	2015-08-03 05:50:57 UTC+00:00	0:00:48
5455	Received	71618082	2015-08-03 05:50:30 UTC+00:00	0:00:18
5383	Received	2529707	2015-07-31 06:23:45 UTC+00:00	0:00:10
5378	Received	2529707	2015-07-31 05:38:54 UTC+00:00	0:00:15
5377	Missed	2529707	2015-07-31 05:38:52 UTC+00:00	0:00:00
5121	Missed	2171739	2015-07-24 10:03:12 UTC+00:00	0:00:00
4519	Dialled	6738727	2015-07-01 07:09:48 UTC+00:00	0:00:00
4446	Dialled	1358256	2015-06-29 15:20:44 UTC+00:00	0:00:00
4445	Missed	1358256	2015-06-29 15:19:26 UTC+00:00	0:00:00
4202	Missed	6958477	2015-06-20 07:54:00 UTC+00:00	0:00:00
3755	Received	5953366	2015-06-04 04:22:12 UTC+00:00	0:00:14
3754	Dialled	5953366	2015-06-04 03:39:22 UTC+00:00	0:00:00
3717	Received	6604821	2015-05-31 10:53:54 UTC+00:00	0:00:17
3714	Missed	2529707	2015-05-31 02:51:19 UTC+00:00	0:00:00
3499	Dialled	3809280	2015-05-23 12:36:23 UTC+00:00	0:00:00
3498	Dialled	2529707	2015-05-23 11:48:46 UTC+00:00	0:00:00
3497	Dialled	2529707	2015-05-23 11:48:37 UTC+00:00	0:00:00
3496	Dialled	2529707	2015-05-23 11:48:30 UTC+00:00	0:00:00
3495	Dialled	3809280	2015-05-23 11:43:02 UTC+00:00	0:00:00
2838	Missed	5092295	2015-05-03 16:10:12 UTC+00:00	0:00:00

Fig. 13.16 WhatsApp calls artifacts using Andriller

13.5 Results of Research Findings of Forensic Evidence for Google Android Device

In this section, we present our research findings of forensic evidence for Android device based on investigation. Table 13.5 presents the examination evidence of forensic investigation for SMS, browser history, contact logs, Email, etc. Listed in this table are the different artifacts such as contact number, messages, SIM-stored message, contacts, etc.

13.6 Conclusion

In this paper, we have presented forensic evidence for Android device. This research paper has focused to extract information stored on Android device. The different software tools are used for extracting the data in Android device. The aim was to determine key artifacts present in the memory of Android device using accessible devices and software. The research finding evidence include artifacts that help forensic examiners and investigation agencies during any criminal occurrence and can be used as evidence in the court of law. In the future, the recovery of artifacts of Android device residing on RAM of Android device can be a part of our research scope.

Table 13.5 Research findings of forensic analysis for android device

	Test	Rooted android device	
		Found	Not found
SMS	SMS collected	✓	
	View deleted SMS	✓	
	SMS storage limits		✓
	View SMS contact name		
	View SMS contact number	✓	
	View SMS body	✓	
	View SMS time stamp	✓	
Call logs	Call logs collected regularly		
	View deleted calls	✓	
	View call time stamp	✓	
	View call contact number	✓	
	View call contact name	✓	
	View call duration		✓
Gallery	Images/pics collected	✓	
	View deleted images	✓	
Apps	View app installed	✓	
	View app package name	✓	
Contacts	Contacts collected regularly		
	View deleted contacts	✓	
	View contact name	✓	
	View contact number	✓	
	View contact date added		✓
SIM	View SIM number	✓	
	View stored contacts	✓	
	View stored SMS/MMS	✓	
	View no. of contacts saved	✓	
Bluetooth	Number of other devices connected	✓	
	View MAC address	✓	
	Last seen date and time stamp		✓
	Device name	✓	
	Phone class	✓	
Wi-Fi	View SSID	✓	
	Display network name	✓	
	Security mode WEP	✓	
	Show password		✓
	MAC address of router	✓	
Calendar	Events and reminders	✓	
Device account and ID	Information about device	✓	
Email		✓	

References

1. Zhao, X., Tian, D.: The architecture design of streaming media applications for Android OS. 2012 I.E. international conference on computer science and automation engineering. Beijing, pp. 280–283, (2012)
2. Alec, Y., et al.: Computer forensics education. s.l. IEEE Secur. Priv. **1**, 15–23 (2003)
3. Kaur, H., Xiaohui, T.: ICT and millennium development goals: a united nations perspective, p. 271. Springer Publishers, New York (2014)
4. Li, Q., Hu, X., Wu, H.: Database management strategy and recovery methods of Android. Software engineering and service science (ICSESS), 2014 5th IEEE International Conference on, Beijing, pp. 727–730 (2014)
5. Chen, S.W., Yang, C.H., Liu, C.T.: Design and implementation of live SD acquisition tool in android smart phone. Genetic and Evolutionary Computing (ICGEC), 2011 Fifth International Conference on, Xiamen, pp. 157–162 (2011)
6. Simao, F., Sicoli, L., Melo, F., Deus, de Sousa, R.: Acquisition and analysis of digital evidence in Android smartphones, Int J FORENSIC Comput Sci, **1** 28–43, (2011)
7. Chung, H., Park, J., Lee, S., Kang, C.: Digital forensic investigation of cloud storage services. Digit Invest. **9**, 81–95 (2012)
8. [Online] www.sqlitebrowser.org/
9. SQLite Consortium.: SQLite - Write-Ahead Logging, 2014. https://www.sqlite.org/draft/wal.html (2014)
10. Thakur, S.N.: Forensic analysis of WhatsApp on Android smartphones. University Of New Orleans, New Orleans (2013)
11. Al Barghouthy, N., Marrington, A.: A comparison of forensic acquisition techniques for android devices: a case study investigation of orweb browsing sessions. 2014 6th International Conference on New Technologies, Mobility and Security (NTMS), Dubai, 2014, pp. 1–4 (2014)
12. [Online] http://www.framaroot.net/
13. Guido, M., Ondricek, J., Grover, J., Wilburn, D., Nguyen, T., Hunt, A.: Automated identification of installed malicious Android applications, digital investigation. Int. J. Digital Forensics Incident Response **10**, S96–S104 (2013)
14. Justin,G.: Android forensics: automated data collection and reporting from a mobile device. Thesis. Rochester Institute of Technology (2013)
15. Awan, F.A.: Forensic examination of social networking applications on smartphones. 2015 Conference on information assurance and cyber security (CIACS). Rawalpindi, 2015, pp. 36–43 (2015)
16. [Online] http://accessdata.com/
17. Zhao, H., Yang, F., Zhang, N., Zhang, J.: MMSnap: An MMS based forensic system for recovering stolen phones. Electronics information and emergency communication (ICEIEC), 2015 5th International conference on, Beijing, 2015, pp. 150–154 (2015)
18. [Online] http://www.mobiledit.com/forensic
19. Dibb, P., Hammoudeh, M.: Forensic data recovery from android os devices: an open source toolkit. Intelligence and security informatics conference (EISIC), 2013 European, Uppsala, 2013, pp. 226–226 (2013)
20. Shortall, A., Azhar, M.A.H.B.: Forensic acquisitions of WhatsApp data on popular mobile platforms. 2015 Sixth International Conference on Emerging Security Technologies (EST), Braunschweig, 2015, pp. 13–17 (2015)
21. [Online] http://andreas-mausch.github.io/whatsapp-viewer/
22. http://www.andriller.com/
23. Zhou, F., Yang, Y., Ding, Z., Sun, G.: Dump and analysis of Android volatile memory on WeChat. 2015 I.E. international Conference on communications (ICC), London, 2015, pp. 7151–7156 (2015)

Chapter 14
A Systematic Literature Review on Software-Defined Networking

Izzat M. Alsmadi, Iyad AlAzzam, and Mohammed Akour

Abstract Software-Defined Networking (SDN) is a recently evolving networking architecture that focuses on the separation of control and data planes. Unlike traditional switches, SDN switches include flow tables that are remotely controlled by a separate software application, the controller. SDN is not completely new; it formulates an architecture on top of several good practices. In this paper, we examined the obtainable knowledge about SDN through conducting a systematic literature review (SLR) to evaluate the current SDN state of the art in terms of research tracks, publications, trends, etc. We systematically evaluate research in SDN based on questions formulated for this purpose. The results present outline information about the most active research areas, tools, security issues, obstacles, limitations, strengths, and opportunities in SDN.

14.1 Introduction

The continuous growth of the Internet, smart applications, e-commerce, multimedia applications, social networks, etc. poses a continuous challenge for networks on keeping up with such evolution in terms of bandwidth, information overload, complexity, etc. Enterprises such as Google, Facebook, Microsoft, eBay, and Amazon use a very large number of data centers. A huge volume of data is exchanged in those centers. Data centers include tenants or virtual machines (VMs) to divide virtually or logically the network into different nodes, clusters, or slices. New services based on user or customer demand may cause new VMs to be created. For each newly created VM or tenant, resources, management, control,

I.M. Alsmadi (✉)
Department of Computing and Cyber Security, University of Texas A&M, One University Way, San Antonio 78224, TX, USA
e-mail: ialsmadi@tamusa.edu

I. AlAzzam • M. Akour
Computer Information Systems Department, Yarmouk University, Irbid, Jordan
e-mail: eyadh@yu.edu.jo; mohammed.akour@yu.edu.jo

I.M. Alsmadi et al. (eds.), *Information Fusion for Cyber-Security Analytics*, Studies in Computational Intelligence 691, DOI 10.1007/978-3-319-44257-0_14

security, etc. are all should be allocated dynamically to accommodate that particular VM needs.

One of the serious challenges in cloud computing or the Internet traffic is that demand varies widely from day to day or even from hour to hour. This fluctuation makes it very hard to manage this process manually. On the other hand, traditional switches are vendor specific and the administration and configuration/reconfiguration of those switches are labor intensive. Similarly, the management of security controls such as firewalls is labor intensive as the process to add, update, or maintain access control lists (ACLs) in those firewalls is accomplished manually by network administrators.

OpenFlow is an algorithm developed to define interaction between controller and switches. Specifically, OpenFlow (OF) includes detailed specifications on how the controller should communicate with its OF switches. OF switches are different from traditional switches in that they are built to be very basic with no control functions and include only data or forwarding elements. Most newly designed switches start supporting both modes: traditional and OF. Controller is a software program that acts as a networking operating system (NOS) for the control and administration of OpenFlow switches.

SDN has several initiatives that came to solve specific problems in networks. Switches, routers, or other network components are vendor specific. Networking companies, for business not technical purposes, do not allow users to program applications on top of those networking components. One of the main goals of SDN is to have an open networking architecture that is not vendor locked-in or specific. Further, this network architecture should also be developers or network administrators to interact with the switches and, for example, use or customize flow or access control algorithms.

In this paper, we conducted an SLR on SDN. We followed SLR systematic research investigation process. Key terms that can distinguish publications in SDN are formulated. Key questions that can best extract recent research trends in SDN are formulated. To the best of our knowledge, there is no research that discusses SLR in SDN. The closest to the scope of SLR in SDN will be survey papers. There are some papers that conducted surveys in SDN (e.g., [10, 48, 207–211, 280]).

14.2 SDN Road Map

While SDN is new as an architecture, no new networking technologies were invented, and the architecture coined old concepts and combined some new practices. Most references considered the work of Casado, his PhD thesis, Nicira networks' establishment, SANE, and ethane papers [212, 214] as the starting hype. However, existing research papers before that (e.g., [213]) discussed the core idea in SDN which is to split the routing or the intelligence knowledge from router and switches and include it in a separate control unit. What was interesting in SDN story is that its advances accelerated almost in parallel in both the academia and the industry. Similar to Google story, researchers in SDN and graduate students from Stanford established new startups, Casado: Nicira networks (2007) with his two advisors, Nick McKeown and Scott Shenker and Kyle Forster and Guido Appenzeller, BigSwitch networks (2010). In this road map, however, we will focus only on the advance at the research and publications' level. Members from the University of California, Berkeley, such as Scott Shenker, were also early contributors in SDN. OpenDaylight is currently an open source project to build the controller and SDN architecture around in which most vendors come together to support it.

Nicira later on introduced OpenFlow as an instance of SDN and a protocol to regulate the communication between SDN controller and switches. Nicira (later became part of VMware) contributes also to the development of the Open Virtual Switch project (Open vs. Switch, 2008), an open source virtual switch that enables software applications to interact with switches. While protocols other than OpenFlow can be used in SDN, however, currently OpenFlow is associated with SDN. We showed that in our string search, the two words that mostly define SDN in literature related to SDN are the abbreviation SDN and OpenFlow. GENI is established in the USA in 2009 as a national project to promote SDN research through an SDN-based networking lab that is open for all researchers. Open Networking Foundation (ONF), the company behind OpenFlow standards, is established in 2010.

Tables 14.1 and 14.2 show the top 22 papers published in SDN based on citation. In order to collect the top ten papers in terms of citation, we used the same two terms that most distinguish research papers related to SDN: SDN and OpenFlow.

Table 14.1 Top 1–11 SDN/OpenFlow published papers (based on citation)

No.	Authors/year	General description
1	McKeown et al. (2008) [215]	A very early paper in OpenFlow Stanford team about the initial goal of OpenFlow to manage university campus network
2	Gude et al. (2008) [216]	Another early paper from Stanford team about SDN network operating system or controller (NOX)
3	Benson et al. (2010) [217]	Using OpenFlow architecture for cloud data centers
4	Casado et al. (2007) [214]	Ethane: network access control based on OpenFlow, from the first team of SDN, Casado, and advisors
5	Mysore et al. (2009) [218]	PortLand, an SDN-based solution for scalable fault-tolerant cloud data centers
6	Heller et al. (2010) [219]	An SDN-based solution for cloud data centers, energy-saving tree architecture
7	Koponen et al. (2010) [220]	Onix, first SDN distributed controller
8	Dobrescu et al. (2009) [221]	Distribution, parallelism, and scalability issues-routers
9	Han et al. (2010) [222]	Distribution, parallelism, and scalability issues-routers
10	Curtis et al. (2011) [223]	SDN-scalability issues, distribution
11	Farrington et al. (2010) [224]	Optical switching/data centers

Table 14.2 Top 12–22 SDN published papers (based on citation)

No.	Authors/year	General description
12	Lantz et al. (2010) [225]	An early contribution from Stanford research team about using SDN for home or campus networks
13	Sherwood et al. (2009) [226]	SDN-scalability issues, distribution
14	Guo et al. (2010) [227]	SDN-based cloud data center virtualization
15	Sherwood et al. (2010a, b) [228, 233]	An early paper contribution, allowing same production network to be used for testing based on SDN
16	Ganjali et al. (2008) [229]	SDN-scalability issues, distribution
17	Casado et al. (2006) [212]	Early paper contribution. SDN-based security policy implementation
18	Foster et al. (2011) [230]	Network or policy programming language
19	Reitblatt et al. (2012) [231]	Network configuration/reconfiguration/SDN design enhancements
20	Kazemian et al. (2012) [232]	SDN-based network testing and QA issues
21	Sherwood et al. (2010a, b) [228, 233]	SDN distribution/testing and experimentations
22	Khurshid et al. (2013) [234]	SDN-based network testing and QA issues

Those two terms were selected after several trials of combinations between different key terms. Results, in terms of the number of citations, vary from one website to another. We focused on the most agreeable ones between the five research indexing websites: IEEE Xplore, ACM Digital Library, Google Scholar, Microsoft Academic Search, ScienceDirect, and CiteSeerX.

The top cited papers can give us indication what are the top research trends in SDN. We can see that the first 11 most cited papers can be classified into:

1. Early contributions by Stanford team [214–216].
2. Cloud data center-related issues (e.g., scalability, fault tolerance) [217–219].
3. SDN scalability and distribution issues [220, 223].
4. Other trends: optical/firmware [221, 222, 224].

One more notice is that there are some other trends that have been evolving more recently. However, they are not getting yet the size of research as those earlier subjects. The new most recent subjects shown in Table 14.2 include SDN security issues, SDN testing and QA, SDN wireless, etc.

Table 14.2 shows the next 11 research papers in terms of citation count.

We think that while research focuses in Table 14.1 will continue to exist in the future, we think that security and testing issues in particular will get more research focus in the new future especially as those two areas include both most significant SDN opportunities and challenges.

14.3 Goals, Research Questions, and Metrics

In conducting this SLR about SDN, we aimed at achieving the following goals:

1. To classify the research papers published in SDN and be able to summarize research trends in SDN.
2. To show the different challenges and opportunities that are posed or opened based on this new networking paradigm.
3. To show how SDN evolves and how can new researchers find open research areas in SDN.
4. To identify, for SDN, most active researchers, teams or groups, conferences, workshops, and journals.

Based on the previously defined goals, we formulated the following questions that our SLR investigated. We divided some research questions into further sub-questions:

R1: What are the main SDN research areas investigated in published papers?
R2: What tools have been used or developed in SDN? How can those tools be classified?
R3: What are the current investigated security issues in SDN?

- R3.1: What are the security problems related to SDN architecture?
- R3.2: What are the security opportunities SDN can bring to networking, cloud computing, etc?

R4: What are the obstacles and limitations of SDN?
R5: What are the strengths and opportunities in SDN?
R6: Research dissemination and trend issues:

- R6.1: What are the most cited papers, authors, popular conferences, and journals publishing about SDN?
- R6.2: Who are the top ten authors in terms of the number of publications?
- R6.3: Who are the top ten authors in terms of citation counts?
- R6.4: Where are the top ten most active teams located?
- R6.5: What are the top ten conferences in terms of SDN publications?
- R6.6: What are the top ten journals in terms of journal publications?

14.4 Article Selection

Selecting the right articles based on the research questions is a major step in SLR. The following steps summarize article-selection stage.

14.4.1 Step 1: Article Identification

We started the process by conducting several combinations of search for SDN key terms in the following academic research libraries: IEEE Xplore, ACM Digital Library, Google Scholar, Microsoft Academic Search, ScienceDirect, and CiteSeerX. Initially we tried different combinations of the following SDN key terms: SDNs, SDN, SDN, SDNs, OpenFlow, OpenFlow, and SDN. In each combination, we compared results in terms of the percentage of relevant papers to the total number of papers. Finally, we noticed that the best combination that retrieved all papers that are relevant to the subject is when using SDN as an abbreviation together with OpenFlow as one term. Initial results retrieve (208 articles in IEEE Xplore, 236 in ACM, 3030 in Google Scholar, 16 MS Academic Research, and 223 in CiteSeerX).

14.4.2 Step 2: Exclusion Criteria

We defined several exclusion criteria including:

1. An article paper published in a language other than English.
2. Our intention was to exclude articles published before 2006 with our assumption that Casado et al. paper 2006 can be considered as the start of coining SDN architecture. As we mentioned earlier, SDN is not new in terms of technology or invention, it is rather a new way of designing network architecture. However, it should be mentioned that there are some important papers before Casado et al. paper in 2006 (e.g., [213]) that are considered significant in the road map of SDN. In our search collection, OpenFlow protocol focused the search for the most recent publications in SDN after adopting OpenFlow protocol (i.e., after 2010). We accepted this assumption to focus our search to the most relevant research publications to the current SDN architecture. We will have a separate section for SDN road map focusing on papers published between 2004 (Feamster paper till 2011).
3. We excluded technical reports and selected only papers published in conferences, journals, or workshops. We excluded also articles, presentations, etc., although some of those included significant information and contribution. SDN has a unique research stand. This is since it is one of those few research fields that is growing almost in parallel between the academia and the industry. Both sides are trying to get a share in this new field.
4. Research papers indexing websites may include also indexed references to editorial introductions or prefaces. We excluded those also from the retrieved results.

14.4.3 Step 3: Inclusion Criteria

In the selection of proper search terms we described earlier, we ended up with two specific terms that we thought that they can best include the most current relevant papers to SDN. Those were SDN and OpenFlow. We also noticed that since OpenFlow protocol was proposed years after embracing SDN, there are publications between the years 2007 and 2011 that were discussing SDN without including any reference to OpenFlow. Hence we decided manually to include those papers after investigating them and their relevancy to our paper subject. The final number of papers included in our literature survey ended up to be 237 papers. ACM and IEEE included the largest percentage of relevant papers from the general retrieved results. This is based on the inclusion/exclusion criteria we described earlier.

14.4.4 Step 4: Final Article Set

In evaluating the difference in publications and statistics between the different websites, we noticed several issues. We tried to combine or aggregate results from the different websites, for example, when considering top papers, authors, publications, etc. We noticed that websites are indexing different papers. Hence we combined all articles from all different websites to get the top counts based on the five indexing websites that we used. As described earlier, based on the inclusion/exclusion process described earlier, many retrieved results were eliminated.

14.4.5 Iterative Development of Literature Mapping

The process of evaluating the different statistics is an iterative one. If we find a problem in the selection in one website, we will modify it and repeat the new process across all websites. Results published in this paper are according to the last process of gathering data collected from the different websites before the submission of this paper for publication. We acknowledge, however, this is a very recent evolving field where even a very short period such as a month can possibly change the statistics related to this subject.

14.5 Mapping Research and Evaluation

In this section, we will answer research questions based on the collected data.

R1: What are the main SDN research areas investigated in published papers?
We made our own classification of the collected papers and their classification. Table 14.3 shows examples of papers that discussed SDN and network security attacks.

Table 14.3 Security attacks/vulnerabilities

Spoofing	DoS/flooding/DNS amplification	Information disclosure, worms/scanners/sniffers/ MIM/botnets
Yao et al. (2011), Li et al. (2011), Li and Hong (2011), Jafarian et al. (2012) Jafarian et al. [260], Braga et al. (2010), Zaalouk et al. (2014)	Suh et al. (2010), Chu et al. (2010), Koponen et al. (2011), Choi et al. (2010), Shin et al. (2013) (2), Braga et al. (2010), Yu et al. (2014) [262], YuHunag et al. (2010), Benton et al. (2013), Chung et al. (2013), Shin and Gu (2013), Popa et al. (2010), Karame (2013), Kotani, Yasuo Okabe (2012), Lu et al. (2012), Schehlmann and Baier (2013), Zaalouk et al. (2014)	Jafarian et al. (2012) [260], Li et al. (2011), Li and Hong (2011), Benton et al. (2013), Mehdi et al. (2011), Mendonca et al. (2012), Song et al. (2014)
Tampering/dynamic flow tunneling	Fingerprinting	Insiders/security aware routing
Shin and Gu (2013), Shalimov et al. (2013)	Shin and Gu (2013)	Popa et al. (2010), Shin and Gu (2012)

Table 14.3 showed that there are some subjects such as DoS, flooding, or DNS amplification that have a significant amount of publications. In those papers, researchers showed challenges and opportunities that SDN can present in terms of those attack detections and preventions. In comparison with traditional networks that can be considered IP-based networks, SDN can be considered as flow-based networks where programs can be developed on top of the network to customize collecting flow-based statistics that can help detect and deal with those attacks at finder details' levels.

In terms of security applications, we classified those security controls into firewalls, access controls, IDS/IPS, provisioning, load balancing, policy management, traffic monitoring, wireless or mobile networks, and home or Wi-Fi networks. Some of those types can be classified as indirect security controls or security controls supporting tasks. For visibility purposes, we divided those controls among two tables: Table 14.4 and Table 14.5.

Our own classification of security controls shown in Tables 14.4 and 14.5 is proposed based on SDN literature as well as predicting future security controls and services with the evolution of SDN in particular and programmable networks in general.

Table 14.6 shows research publications related to SDN-cloud-security issues. We further classified this area into general, data centers and visualization, monitoring, orchestration, control, and migration.

R2: What tools have been used or developed in SDN? Table 14.7 shows the number of papers about used proposed and implemented tools in the SDN area. We have

Table 14.4 Security controls/applications (1)

Firewalls	Access control/VLAN/slicing/ virtualization	IDS/IPS/NIDS/NIPS/SDP
Casado et al. (2006) [212]	Nayak et al. (2009), Casado et al. (2009), Yamasaki et al. (2011), Sherwood et al. (2009)(1) [226], Sherwood et al. (2009)(2) [226], Sherwood et al. (2010a, b) [228, 233], Tootoonchian and Ganjali (2010)	Goodney et al. (2010)
Song et al. (2013a, b) [83, 235], Hu et al. (2014) (2) [211], Katta et al. (2012), Hand et al. (2013), Jia and Wang (2013), Suh et al. (2014) [18], Zhu et al. (2014), Fayaz and Sekar (2014)	Dixit et al. (2013) [261], Yazici et al. (2014), Banjar et al. (2014), Dixit et al. (2013) [261], Gutz et al. (2012), Yong-Juan et al. (2013), Kinoshita et al. (2012), Hideki et al. (2014), Dangovas and Kuliesius (2014), Wen et al. (2013)	Yu et al. (2014) [262], Kerner (2012), Hand et al. (2013), Skowyra et al. (2013)(1), Chung et al. (2013)(1,2), Yi and Zhigang (2013), Heorhiadi et al. (2012), Giotis et al. (2013)
NAT/privacy protection/ anonymity	**Provisioning/migration/hybrid networks**	**Distribution, load balancing/ scalability/fault tolerance**
Mendonca et al. (2012), Kopsel and Woesner (2011), Kotronis et al. (2013), Suñé et al. (2014), Paterson (2014), Thuemmler et al. (2013)	Bari et al. (2013)(1) [76], Levin et al. (2013), Vissicchio et al. (2014), Vanbever and Vissicchio (2014), Vanbever et al. (2013) [256], Zhang et al. (2014), Kang et al. (2012) [204]	Sharma et al. (2011), Handigol et al. (2009), Wang et al. (2011), Dixon et al. (2011), Schmid and Suomela (2013), Heorhiadi et al. (2012), Yeganeh et al. (2013) [270], Laurent et al. (2014), Reitblatt et al. (2013) [251]

divided the tools into ten types (protocol, control architecture and platform, middle box, simulation, testing, framework, programming and debugging, visualization, security, and system). Programming and debugging seem to be the hottest area in the SDNs' tool. Control architecture and platform and framework can be ranked as second as an attractive research area. Table 14.7 shows publications in SDN tools.

Cabral et al. [236] propose a protocol and technique to enhance the forwarding plane called Protocol-Oblivious Forwarding (POF). This protocol assists in reducing the network cost through employing commodity forwarding element.

An experimentation tool is presented in Voellmy et al. [237] called Mini-CCNx for the Named Data Networking (NDN). This tool is able to reproduce the experiments on the test bed for NDN through using dynamic routing protocol and multicast content delivery. SDN control architecture called Procera is expressed and explained in Qazi et al. [238, 258].

Table 14.5 Security controls/applications (2)

Policy languages and management	Traffic/BW monitoring, management/DPI
Feamster et al. (2010), Wang et al. (2012) [179], Hinrichs et al. (2008), Voellmy et al. (2012) [237], Son et al. (2013), Nayak et al. (2009), Monsanto et al. (2013), Fayazbakhsh et al. (2013) [239], Voellmy and Hudak (2011), Foster et al. (2011) [230], Foster et al. (2013) [98], Katta et al. (2012), Voellmy et al. (2013) Voellmy et al. [246], Qazi et al. (2013a, b) [238, 258], Ferguson et al. (2012), Ferguson et al. (2013), Kazemian et al. (2013), Yu et al. (2010) [272], Anderson et al. (2014) [257], Bari et al. (2013)(2) [108], Kim and Feamster (2013), Gibb et al. (2012)	Jain et al. (2013), Zaalouk et al. (2014), Wang et al. (2013a, b) (Wang et al. [139, 141]), Ballard et al. (2010), Nayak et al. (2009), Curtis et al. (2011)(2) [223], Qazi et al. (2013a, b) ([238, 258], Sun et al. (2014), Choi et al. (2014)(1) [247], Choi et al. (2014) (2) [247], Chowdhury et al. (2014) [59], Jose et al. (2011), Yu et al. (2013), Shin et al. (2012), Karame (2013), Shirali-Shahreza and Ganjali (2013)(1,2) [67], Argyropoulos et al. (2012), Giotis et al. (2013), Huang et al. (2011), Rasley et al. (2014), Raumer et al. (2014)
Wireless/mobile	Wi-Fi, home networking
Ding et al. (2014), Baldini et al. (2012), Jin and Wang (2013) [78], Hurel et al. (2014), Gember et al. (2012)(2), Staessens et al. (2011), Basta et al. (2013) [116], Namal et al. (2013) [168], Katti and Li (2014), Moradi et al. (2014), Liyanage et al. (2014), Hampel et al. (2013), Skowyra et al. (2013)(1)	McKeown et al. (2008) [215], Yap et al. (2011), Clark et al. (2009), Mehdi et al. (2011), Feamster et al. (2010), Yap et al. (2009)(1), Yap et al. (2009)(2), Schulz-Zander et al. (2014)

Table 14.6 SDN-cloud security

General	Data centers/virtualization	Monitoring
Popa et al. (2010), Benson et al. (2011), Pitt (2013), Miao et al. (2014), Wailly et al. (2011), Hurel et al. (2014), Vaughan-Nichols (2011), Carrozza et al. (2013), Tsugawa et al. (2014)	Bates et al. (2014), Wang et al. (2013a, b) (Wang et al. [139, 141]), Casado and Corn (2014), Tavakoli et al. (2009), Heller et al. (2010) [219], Curtis et al. (2011) (2) [223], Erickson et al. (2011), Moshref et al. (2013), Kang et al. (2013) [111]	Shin and Gu (2012), Wang et al. (2013a, b) [139, 141], Shin et al. (2012)
Orchestration	Controls/access management	Migration/mapping
Gember et al. (2013), Zaalouk et al. (2014)	Chung et al. (2013), Raghavendra et al. (2012), Faraji et al. (2014), Kretzschmar and Golling (2011)	Chryssa et al. (2014)

Procera contains a declarative policy language derived from the information of functional reactive programming. Fayazbakhsh et al. [239] introduce SIMPLE which is an SDN-based strategy enforcement layer intended for enhancing the middle box especially the traffic steering. In Monaco et al. [240] a new architecture is developed in SDN called flow tags. This architecture improved middle box export tags in order to supply the need and essential casual context.

Table 14.7 Software defined network tools

Tool categorization	[References]
Protocol	Cabral et al. (2013) [236], Gupta et al. (2013) [242]
Control architecture and platform	Qazi et al. (2013a, b) [238, 258], Monaco et al. (2013) [240], McGeer (2013) [241], Haw et al. (2014) [245], Nelson et al. (2014) [248], Dixit et al. (2014), Yu et al. (2014) [262]
Middle box	Fayazbakhsh et al. (2013) [239]
Simulation	Kuzniar et al. (2012) [243]
Testing	Voellmy et al. (2012) [237], Vishnoi et al. (2014) [244]
Framework	Voellmy et al. (2013) [246], Handigol et al. (2012) [255], Khan et al. (2014) [259], Jafarian et al. (2012) [260], Vestin et al. (2013) [264], Hong et al. (2013a, b) [162, 265]
Programming and debugging	Yegulalp (2013) [274], Erickson (2013) [249], Bozakov and Papadimitriou (2012) [250], Porras et al. (2012) [252], Georgopoulos et al. (2013) [254], Vanbever et al. (2013) Vanbever et al. [256], Qazi et al. (2013a, b) [238, 258], Ghobadi et al. (2012) [267]
Visualization	Reitblatt et al. (2013) [251]
Security	Monsanto et al. (2012) [253]
System	Anderson et al. (2014) [257]

In McGeer [241] a controller platform called Yanc is introduced for SDN; it depicts the state and configuration of the network as a file system which allows and permits system and user applications to cooperate and work together via standard and typical file I/Os. In Gupta et al. [242] a protocol for the safe update for OpenFlow network is explained. The protocol meets the weak flow and packet consistency conditions. In Kuzniar et al. [243] a simulation tool called FS-SDN is proposed in order to deal with the problem of evaluating and prototyping applications in SDN precisely and correctly at high level.

In Vishnoi et al. [244] a testing approach called SOFT is proposed to test the interoperability of OpenFlow switches. The main thing about SOFT is to recognize and create input test that makes the various OpenFlow implementation to act and perform in an irregularity way. In Haw et al. [245] a smart flow management policy in an OpenFlow controller system called SmartTime is introduced. It merges the proactive eviction rule flow with adaptive time-out heuristic. A framework for SDN is proposed in Voellmy et al. [246] to decrease the delivery time of the contents through enhancing network control, network management, and content delivery in Long-Term Evolution (LTE). A system called Maple is presented in Choi et al. [247] that makes SDN programming simple through using a standard programming language for determining the whole network behavior.

In Nelson et al. [248] a new architecture is proposed for SDN called software-defined unified virtual monitoring (SuVMF). This framework is used to afford and support the processes of monitoring and controlling management abstraction. An SDN controller programming language is presented in Erickson [249] called FlowLog. The main difference between FlowLog and other languages is that in OpenFlow there is a unified abstraction for the control plane tier, controller state tier, and data plane tier. Beacon is a quick, open source Java-based OpenFlow controller that assists threaded operation and event based together. It has been produced in 2010 and used in research and teaching [250].

In Reitblatt et al. [251] a visualization layer called AutoSlice is presented which computerizes the process of SDN slices and the deployment as well. In Porras et al. [252] a programming language called FatTire is presented. This language is used to write software for fault-tolerant network. FatTire allows programmers to determine the paths through the network with the required level of fault tolerance. Security software called FortNox is presented in Monsanto et al. [253]. This framework offers constraint enforcement and role-based authorization intended for the OpenFlow controller (NOX). In Georgopoulos et al. [254] declarative programming language called NetCore is defined for articulating policies on SDN regarding packet forwarding. This high-level language is compositional and communicative and includes formal semantic.

A framework is presented in Handigol et al. [255] using OpenFlow to increase the QoE fairness through increasing the technology efficiency in SDN. A debugger in SDN is proposed in Vanbever et al. [256] called NDB which is stimulated and encouraged by GDB. Two helpful primitives are implemented in NDB (backtraces and breakpoints) for debugging SDN. Hotsawp is a system used to upgrade the SDN controllers in correct manner and without disruption [257]. Hotswap preserves and retains the network events history. NetKAT is a mathematical base programming language for network presented in Qazi et al. [238, 258]. Atlas is a framework which encompasses the application awareness in SDN. It permits precise and scalable categorization of applications in SDN [259].

An SDN framework called iSDF is introduced in Jafarian et al. [260] to overcome and assist the limitations of service delivery in ISP regarding deployment flexibility, cost, operational ease, and scalability. OpenFlow Random Host Mutation (OF-RHM) is a procedure to mutate the addresses of IP host with excessive randomness and speed, while preserving the integrity of configuration and reducing the overhead operation [261]. ElastiCon is a stretchy disseminated controller architecture wherein the pool of controllers shrinks or expands dynamically according to the traffic circumstances and the load that is moved across controllers dynamically [262]. GatorCloud is an architecture for cloud resource management which facilitates sharing resources among various service models dynamically. It uses balloon abstraction to encapsulate the related resources of the services and the execution context [263]. Snap is a packet processing framework that improves packet processing in comparison with conventional software router through utilizing the available parallelism on modern GPU [264]. An SDN framework called Odin has been introduced in order to present programmability in WLANS through simplifying the client management procedures [162, 265]. In Nelson et al. [266] SWAN system is presented to improve and increase the inter-data center network utilization through directing and managing centrally the traffic of every service and reconfiguring the data plane to accommodate the existing traffic demand frequently. FlowLog is a declarative programming language for developing SDN controller programs [267]:

R3: What are the current investigated security issues in SDN?
Based on our investigation of SDN security subject in SDN, we noticed that this subject should be further divided into two sub-questions:

R3.1. What are the security problems related to SDN architecture?
Several papers discussed security problems related to SDN architecture. As a new architecture, it is expected that such architecture will pose both security challenges and opportunities.

In research papers, there are some focused areas and concerns in this regard. We can summarize them as the following:

1. Several papers showed concerns related to OpenFlow communication protocol security and how much such protocol is secure or vulnerable for external intrusion. In particular, many authors pointed out that encryption method offered in OF protocol for the communication between the control and its switches (TLS) is left optional, and in fact many developed controllers do not use it (Namal et al. 2010; Kloeti et al. 2012; Benton et al. 2013; Meyer and Schwenk 2013). OF manual is further described that users can decide their own encryption method. We think that this is a security concern mentioned in many research papers and should be handled properly in the next OF versions.
2. The controller as a central security and control is another very serious security concern described in many research papers. The concerns are not only from security perspective but also from scalability and fault tolerance perspectives. This is why distributed controllers and load balancing approaches are proposed in many SDN research papers and as we saw in previous statistics where getting a major focus.

   There are some serious concerns that if controller is compromised, then the whole network will be at risk as the controller in SDN contains the complete network picture and intelligence. In addition to distributed controller architecture proposals, there are other proposals to secure the controller and the communication with the controller through much secured encrypted channels.
3. Upper level applications or middle boxes can communicate and interact with the controller. This is another security concern where such applications can be used intentionally or unintentionally to compromise the controller or its modules. Another protocol in addition to OF should be proposed and made standard in this region to regulate the communication between upper level applications and the controller in such manner that prevent exposing controller resources.
4. SDN network includes a large number of traffic that is communicated between the controller and its switches. There are some serious concerns that DoS or flooding attacks can be made easy to flood SDN network with new flows from new sources. This makes all traffic be forwarded to the controller to make decision about which may eventually cause DoS. Effective methods are proposed to allow controller to monitor the possible occurrence of such DoS activities and be able to stop or counterattack it.
5. Middle man attacks or information leak problem is also another security concern especially as the controller sends control messages to OF switches remotely.

If this channel is compromised, controller or legitimate hosts can be impersonated which may lead to serious information leakage.

Those are examples of security concerns listed in surveyed research papers related to SDN architecture.

R3.2. What are the security opportunities SDN can bring to networking, cloud computing, etc.
As we mentioned earlier, as a new architecture, SDN is posing both security concerns and opportunities. In this section, we will focus on some of the opportunities that were mentioned in surveyed research papers:

1. Existing research papers indicated that SDN can offer the ability to deal with security controls in completely different manners in comparison with traditional security controls (e.g., [211, 230, 252]; Clark et al. 2009; Naous et al. 2009; Katta et al. 2012; Wen et al. 2013). For example, SDN programmability feature may help build customized security services on demand. In other words, the same security service can be provided to the different customers or clients differently. Attributes related to this security control can be user defined. For example, an ISP may give home users of the Internet service the ability to run customized firewalls where users can decide bandwidth limitations, websites to filter, number of users to allow, rate limit traffic in a day or a month, and many other parameters that can be customized per user. This is largely possible since SDN architecture is flow based and not IP based. Network administrators can hence have more fine-grained control on traffic compared with traditional security measures.
2. In relation to flow-based management in SDN rather than IP management, SDN can allow network security measures to rely on more specific attributes other than IP, MAC addresses, or ports typically permitted or denied in traditional firewalls or IDSs. OpenFlow earlier versions allow 12 attributes in packet headers, and new OpenFlow protocol versions (i.e., 1.2 and above) have up to 40 different attributes in which flows can be defined, categorized, or filtered. Those extra attributes are related to the exact network protocol, dealing with IP version 6 and many other new attributes that can give network administrators more control on flow management.
3. Research papers indicated that most traditional security controls will need to be revisited based on SDN to evaluate the required changes and how could those security controls be modified to optimize the usage of SDN. SDN programmability and the ability to give users more controls on switches and network traffic seem to receive conflicting opinions from security perspectives. On one side, such control is an important tool with network and security administrators to have ultimate control and management in the network. On the other hand, allowing such information to be exposed may risk such information to be compromised by illegitimate users, and hence risk on the network can be far more serious in comparison with traditional networks that hide control and routing protocols in switches.

4. Insider threats are also getting more focused in SDN security (Juba et al. 2013; Popa et al. 2010). This is since an insider, intentionally or unintentionally, can have more power and control under OpenFlow networks in comparison with traditional networks. As we mentioned earlier, such power can play in both sides, positive and negative impacts.

Those are few of the security concerns and opportunities that are discussed in research papers of SDN security subject in particular.

R4: What are the challenges of SDN?
Although SDN provides evidences in facilitating, developing, maintaining, and providing automation to network management, there are technical challenges that can limit its operation and performance in cloud computing, information technology organizations, and networking enterprises. The following are examples of the barriers mentioned and described about SDN adoption:

- SDN supports both centralized and distributed controller's models. Having both models in SDN is considered as a challenge; several articles argued about the pros and cons of the SDN centralized and distributed models, i.e., the centralized control plane pledges the consistency of network status by offering only one management point. This brings one main limitation; the controller should update OpenFlow switches more than traditional routers, which might incur overload [213, 279].
- Network visibility and management is another challenge that was addressed in the selected papers. In spite of the powerful monitoring tools that are provided by SDN, the debugging, troubleshooting, and enforcing security compliance are still considered hard missions in distributed SDN [269].
- Many research articles explored the most important scalability concerns as a challenge of SDN; they determine and discussed different metrics that can potentially be affected as the network grows [220, 223, 270–272, 275, 277]. The centralized model can increase the cost of control plane scalability. Pooling all the activities in one node requires more computation power, data storage, and throughput to manage the traffic all that could increase the response time.
- Deficiency of standards is another challenge that was addressed in the research articles. Although OpenFlow protocol provides only one specification for each version, still the variety of network hardware and software platforms drives providers and users to implement and deploy compatible OpenFlow libraries for each and every platform of OpenFlow implementation.
- Like any new technology SDN, enterprises' economic and necessary technical experts' issues could be the main limitation of building and deploying it. Robustness, resilience, and scalability are limiting the SDN deployment in terms of logic centralization warrantees. Several articles showed that SDN could reduce reliability. This is largely due to the centralization of control functions into the controller.

Table 14.8 Highlighted SDN challenges and obstacles

No.	Authors/year	Tackled SDN issues
1	Ashton et al. (2013) [298], Yazici et al. (2012), Macapuna et al. (2010) [300]	SDN reliability
2	Marsan (2012) [301]	SDN security
3	Yeganeh et al. (2013) [270], Voellmy et al. (2012) [237], Ashton et al. (2013) [298]	SDN scalability
4	Cai et al. (2010) [302], Cai et al. (2011) [402] [303]	Performance
5	Heller et al. (2012) [403] [304], Hu et al. (2012a, b) [404] [175, 305]	Controller placement

Table 14.8 shows most of the SDN challenges and limitation issue that were addressed and discussed in several surveyed papers.

R5. New Possible Opportunities of SDN
Cloud computing takes an important role in the market. The opportunity of having SDN to support cloud-based networks is investigated by several researches. The papers showed how SDN can be considered as a new supplementary technology for virtualization. In cellular network field, 35 articles out of 200 proposed SDN-based architecture as a solution for several networking issues. SDN is expected to improve how networks are developed, operated, and maintained. After scanning the selected papers, we identified the following five profits which an enterprise can gain by deploying SDN:

1. SDN provides Software-Defined Wireless Networking (SDWN) as a technology to supplement the wireless networks. It offers radio resource and mobility management, routing, and multi-home networking. Employing SDN functionalities to the relay between the home network and edge networks could solve multi-homing in wireless networks.
2. Realizing traffic offloading: Employing SDN architecture allows to aggregate offloading data centers in the mobile network and triggers the chosen traffic to these data centers without modifications to the functionality of network elements in the core mobile network.
3. New services are provided quickly and flexibly: SDN allows creating several VM instances, and the way SDNs can be set up is a far better complement to VMs than plain old physical networks.
4. Flexibility and comprehensive network management: SDN offers network experimentation tolerance. Even if one can exceed the limits forced by SNMP, the experiment can be done along with the new network configurations without being disabled by their consequences. Moreover, SDN divides the control plane (which manages the traffic) from the data plane (which forwards traffic based upon the decisions that the control plane takes).
5. Better and more granular security: VM's management in dynamic and complex environments is very tedious. SDNs can provide the kind of fine-grained security for applications, endpoints, and BYOD devices that a conventional hard-wired network cannot provide.

Table 14.9 SDN opportunities

No.	Authors/year	General description
1	Feamster et al. (2014) [278], Levin et al. (2012) [279], Nunes et al. (2014) [289]	Pros and cons of SDN centralized and distributed control models
2	Kreutz et al. (2013) [280]	Secure and dependable SDN
3	Jammala et al. (2014) [281], Jin et al. (2014), and Farhadi et al. (2014) [295]	The benefit of programmability of SDN network
4	Young-Jin Kim et al. (2014), Yeganeh et al. (2013) [270], Yu et al. (2010) [272], Jin et al. (2013) [277], Jin et al. (2003) [277], Tootoonchian et al. (2012) [271], Voellmy et al. (2012) [237], IBM (2012) [284], Zdravko Bozakov et al. (2012)	SDN scalability issues
5	Lee et al. (2014) [60], aj Jain (2012) Sun et al. (2012) [285], Pries et al. (2012) [286]	Mobile cloud computing
6	Haw et al. (2014) [245], Gember et al. (2012), Arijit Banerjee (2013), Junguk Cho (2014)	Traffic offloading in wireless network
7	HP (2012) [287], Brocade Communication (2012) [288], Bozakov et al. (2012) [250], Scott et al. (2014) [290], Kotronis et al. (2012) [291], Heller et al. (2013) [292], Agarwal et al. (2014) [293], Young-Jin Kim et al. (2014)	Device configuration and troubleshooting
8	Na et al. (2014) [296], Sivaraman et al. (2013) [297], Baker et al. (2012) [310]	SDN agility

Table 14.10 Number of published articles in each year

Year	No. of publications
2012	96
2013	207
2014	122

Table 14.9 shows most of the SDN strengths and opportunities. The table demonstrates some published or preprint articles that addressed and discussed the mentioned opportunities.

R6: What are the most popular conferences and journals publishing about SDN? Table 14.10 shows publications in SDN in the last 3 years (based on our selection, inclusion, and exclusion process).

Table 14.11 shows distribution of publications based on venues. Conferences get the large percentage of publications. As a new field, researchers want to publish their contribution early where, for example, publication in journals and magazines typically takes much longer time in comparison with conferences, workshops, or symposiums.

Table 14.11 Number of articles in each venue

Types of articles	Number of articles
Journal	24
Conference	200
Magazine	2
Symposium	47
Workshop	131
Others	51

Table 14.12 Top authors in SDN by number of publications

Author	Count	Author	Count	Author	Count
J. Rexford	17	R. Casellas	6	S. Shenker	5
N. McKeown	13	R. Martinez	6	J. Mogul	5
N. Foster	9	R. Munoz	6	A. Guha	5
A. Feldmann	8	N. Feamster	6	H. Zeng	5
B. Heller	7	D. Walker	6	V. Jeyakumar	5
M. Canini	7	M. Reitblatt	5	T. Koponen	5
		M. Yu	5		

14.6 Mapping Demographics

Because IEEE Computer Society staff will do the final formatting of your paper, some figures may have to be moved from where they appeared in the original submission. Figures and tables should be sized as they are to appear in print. Figures or tables not correctly sized will be returned to the author for reformatting.

In demographic statistics, we aggregated results from the five indexing websites. Table 14.12 shows top authors published in our specific surveyed area. In all statistics, we did not include the complete counts in the tables, but on those in the top according to a cut off we decided in each table separately.

Interestingly that while Rexford is listed as the top author in this specific area, he is the first or the only author in only two papers out of the 17 included in our collection. Table 14.13 shows the top institutions publishing in SDN area within the list of papers that we collected.

Table 14.13 shows that SDN is getting focused from universities ranked as top-ranked universities in the world. This is a typical trend for such universities focusing on new research areas. Table 14.14 shows top conferences or journals publishing in SDN. ACM SIGCOMM seems to be taking the lead in this category. We noticed however that many conferences and workshops accept papers for work in progress or for very short papers (1–2 pages). Possibly this is the trend given that this is a very new emerging area. We noticed also that publication cycle is very short and many conferences publish their papers before the time of the actual conference or event.

Table 14.13 Top institutions publishing in SDN

Institution	Count	Institution	Count
Princeton University	23	University of Illinois at Urbana-Champaign	7
Stanford University	20	University of Wisconsin Madison	7
Technical University of Berlin	14	Yale University	6
Cornell University	11	Tsinghua University	6
University of California, Berkeley	8	Carnegie Mellon University	5
Georgia Institute of Technology	8	Microsoft Research	5
University of California, San Diego	7	HP Labs	5
University of Southern California	7	Swiss Federal Institute of Technology, Lausanne	5
IBM Thomas J. Watson Research Center	7		

Table 14.14 Top SDN conferences

Publication	Count	Publication	Count
SIGCOMM **HotSDN '13**	31	**ACM CoNEXT '13**	10
EWSDN	22	SDN4FNS	10
SIGCOMM Computer Communication Review 2014	21	**CFI '14 2014**	9
SIGCOMM 2013	20	**HotNets-XII 2013**	9
ACM HotSDN '12	17	IEEE Communications Surveys	8
NOMS	15	ICC	6

14.7 Conclusion

This systematic literature review (SLR) investigated SDN literature and research dissemination. Five indexing agencies are surveyed for SDN research publications. First, we presented different challenges and opportunities that are evolving as a result of SDN emergence. Second, we highlighted active research areas in SDN according to the collected dataset.

Finally, we provided current and future research tracks in SDN. The large amount of publications in SDN given the relatively short amount of lifetime and the extensive industrial support to SDN showed that this area will continue to expand in both the academia and industry in the few coming years. SDN can act as an enabler or a steroid where most applications built on top of networks (e.g., security services, monitoring, distribution services) will have to evolve in response to the evolving architecture.

References

1. Kawashima, R., Matsuo, H.: Performance evaluation of non-tunneling edge-overlay model on 40GbE environment. In: 2014 I.E. 3rd Symposium on Network Cloud Computing and Applications (NCCA), pp. 68–74, 5–7 February 2014
2. Carrozzo, G., Monno, R., Belter, B., Krzywania, R., Pentikousis, K., Broadbent, M., Kudoh, T., Takefusa, A., Vieo-Oton, A., Fernandez, C., Puvpe, B., Tanaka, J.: Large-scale SDN experiments in federated environments. In: 2014 International Conference on Smart Communications in Network Technologies (SaCoNeT), pp. 1–6, 18–20 June 2014
3. Huang, C., Zhu, J., Luo, M., Chou, W.: A new mechanism for SDN network virtualization service. In: 2014 International Conference on Smart Communications in Network Technologies (SaCoNeT), pp. 1–6, 18–20 June 2014
4. Jmal, R., Chaari Fourati, L.: Implementing shortest path routing mechanism using Openflow POX controller. In: The 2014 International Symposium on Networks, Computers and Communications, pp. 1–6, 17–19 June 2014
5. Kawai, Y., Sato, Y., Ata, S., Huang, D., Medhi, D., Oka, I.: A database oriented management for asynchronous and consistent reconfiguration in Software-Defined Networks. In: 2014 I.E. on Network Operations and Management Symposium (NOMS), pp. 1–5, 5–9 May 2014
6. Araniti, G., Cosmas, J., Iera, A., Molinaro, A., Morabito, R., Orsino, A.: OpenFlow over wireless networks: performance analysis. In: 2014 I.E. International Symposium on Broadband Multimedia Systems and Broadcasting (BMSB), pp. 1–5, 25–27 June 2014
7. Huang, W.-Y., Chou, T.-Y., Hu, J.-W., Liu, T.-L.: Automatical end to end topology discovery and flow viewer on SDN. In: 2014 28th International Conference on Advanced Information Networking and Applications Workshops (WAINA), pp. 910–915, 13–16 May 2014
8. Gorja, P., Kurapati, R.: Extending open vSwitch to L4-L7 service aware OpenFlow switch. In: 2014 I.E. International on Advance Computing Conference (IACC), pp. 343–347, 21–22 February 2014
9. Cvijetic, N., Tanaka, A., Ji, P.N., Sethuraman, K., Murakami, S., Wang, T.: SDN and OpenFlow for dynamic flex-grid optical access and aggregation networks. J. Lightwave Technol. **32**(4), 864–870 (2014)
10. Lara, A., Kolasani, A., Ramamurthy, B.: Network innovation using OpenFlow: a survey. IEEE Commun. Surv. Tut. **16**(1), 493–512 (2014). First Quarter
11. Sato, G., Uchida, N., Shibata, Y.: Performance evaluation of PC router based cognitive wireless network for disaster-resilient WANs. In: 2014 28th International Conference on Advanced Information Networking and Applications Workshops (WAINA), pp. 611–616, 13–16 May 2014
12. Javid, T., Riaz, T., Rasheed, A.: A layer2 firewall for software defined network. In: 2014 Conference on Information Assurance and Cyber Security (CIACS), pp. 39–42, 12–13 June 2014
13. Broadbent, M., Georgopoulos, P., Kotronis, V., Plattner, B., Race, N.: OpenCache: leveraging SDN to demonstrate a customisable and configurable cache. In: 2014 I.E. Conference on Computer Communications Workshops (INFOCOM WKSHPS), pp. 151–152, 27 April 2014–2 May 2014
14. Nunes, B., Mendonca, M., Nguyen, X., Obraczka, K., Turletti, T.: A survey of Software-Defined Networking: past, present, and future of programmable networks. IEEE Commun. Surv. Tut. **PP**(99), 1–18
15. Kim, H., Kim, J., Ko, Y.-B.: Developing a cost-effective OpenFlow testbed for small-scale Software Defined Networking. In: 2014 16th International Conference on Advanced Communication Technology (ICACT), pp. 758–761, 16–19 February 2014
16. de Oliveira, R.L.S., Shinoda, A.A., Schweitzer, C.M., Rodrigues Prete, L.: Using Mininet for emulation and prototyping software-defined networks. In: 2014 I.E. Colombian Conference on Communications and Computing (COLCOM), pp. 1–6, 4–6 June 2014

17. Phemius, K., Bouet, M., Leguay, J.: DISCO: distributed multi-domain SDN controllers. In: 2014 I.E. Network Operations and Management Symposium (NOMS), pp. 1–4, 5–9 May 2014
18. Suh, M., Park, S.H., Lee, B., Yang, S.: Building firewall over the software-defined network controller. In: 2014 16th International Conference on Advanced Communication Technology (ICACT), pp. 744–748, 16–19 February 2014
19. Sher-DeCusatis, C.J., DeCusatis, C.: Developing a software defined networking curriculum through industry partnerships. In: 2014 Zone 1 Conference of the American Society for Engineering Education (ASEE Zone 1), pp. 1–7, 3–5 April 2014
20. Camillo Penna, M., Jamhour, E., Miguel, M.L.F.: A clustered SDN architecture for large scale WSON. In: 2014 I.E. 28th International Conference on Advanced Information Networking and Applications (AINA), pp. 374–381, 13–16 May 2014
21. Smith, P., Schaeffer-Filho, A., Hutchison, D., Mauthe, A.: Management patterns: SDN-enabled network resilience management. In: 2014 I.E. Network Operations and Management Symposium (NOMS), pp. 1–9, 5–9 May 2014
22. Izquierdo-Zaragoza, J.-L., Fernandez-Gambin, A., Pedreno-Manresa, J.-J., Pavon-Marino, P.: Leveraging Net2Plan planning tool for network orchestration in OpenDaylight. In: 2014 International Conference on Smart Communications in Network Technologies (SaCoNeT), pp. 1–6, 18–20 June 2014
23. Kim, E.-D., Lee, S.-I., Choi, Y., Shin, M.-K., Kim, H.-J.: A flow entry management scheme for reducing controller overhead. In: 2014 16th International Conference on Advanced Communication Technology (ICACT), pp. 754–757, 16–19 February 2014
24. Karl, M., Herfet, T.: Transparent multi-hop protocol termination. In: 2014 I.E. 28th International Conference on Advanced Information Networking and Applications (AINA), pp. 253–259, 13–16 May 2014
25. Malacarne, A., Paolucci, F., Cugini, F., Mastropaolo, A., Bottari, G., Poti, L.: Multiplexing of asynchronous and independent ASK and PSK transmissions in SDN-controlled intra-data center network. J. Lightwave Technol. **32**(9), 1794–1800 (2014)
26. Rodrigues Prete, L., Schweitzer, C.M., Shinoda, A.A., Santos de Oliveira, R.L.: Simulation in an SDN network scenario using the POX Controller. In: 2014 I.E. Colombian Conference on Communications and Computing (COLCOM), pp. 1–6, 4–6 June 2014
27. Kim, S.-M., Choi, H.-Y., Park, P.-W., Min, S.-G., Han, Y.-H.: OpenFlow-based Proxy mobile IPv6 over software defined network (SDN). In: 2014 I.E. 11th Consumer Communications and Networking Conference (CCNC), pp. 119–125, 10–13 January 2014
28. Xu, X., Zhang, H., Dai, X., Hou, Y., Tao, X., Zhang, P.: SDN based next generation mobile network with service slicing and trials. In: Communications, China, vol. 11, no. 2, pp. 65–77, February 2014
29. Dotcenko, S., Vladyko, A.; Letenko, I.: A fuzzy logic-based information security management for software-defined networks. In: 2014 16th International Conference on Advanced Communication Technology (ICACT), pp. 167–171, 16–19 February 2014
30. Malboubi, M., Wang, L., Chuah, C.-N., Sharma, P.: Intelligent SDN based traffic (de) Aggregation and Measurement Paradigm (iSTAMP). In: 2014 Proceedings IEEE INFOCOM, pp. 934–942, 27 April 2014–2 May 2014
31. Masutani, H., Nakajima, Y., Kinoshita, T., Hibi, T., Takahashi, H., Obana, K., Shimano, K., Fukui, M.: Requirements and design of flexible NFV network infrastructure node leveraging SDN/OpenFlow. In: 2014 International Conference on Optical Network Design and Modeling, pp. 258–263, 19–22 May 2014
32. Cleder Machado, C., Zambenedetti Granville, L., Schaeffer-Filho, A., Araujo Wickboldt, J.: Towards SLA policy refinement for QoS management in Software-Defined Networking. In: 2014 I.E. 28th International Conference on Advanced Information Networking and Applications (AINA), pp. 397–404, 13–16 May 2014
33. Zuo Q., Chen, M., Ding K., Xu B.: On generality of the data plane and scalability of the control plane in Software-Defined Networking. In: Communications, China, vol. 11, no. 2, pp. 55–64, February 2014

34. Bozakov, Z., Papadimitriou, P.: Towards a scalable software-defined network virtualization platform. In: 2014 I.E. Network Operations and Management Symposium (NOMS), pp. 1–8, 5–9 May 2014
35. Laga, S., Van Cleemput, T., Van Raemdonck, F., Vanhoutte, F., Bouten, N., Claeys, M., De Turck, F.: Optimizing scalable video delivery through OpenFlow layer-based routing. In: 2014 I.E. Network Operations and Management Symposium (NOMS), pp. 1–4, 5–9 May 2014
36. Munoz, R., Casellas, R., Martinez, R., Vilalta, R.: PCE: what is it, how does it work and what are its limitations? J. Lightwave Technol. **32**(4), 528–543 (2014)
37. Iyer, A.; Kumar, P., Mann, V.: Avalanche: data center Multicast using software defined networking. In: 2014 Sixth International Conference on Communication Systems and Networks (COMSNETS), pp. 1–8, 6–10 January 2014
38. Sambo, N., Meloni, G., Paolucci, F., Cugini, F., Secondini, M., Fresi, F., Poti, L., Castoldi, P.: Programmable transponder, code and differentiated filter configuration in elastic optical networks. J. Lightwave Technol. **32**(11), 2079–2086 (2014)
39. Zinner, T., Jarschel, M., Blenk, A., Wamser, F., Kellerer, W.: Dynamic application-aware resource management using Software-Defined Networking: implementation prospects and challenges. In: 2014 I.E. Network Operations and Management Symposium (NOMS), pp. 1–6, 5–9 May 2014
40. Martinello, M., Ribeiro, M.R.N., de Oliveira, R.E.Z., de Angelis Vitoi, R.: Keyflow: a prototype for evolving SDN toward core network fabrics. IEEE Netw. **28**(2), 12–19 (2014)
41. Mueller, J., Chen, Y., Reichel, B., Vlad, V., Magedanz, T.: Design and implementation of a Carrier Grade Software Defined Telecommunication Switch and Controller. In: 2014 I.E. Network Operations and Management Symposium (NOMS), pp. 1–7, 5–9 May 2014
42. Hu, F., Hao, Q., Bao, K.: A survey on Software Defined Networking (SDN) and OpenFlow: from concept to implementation. IEEE Commun. Surv. Tut. **PP**(99), 1
43. Autenrieth, A.; Szyrkowiec, T., Grobe, K., Elbers, J.-P., Kaczmarek, P., Kostecki, P., Kellerer, W.: Evaluation of virtualization models for optical connectivity service providers. In: 2014 International Conference on Optical Network Design and Modeling, pp. 264–268, 19–22 May 2014
44. Kim, W.-S., Chung, S.-H., Ahn, C.-W., Do, M.-R.: Seamless handoff and performance anomaly reduction schemes based on OpenFlow access points. In: 2014 28th International Conference on Advanced Information Networking and Applications Workshops (WAINA), pp. 316–321, 13–16 May 2014
45. Alvizu, R., Maier, G.: Can open flow make transport networks smarter and dynamic? An overview on transport SDN. In: 2014 International Conference on Smart Communications in Network Technologies (SaCoNeT), pp. 1–6, 18–20 June 2014
46. Sharma, S., Staessens, D., Colle, D., Palma, D., Goncalves, J., Pickavet, M., Cordeiro, L., Demeester, P.: Demonstrating resilient quality of service in Software Defined Networking. In: 2014 I.E. Conference on Computer Communications Workshops (INFOCOM WKSHPS), pp. 133–134, 27 April 2014–2 May 2014
47. Soltani, A.; Bazlamacci, C.F.: HyFI: hybrid flow initiation in software defined networks. In: 2014 5th International Conference on Information and Communication Systems (ICICS), pp. 1–6, 1–3 April 2014
48. Xia, W., Wen, Y., Foh, C.H., Niyato, D., Xie, H.: A survey on Software-Defined Networking. IEEE Commun. Surv. Tut. **PP**(99), 1 (2014)
49. Munoz, R., Casellas, R., Vilalta, R., Martinez, R.: Dynamic and adaptive control plane solutions for flexi-grid optical networks based on stateful PCE. J. Lightwave Technol. **32** (16), 2703–2715 (2014)
50. Awobuluyi, O.: Periodic control update overheads in OpenFlow-based enterprise networks. In: 2014 I.E. 28th International Conference on Advanced Information Networking and Applications (AINA), pp. 390–396, 13–16 May 2014
51. Chang, D., Kwak, M., Choi, N., Kwon, T., Choi, Y.: C-flow: an efficient content delivery framework with OpenFlow. In: 2014 International Conference on Information Networking (ICOIN), pp. 270–275, 10–12 February 2014

52. Sama, M.R., Ben Hadj Said, S., Guillouard, K., Suciu, L.: Enabling network programmability in LTE/EPC architecture using OpenFlow. In: 2014 12th International Symposium on Modeling and Optimization in Mobile, Ad Hoc, and Wireless Networks (WiOpt), pp. 389–396, 12–16 May 2014
53. Latifi, S., Durresi, A., Cico, B.: Emulating enterprise network environments for fast transition to software-defined networking. In: 2014 3rd Mediterranean Conference on Embedded Computing (MECO), pp. 294–297, 15–19 June 2014
54. Sgambelluri, A., Adami, D., Donatini, L., Gharbaoui, M., Martini, B., Giordano, S., Castoldi, P.: IT and network SDN orchestrator for Cloud Data Center. In: 2014 I.E. Network Operations and Management Symposium (NOMS), pp. 1–2, 5–9 May 2014
55. Roy, A.R., Bari, M.F., Zhani, M.F., Ahmed, R., Boutaba, R.: Design and management of DOT: a distributed OpenFlow Testbed. In: 2014 I.E. Network Operations and Management Symposium (NOMS), pp. 1–9, 5–9 May 2014
56. Shin, Y.Y., Kang, S.H., Kwak, J.Y., Lee, B.Y., Hyang, S.: The study on configuration of multi-tenant networks in SDN controller. In: 2014 16th International Conference on Advanced Communication Technology (ICACT), pp. 1223–1226, 16–19 February 2014
57. Ji, Y., Zhang, J., Zhao, Y., Li, H., Yang, Q., Ge, C., Xiong, Q., Xue, D., Yu, J., Qiu, S.: All Optical switching networks with energy-efficient technologies from components level to network level. IEEE J. Sel. Areas Commun. **PP**(99), 1
58. Papagianni, C., Androulidakis, G., Papavassiliou, S.: Virtual topology mapping in SDN-enabled clouds. In: 2014 I.E. 3rd Symposium on Network Cloud Computing and Applications (NCCA), pp. 62–67, 5–7 February 2014
59. Chowdhury, S.R., Bari, M.F., Ahmed, R., Boutaba, R.: PayLess: a low cost network monitoring framework for Software Defined Networks. In: 2014 I.E. Network Operations and Management Symposium (NOMS), pp. 1–9, 5–9 May 2014
60. Lee, B., Park, S.H., Shin, J., Yang, S.: IRIS: the Openflow-based recursive SDN controller. In: 2014 16th International Conference on Advanced Communication Technology (ICACT), pp. 1227–1231, 16–19 February 2014
61. Rosa, R.V., Esteve Rothenberg, C., Madeira, E.: Virtual data center networks embedding through Software Defined Networking. In: 2014 I.E. Network Operations and Management Symposium (NOMS), pp. 1–5, 5–9 May 2014
62. Zhao, Y., Zhang, J., Yang, H., Yu, X.: Data center optical networks (DCON) with OpenFlow based Software Defined Networking (SDN). In: 2013 8th International ICST Conference on Communications and Networking in China (CHINACOM), pp. 771–775, 14–16 August 2013
63. Guimaraes, C., Corujo, D., Aguiar, R.L., Silva, F., Frosi, P.: Empowering software defined wireless Networks through Media Independent Handover management. In: 2013 I.E. Global Communications Conference (GLOBECOM), pp. 2204–2209, 9–13 December 2013
64. Binczewski, A., Bogacki, W., Dolata, L., Lechert, L., Podleski, L., Przywecki, M., Oehlschlaeger, A., Dunne, J., Simeonidou, D., Zervas, G., Rofoee, B.R.: Enabling service market in metro and access networks—the ADDONAS project. In: 2013 Second European Workshop on Software Defined Networks (EWSDN), pp. 19–24, 10–11 October 2013
65. Mann, V., Kannan, K., Vishnoi, A., Iyer, A.S.: NCP: Service replication in data centers through software defined networking. In: 2013 IFIP/IEEE International Symposium on Integrated Network Management (IM 2013), pp. 561–567, 27–31 May 2013
66. Fratczak, T., Broadbent, M., Georgopoulos, P., Race, N.: HomeVisor: adapting home network environments. In: 2013 Second European Workshop on Software Defined Networks (EWSDN), pp. 32–37, 10–11 October 2013
67. Shirali-Shahreza, S., Ganjali, Y.: Empowering Software Defined Network controller with packet-level information. In: 2013 I.E. International Conference on Communications Workshops (ICC), pp. 1335–1339, 9–13 June 2013
68. Thanh, N.H., Cuong, B.D., Thien, T.D., Nam, P.N., Thu, N.Q., Huong, T.T., Nam, T.M.: ECODANE: a customizable hybrid testbed for green data center networks. In: 2013 International Conference on Advanced Technologies for Communications (ATC), pp. 312–317, 16–18 October 2013

69. Crowcroft, J., Oliver, H., Bar-Geva, Y.: Tearing down the Protocol Wall with Software Defined Networking. In: 2013 I.E. SDN for Future Networks and Services (SDN4FNS), pp. 1–9, 11–13 November 2013
70. Ben Hadj Said, S., Sama, M.R., Guillouard, K., Suciu, L., Simon, G., Lagrange, X., Bonnin, J.-M.: New control plane in 3GPP LTE/EPC architecture for on-demand connectivity service. In: 2013 I.E. 2nd International Conference on Cloud Networking (CloudNet), pp. 205–209, 11–13 November 2013
71. Gurusanthosh, P., Rostami, A., Manivasakan, R.: SDMA: a semi-distributed mobility anchoring in LTE networks. In: 2013 International Conference on Selected Topics in Mobile and Wireless Networking (MoWNeT), pp. 133–139, 19–21 August 2013
72. Lee, B.-S., Kanagavelu, R., Aung, K.M.M.: An efficient flow cache algorithm with improved fairness in Software-Defined Data Center Networks. In: 2013 I.E. 2nd International Conference on Cloud Networking (CloudNet), pp. 18–24, 11–13 November 2013
73. Lin, W.-C., Liu, G.-H., Kuo, K.-T., Wen, C.H.-P.: D2ENDIST-FM: flow migration in routing of OpenFlow-based cloud networks. In: 2013 I.E. 2nd International Conference on Cloud Networking (CloudNet), pp. 170–174, 11–13 November 2013
74. Zhao, Y., Zhang, J., Gao, L., Yang, H.: Unified control system for heterogeneous networks with Software Defined Networking (SDN). In: 2013 8th International ICST Conference on Communications and Networking in China (CHINACOM), pp. 781–784, 14–16 August 2013
75. Jarschel, M., Zinner, T., Hohn, T., Phuoc, T.-G.: On the accuracy of leveraging SDN for passive network measurements. In: 2013 Australasian Telecommunication Networks and Applications Conference (ATNAC), pp. 41–46, 20–22 November 2013
76. Bari, M.F., Chowdhury, S.R., Ahmed, R., Boutaba, R.: PolicyCop: an autonomic QoS policy enforcement framework for Software Defined Networks. In: 2013 I.E. SDN for Future Networks and Services (SDN4FNS), pp. 1–7, 11–13 November 2013
77. Shimamura, M., Yamanaka, H., Uratani, Y., Nagata, A., Ishii, S., Iida, K., Kawai, E., Tsuru, M.: Architecture for resource controllable NVE to meet service providers' dynamic QoS demands. In: 2013 Fourth International Conference on the Network of the Future (NOF), pp. 1–6, 23–25 October 2013
78. Jin, R., Wang, B.: Malware detection for mobile devices using Software-Defined Networking. In: 2013 Second GENI Research and Educational Experiment Workshop (GREE), pp. 81–88, 20–22 March 2013
79. Kanagavelu, R., Lee, B.S., Felipe Miguel, R., Dat, L.N.T., Mingjie, L.N.: Software defined network based adaptive routing for data replication in Data Centers. In: 2013 19th IEEE International Conference on Networks (ICON), pp. 1–6, 11–13 December 2013
80. Shang, Z., Chen, W., Ma, Q., Wu, B.: Design and implementation of server cluster dynamic load balancing based on OpenFlow. In: 2013 International Joint Conference on Awareness Science and Technology and Ubi-Media Computing (iCAST-UMEDIA), pp. 691–697, 2–4 November 2013
81. Zhang, Y., Beheshti, N., Beliveau, L., Lefebvre, G., Manghirmalani, R., Mishra, R., Patneyt, R., Shirazipour, M., Subrahmaniam, R., Truchan, C., Tatipamula, M.: StEERING: a software-defined networking for inline service chaining. In: 2013 21st IEEE International Conference on Network Protocols (ICNP), pp. 1–10, 7–10 October 2013
82. Sugiki, A.: An integrated management framework for virtual machines, switches, and their SDNs. In: 2013 19th IEEE International Conference on Networks (ICON), pp. 1–6, 11–13 December 2013
83. Song, S., Hong, S., Guan, X., Choi, B.-Y., Choi, C.: NEOD: Network Embedded On-line Disaster management framework for Software Defined Networking. In: 2013 IFIP/IEEE International Symposium on Integrated Network Management (IM 2013), pp. 492–498, 27–31 May 2013
84. Farias, F., Salvatti, J., Victor, P., Abelem, A.: Integrating legacy forwarding environment to OpenFlow/SDN control plane. In: 2013 15th Asia-Pacific Network Operations and Management Symposium (APNOMS), pp. 1–3, 25–27 September 2013

85. Jivorasetkul, S., Shimamura, M., Iida, K.: Better network latency with end-to-end header compression in SDN architecture. In: 2013 I.E. Pacific Rim Conference on Communications, Computers and Signal Processing (PACRIM), pp. 183–188, 27–29 August 2013
86. Nguyen, K., Minh, Q.T., Yamada, S.: Towards optimal disaster recovery in backbone networks. In: 2013 I.E. 37th Annual Computer Software and Applications Conference (COMPSAC), pp. 826–827, 22–26 July 2013
87. Yamashita, S., Tanaka, H., Hori, Y., Otani, M., Watanabe, K.: Development of network user authentication system using OpenFlow. In: 2013 Eighth International Conference on Broadband and Wireless Computing, Communication and Applications (BWCCA), pp. 566–569, 28–30 October 2013
88. Shimonishi, H., Shinohara, Y., Chiba, Y.: Vitalizing data-center networks using OpenFlow. In: 2013 I.E. Photonics Society Summer Topical Meeting Series, pp. 250–251, 8–10 July 2013
89. Simeonidou, D., Nejabati, R., Channegowda, M.P.: Software defined optical networks technology and infrastructure: enabling software-defined optical network operations. In: 2013 Optical Fiber Communication Conference and Exposition and the National Fiber Optic Engineers Conference (OFC/NFOEC), pp. 1–3, 17–21 March 2013
90. Kawasumi, R., Hirota, Y., Murakami, K., Tode, H.: Multicast distribution system with functions of time-shift and loss-recovery based on in-network caching and OpenFlow control. In: 2013 Eighth International Conference on P2P, Parallel, Grid, Cloud and Internet Computing (3PGCIC), pp. 641–646, 28–30 October 2013
91. Sgambelluri, A., Paolucci, F., Cugini, F., Valcarenghi, L., Castoldi, P.: Generalized SDN control for access/metro/core integration in the framework of the interface to the Routing System (I2RS). In: 2013 I.E. Globecom Workshops (GC Wkshps), pp. 1216–1220, 9–13 December 2013
92. Nguyen, K., Minh, Q.T., Yamada, S.: Increasing resilience of OpenFlow WANs using multipath communication. In: 2013 International Conference on IT Convergence and Security (ICITCS), pp. 1–2, 16–18 December 2013
93. Sadasivarao, A., Syed, S., Pan, P., Liou, C., Monga, I., Guok, C., Lake, A.: Bursting data between data centers: case for transport SDN. In: 2013 I.E. 21st Annual Symposium on High-Performance Interconnects (HOTI), pp. 87–90, 21–23 August 2013
94. Woesner, H., Fritzsche, D.: SDN and OpenFlow for converged access/aggregation networks. In: 2013 Optical Fiber Communication Conference and Exposition and the National Fiber Optic Engineers Conference (OFC/NFOEC), pp. 1–3, 17–21 March 2013
95. Munoz, R., Casellas, R., Martinez, R.: PCE: what is it, how does it work and what are its limitations? In: 2013 Optical Fiber Communication Conference and Exposition and the National Fiber Optic Engineers Conference (OFC/NFOEC), pp. 1–55, 17–21 March 2013
96. Kloti, R., Kotronis, V., Smith, P.: OpenFlow: a security analysis. In: 2013 21st IEEE International Conference on Network Protocols (ICNP), pp. 1–6, 7–10 October 2013
97. Vahlenkamp, M., Schneider, F., Kutscher, D., Seedorf, J.: Enabling information centric networking in IP networks using SDN. In: 2013 I.E. SDN for Future Networks and Services (SDN4FNS), pp. 1–6, 11–13 November 2013
98. Foster, N., Guha, A., Reitblatt, M., Story, A., Freedman, M.J., Katta, N.P., Monsanto, C., Reich, J., Rexford, J., Schlesinger, C., Walker, D., Harrison, R.: Languages for software-defined networks. IEEE Commun. Mag. **51**(2), 128–134 (2013)
99. Othman, M.M.O., Okamura, K.: Hybrid control model for flow-based networks. In: 2013 I.E. 37th Annual Computer Software and Applications Conference Workshops (COMPSACW), pp. 765–770, 22–26 July 2013
100. Ortiz, S.: Software-defined networking: on the verge of a breakthrough? Computer **46**(7), 10–12 (2013)
101. Katrinis, K., Wang, G., Schares, L.: SDN control for hybrid OCS/electrical datacenter networks: an enabler or just a convenience? In: 2013 I.E. Photonics Society Summer Topical Meeting Series, pp. 242–243, 8–10 July 2013

102. Ishimori, A., Farias, F., Cerqueira, E., Abelem, A.: Control of multiple packet schedulers for improving QoS on OpenFlow/SDN networking. In: 2013 Second European Workshop on Software Defined Networks (EWSDN), pp. 81–86, 10–11 October 2013
103. Meyer, D.: The software-defined-networking research group. IEEE Internet Comput. **17**(6), 84–87 (2013)
104. Pongracz, G., Molnar, L., Kis, Z.L.: Removing roadblocks from SDN: OpenFlow software switch performance on Intel DPDK. In: 2013 Second European Workshop on Software Defined Networks (EWSDN), pp. 62–67, 10–11 October 2013
105. Sanchez, R., Hernandez, J.A., Larrabeiti, D.: Using transparent WDM metro rings to provide an out-of-band control network for OpenFlow in MAN. In: 2013 15th International Conference on Transparent Optical Networks (ICTON), pp. 1–4, 23–27 June 2013
106. Park, S.M., Ju, S., Kim, J., Lee, J.: Software-defined-networking for M2M services. In: 2013 International Conference on ICT Convergence (ICTC), pp. 50–51, 14–16 October 2013
107. Sun, G., Liu, G., Zhang, H., Tan, W.: Architecture on mobility management in OpenFlow-based radio access networks. In: 2013 I.E. Global High Tech Congress on Electronics (GHTCE), pp. 88–92, 17–19 November 2013
108. Bari, M.F., Roy, A.R., Chowdhury, S.R., Zhang, Q., Zhani, M.F., Ahmed, R., Boutaba, R.: Dynamic controller provisioning in Software Defined Networks. In: 2013 9th International Conference on Network and Service Management (CNSM), pp. 18–25, 14–18 October 2013
109. Bakshi, K.: Considerations for Software Defined Networking (SDN): approaches and use cases. In: 2013 I.E. Aerospace Conference, pp. 1–9, 2–9 March 2013
110. Skoldstrom, P., John, W.: Implementation and evaluation of a carrier-grade OpenFlow virtualization scheme. In: 2013 Second European Workshop on Software Defined Networks (EWSDN), pp. 75–80, 10–11 October 2013
111. Kang, M., Kang, E.-Y., Hwang, D.-Y., Kim, B.-J., Nam, K.-H., Shin, M.-K., Choi, J.-Y.: Formal modeling and verification of SDN-OpenFlow. In: 2013 I.E. Sixth International Conference on Software Testing, Verification and Validation (ICST), pp. 481–482, 18–22 March 2013
112. Tajik, S., Rostami, A.: MultiFlow: enhancing IP multicast over IEEE 802.11 WLAN. In: 2013 IFIP Wireless Days (WD), pp. 1–8, 13–15 November 2013
113. Liu, L., Choi, H.Y., Casellas, R., Tsuritani, T., Morita, I., Martinez, R., Munoz, R.: Demonstration of a dynamic transparent optical network employing flexible transmitters/receivers controlled by an OpenFlow-stateless PCE integrated control plane [invited]. IEEE/OSA J. Opt. Commun. Networking **5**(10), A66–A75 (2013)
114. Othman, M.M.O., Okamura, K.: Enhancing control model to ease off centralized control of flow-based SDNs. In: 2013 I.E. 37th Annual Computer Software and Applications Conference (COMPSAC), pp. 467–470, 22–26 July 2013
115. Patel, A.N., Ji, P.N., Wang, T.: QoS-aware optical burst switching in OpenFlow based Software-Defined Optical Networks. In: 2013 17th International Conference on Optical Network Design and Modeling (ONDM), pp. 275–280, 16–19 April 2013
116. Basta, A., Kellerer, W., Hoffmann, M., Hoffmann, K., Schmidt, E.-D.: A virtual SDN-enabled LTE EPC architecture: a case study for S-/P-Gateways functions. In: 2013 I. E. SDN for Future Networks and Services (SDN4FNS), pp. 1–7, 11–13 November 2013
117. Natarajan, S., Ramaiah, A., Mathen, M.: A software defined Cloud-Gateway automation system using OpenFlow. In: 2013 I.E. 2nd International Conference on Cloud Networking (CloudNet), pp. 219–226, 11–13 November 2013
118. Huong, T.T., Thanh, N.H., Hung, N.T., Mueller, J., Magedanz, T.: QoE-aware resource provisioning and adaptation in IMS-based IPTV using OpenFlow. In: 2013 19th IEEE Workshop on Local & Metropolitan Area Networks (LANMAN), pp. 1–3, 10–12 April 2013
119. Zhao, Y., Zhang, J., Zhou, T., Yang, H., Gu, W., Lin, Y., Han, J., Li, G., Xu, H.: Time-aware software defined networking (Ta-SDN) for flexi-grid optical networks supporting data center application. In: 2013 I.E. Globecom Workshops (GC Wkshps), pp. 1221–1226, 9–13 December 2013

120. Ramos, R.M., Martinello, M., Esteve Rothenberg, C.: SlickFlow: resilient source routing in Data Center Networks unlocked by OpenFlow. In: 2013 I.E. 38th Conference on Local Computer Networks (LCN), pp. 606–613, 21–24 October 2013
121. Dimogerontakis, E., Vilata, I., Navarro, L.: Software Defined Networking for community network testbeds. In: 2013 I.E. 9th International Conference on Wireless and Mobile Computing, Networking and Communications (WiMob), pp. 111–118, 7–9 October 2013
122. Arefin, A., Nahrstedt, K.: Multi-stream frame rate guarantee using cross-layer synergy. In: 2013 21st IEEE International Conference on Network Protocols (ICNP), pp. 1–2, 7–10 October 2013
123. Vahlenkamp, M., Schneider, F., Kutscher, D., Seedorf, J.: Enabling ICN in IP networks using SDN. In: 2013 21st IEEE International Conference on Network Protocols (ICNP), pp. 1–2, 7–10 October 2013
124. Takagiwa, K., Ishida, S., Nishi, H.: SoR-based programmable network for future software-Defined Network. In: 2013 I.E. 37th Annual Computer Software and Applications Conference (COMPSAC), pp. 165–166, 22–26 July 2013
125. Bifulco, R., Schneider, F.: OpenFlow rules interactions: definition and detection. In: 2013 I.E. SDN for Future Networks and Services (SDN4FNS), pp. 1–6, 11–13 November 2013
126. Hu, G., Xu, K., Wu, J.: SuperFlow: a reliable, controllable and scalable architecture for large-scale enterprise networks. In: 2013 I.E. 10th International Conference on High Performance Computing and Communications & 2013 I.E. International Conference on Embedded and Ubiquitous Computing (HPCC_EUC), pp. 1195–1202, 13–15 November 2013
127. Casellas, R., Martinez, R., Munoz, R., Vilalta, R., Liu, L., Tsuritani, T., Morita, I.: Control and management of flexi-grid optical networks with an integrated stateful path computation element and OpenFlow controller [invited]. IEEE/OSA J. Opt. Commun. Networking **5**(10), A57–A65 (2013)
128. Valdivieso Caraguay, A.L., Barona Lopez, L.I., Garcia Villalba, L.J.: Evolution and challenges of Software Defined Networking. In: 2013 I.E. SDN for Future Networks and Services (SDN4FNS), pp. 1–7, 11–13 November 2013
129. Shah, S.A., Faiz, J., Farooq, M., Shafi, A., Mehdi, S.A.: An architectural evaluation of SDN controllers. In: 2013 I.E. International Conference on Communications (ICC), pp. 3504–3508, 9–13 June 2013
130. Kim, T., Lee, T., Kim, K.-H., Yeh, H., Hong, M.: An efficient packet processing protocol based on exchanging messages between switches and controller in OpenFlow networks. In: 2013 10th International Conference and Expo on Emerging Technologies for a Smarter World (CEWIT), pp. 1–5, 21–22 October 2013
131. Zhou, T., Gong X., Hu, Y., Que, X., Wang, W.: PindSwitch: a SDN-based protocol-independent autonomic flow processing platform. In: 2013 I.E. Globecom Workshops (GC Wkshps), pp. 842–847, 9–13 December 2013
132. Liu, L., Zhang, D., Tsuritani, T., Vilalta, R., Casellas, R., Hong, L., Morita, I., Guo, H., Wu, J., Martinez, R., Munoz, R.: Field trial of an OpenFlow-based unified control plane for multilayer multigranularity optical switching networks. J. Lightwave Technol. **31**(4), 506–514 (2013)
133. Rendon, O.M.C., Estrada-Solano, F., Granville, L.Z.: A mashup-based approach for virtual SDN management. In: 2013 I.E. 37th Annual Computer Software and Applications Conference (COMPSAC), pp. 143–152, 22–26 July 2013
134. Kawashima, R., Matsuo, H.: Non-tunneling edge-overlay model using OpenFlow for cloud datacenter networks. In: 2013 I.E. 5th International Conference on Cloud Computing Technology and Science (CloudCom), vol. 2, pp. 176–181, 2–5 December 2013
135. Kong, X., Wang, Z., Shi, X., Yin, X., Li, D.: Performance evaluation of software-defined networking with real-life ISP traffic. In: 2013 I.E. Symposium on Computers and Communications (ISCC), pp. 000541–000547, 7–10 July 2013
136. Channegowda, M., Nejabati, R., Simeonidou, D.: Software-defined optical networks technology and infrastructure: enabling software-defined optical network operations [invited]. IEEE/OSA J. Opt. Commun. Networking **5**(10), A274–A282 (2013)

137. Narayanan, R., Lin, G., Syed, A.A.; Shafiq, S., Gilani, F.: A framework to rapidly test SDN use-cases and accelerate middlebox applications. In: 2013 I.E. 38th Conference on Local Computer Networks (LCN), pp. 763–770, 21–24 October 2013
138. Shiraki, O., Nakagawa, Y., Hyoudou, K., Kobayashi, S., Shimizu, T.: Managing storage flows with SDN approach in I/O converged networks. In: 2013 I.E. Globecom Workshops (GC Wkshps), pp. 890–895, 9–13 December 2013
139. Wang, S.-Y., Chou, C.-L., Yang, C.-M.: EstiNet openflow network simulator and emulator. IEEE Commun. Mag. **51**(9), 110–117 (2013)
140. Hata, H.: A study of requirements for SDN switch platform. In: 2013 International Symposium on Intelligent Signal Processing and Communications Systems (ISPACS), pp. 79–84, 12–15 November 2013
141. Wang, J.-Q., Fu, H., Cao, C.: Software defined networking for telecom operators: architecture and applications. In: 2013 8th International ICST Conference on Communications and Networking in China (CHINACOM), pp. 828–833, 14–16 August 2013
142. Teixeira, J., Antichi, G., Adami, D., Del Chiaro, A., Giordano, S., Santos, A.: Datacenter in a box: test your SDN cloud-datacenter controller at home. In: 2013 Second European Workshop on Software Defined Networks (EWSDN), pp. 99–104, 10–11 October 2013
143. Kempf, J., Zhang, Y., Mishra, R., Beheshti, N.: Zeppelin—a third generation data center network virtualization technology based on SDN and MPLS. In: 2013 I.E. 2nd International Conference on Cloud Networking (CloudNet), pp. 1–9, 11–13 November 2013
144. Watashiba, Y., Hirabara, S., Date, S., Abe, H., Ichikawa, K., Kido, Y., Shimojo, S., Takemura, H.: OpenFlow network visualization software with flow control interface. In: 2013 I.E. 37th Annual Computer Software and Applications Conference (COMPSAC), pp. 475–477, 22–26 July 2013
145. Shimizu, T., Nakamura, T., Iwashina, S., Takita, W., Iwata, A.; Kiuchi, M., Kubota, Y., Ohhashi, M.: An experimental evaluation of dynamic virtualized networking resource control on an Evolved mobile core network: a new approach to reducing massive traffic congestion after a devastating disaster. In: 2013 I.E. Region 10 Humanitarian Technology Conference (R10-HTC), pp. 270–275, 26–29 August 2013
146. Azodolmolky, S., Wieder, P., Yahyapour, R.: Performance evaluation of a scalable software-defined networking deployment. In: 2013 Second European Workshop on Software Defined Networks (EWSDN), pp. 68–74, 10–11 October 2013
147. Sgambelluri, A., Giorgetti, A., Cugini, F., Paolucci, F., Castoldi, P.: OpenFlow-based segment protection in Ethernet networks. IEEE/OSA J. Opt. Commun. Networking **5**(9), 1066–1075 (2013)
148. Azodolmolky, S., Nejabati, R., Pazouki, M., Wieder, P., Yahyapour, R., Simeonidou, D.: An analytical model for software defined networking: a network calculus-based approach. In: 2013 I.E. Global Communications Conference (GLOBECOM), pp. 1397–1402, 9–13 December 2013
149. Bhattacharya, B., Das, D.: SDN based architecture for QoS enabled services across networks with dynamic service level agreement. In: 2013 I.E. International Conference on Advanced Networks and Telecommuncations Systems (ANTS), pp. 1–6, 15–18 December 2013
150. Jarschel, M., Wamser, F., Hohn, T., Zinner, T., Tran-Gia, P.: SDN-based application-aware networking on the example of YouTube video streaming. In: 2013 Second European Workshop on Software Defined Networks (EWSDN), pp. 87–92, 10–11 October 2013
151. Hock, D., Hartmann, M., Gebert, S., Jarschel, M., Zinner, T., Tran-Gia, P.: Pareto-optimal resilient controller placement in SDN-based core networks. In: 2013 25th International Teletraffic Congress (ITC), pp. 1–9, 10–12 September 2013
152. Choumas, K., Makris, N., Korakis, T., Tassiulas, L., Ott, M.: Exploiting OpenFlow resources towards a content-centric LAN. In; 2013 Second European Workshop on Software Defined Networks (EWSDN), pp. 93–98, 10–11 October 2013
153. da Cruz, M.A., Castro e Silva, L., Correa, S., Cardoso, K.V.: Accurate online detection of bidimensional Hierarchical Heavy Hitters in software-defined networks. In: 2013 I.E. Latin-America Conference on Communications (LATINCOM), pp. 1–6, 24–26 November 2013

154. Batalle, J., Ferrer Riera, J., Escalona, E., Garcia-Espin, J.A.: On the implementation of NFV over an OpenFlow infrastructure: routing function virtualization. In: 2013 I.E. SDN for Future Networks and Services (SDN4FNS), pp. 1–6, 11–13 November 2013
155. Detti, A., Pisa, C., Salsano, S., Blefari-Melazzi, N.: Wireless mesh Software Defined Networks (wmSDN). In: 2013 I.E. 9th International Conference on Wireless and Mobile Computing, Networking and Communications (WiMob), pp. 89–95, 7–9 October 2013
156. Tseng, C.-W., Yang, Y.-T., Chou, L.-D.: An IPv6-enabled Software-Defined Networking architecture. In: 2013 15th Asia-Pacific Network Operations and Management Symposium (APNOMS), pp. 1–3, 25–27 September 2013
157. Skoldstrom, P., Chial Sanchez, B.: Virtual aggregation using SDN. In: 2013 Second European Workshop on Software Defined Networks (EWSDN), pp. 56–61, 10–11 October 2013
158. Guan, X., Choi, B.-Y., Song, S.: Reliability and scalability issues in Software Defined Network frameworks. In: 2013 Second GENI Research and Educational Experiment Workshop (GREE), pp. 102–103, 20–22 March 2013
159. Arefin, A., Rivas, R., Tabassum, R., Nahrstedt, K.: OpenSession: SDN-based cross-layer multi-stream management protocol for 3D teleimmersion. In: 2013 21st IEEE International Conference on Network Protocols (ICNP), pp. 1–10, 7–10 October 2013
160. Pereini, P., Kuzniar, M., Kostic, D.: OpenFlow needs you! A call for a discussion about a cleaner OpenFlow API. In: 2013 Second European Workshop on Software Defined Networks (EWSDN), pp. 44–49, 10–11 October 2013
161. Luo, M.-Y., Chen, J.-Y.: Software Defined Networking across distributed datacenters over cloud. In: 2013 I.E. 5th International Conference on Cloud Computing Technology and Science (CloudCom), vol. 1, pp. 615–622, 2–5 December 2013
162. Hong, W., Wang, K., Hsu, Y.-H.: Application-aware resource allocation for SDN-based cloud datacenters. In: 2013 International Conference on Cloud Computing and Big Data (CloudCom-Asia), pp. 106–110, 16–19 December 2013
163. Autenrieth, A., Elbers, J.-P., Kaczmarek, P., Kostecki, P.: Cloud orchestration with SDN/OpenFlow in carrier transport networks. In: 2013 15th International Conference on Transparent Optical Networks (ICTON), pp. 1–4, 23–27 June 2013
164. Choi, H.Y., Liu, L., Tsuritani, T., Morita, I.: Demonstration of BER-adaptive WSON employing flexible transmitter/receiver with an extended OpenFlow-based control plane. IEEE Photon. Technol. Lett. **25**(2), 119–121 (2013)
165. Schwarz, M.F., Rojas, M.A.T., Miers, C.C., Redigolo, F.F., Carvalho, T.C.M.B.: Emulated and software defined networking convergence. In: 2013 IFIP/IEEE International Symposium on Integrated Network Management (IM 2013), pp. 700–703, 27–31 May 2013
166. Todorovic, M.J., Krajnovic, N.D.: Simulation analysis of SDN network capabilities. In: 2013 21st Telecommunications Forum (TELFOR), pp. 38–41, 26–28 November 2013
167. Bueno, I., Aznar, J.I., Escalona, E., Ferrer, J., Antoni Garcia-Espin, J.: An OpenNaaS based SDN framework for dynamic QoS control. In: 2013 I.E. SDN for Future Networks and Services (SDN4FNS), pp. 1–7, 11–13 November 2013
168. Namal, S., Ahmad, I., Gurtov, A., Ylianttila, M.: Enabling secure mobility with OpenFlow. In: 2013 I.E. SDN for Future Networks and Services (SDN4FNS), pp. 1–5, 11–13 November 2013
169. Kim, N., Yoo, J.-Y., Kim, N.L., Kim, J.: A programmable networking switch node with in-network processing support. In: 2012 I.E. International Conference on Communications (ICC), pp. 6611–6615, 10–15 June 2012
170. Haleplidis, E., Denazis, S., Koufopavlou, O., Halpern, J., Salim, J.H.: Software-Defined Networking: experimenting with the control to forwarding plane interface. In: 2012 European Workshop on Software Defined Networking (EWSDN), pp. 91–96, 25–26 October 2012
171. Gunter, D., Kettimuthu, R., Kissel, E., Swany, M., Yi, J., Zurawski, J.: Exploiting network parallelism for improving data transfer performance. In: 2012 SC Companion High Performance Computing, Networking, Storage and Analysis (SCC), pp. 1600–1606, 10–16 November 2012

172. Kissel, E., Fernandes, G., Jaffee, M., Swany, M., Zhang, M.: Driving Software Defined Networks with XSP. In: 2012 I.E. International Conference on Communications (ICC), pp. 6616–6621, 10–15 June 2012
173. Shirazipour, M., Zhang, Y., Beheshti, N., Lefebvre, G., Tatipamula, M.: OpenFlow and multi-layer extensions: overview and next steps. In: 2012 European Workshop on Software Defined Networking (EWSDN), pp. 13–17, 25–26 October 2012
174. Syrivelis, D., Parisis, G., Trossen, D., Flegkas, P., Sourlas, V., Korakis, T., Tassiulas, L.: Pursuing a Software Defined Information-centric network. In: 2012 European Workshop on Software Defined Networking (EWSDN), pp. 103–108, 25–26 October 2012
175. Hu, Y., Wang, W., Gong, X., Que, X., Cheng, S.: BalanceFlow: controller load balancing for OpenFlow networks. In: 2012 I.E. 2nd International Conference on Cloud Computing and Intelligent Systems (CCIS), vol. 02, pp. 780–785, 30 October 2012–1 November 2012
176. Kawashima, R.: vNFC: a virtual networking function container for SDN-enabled virtual networks. In: 2012 Second Symposium on Network Cloud Computing and Applications (NCCA), pp. 124–129, 3–4 December 2012
177. de Oliveira Silva, F., de Souza Pereira, J.H., Rosa, P.F., Kofuji, S.T.: Enabling future internet architecture research and experimentation by using Software Defined Networking. In: 2012 European Workshop on Software Defined Networking (EWSDN), pp. 73–78, 25–26 October 2012
178. Ujcich, B., Wang, K.-C., Parker, B., Schmiedt, D.: Thoughts on the Internet architecture from a modern enterprise network outage. In: 2012 I.E. Network Operations and Management Symposium (NOMS), pp. 494–497, 16–20 April 2012
179. Wang, W., Hu, Y., Que, X., Gong, X.: Autonomicity design in OpenFlow based Software Defined Networking. In: 2012 I.E. Globecom Workshops (GC Wkshps), pp. 818–823, 3–7 December 2012
180. Lara, A., Kolasani, A., Ramamurthy, B.: Simplifying network management using Software Defined Networking and OpenFlow. In: 2012 I.E. International Conference on Advanced Networks and Telecommuncations Systems (ANTS), pp. 24–29, 16–19 December 2012
181. Narayan, S., Bailey, S., Daga, A: Hadoop acceleration in an OpenFlow-based cluster. In: 2012 SC Companion High Performance Computing, Networking, Storage and Analysis (SCC), pp. 535–538, 10–16 November 2012
182. Narayanan, R., Kotha, S., Lin, G., Khan, A., Rizvi, S., Javed, W., Khan, H., Khayam, S.A.: Macroflows and microflows: enabling rapid network innovation through a split SDN data plane. In: 2012 European Workshop on Software Defined Networking (EWSDN), pp. 79–84, 25–26 October 2012
183. Jarschel, M., Lehrieder, F., Magyari, Z., Pries, R.: A flexible OpenFlow-controller benchmark. In: 2012 European Workshop on Software Defined Networking (EWSDN), pp. 48–53, 25–26 October 2012
184. Shirazipour, M., John, W., Kempf, J., Green, H., Tatipamula, M.: Realizing packet-optical integration with SDN and OpenFlow 1.1 extensions. In: 2012 I.E. International Conference on Communications (ICC), , pp. 6633–6637, 10–15 June 2012
185. Risdianto, A.C., Mulyana, E.: Implementation and analysis of control and forwarding plane for SDN. In: 2012 7th International Conference on Telecommunication Systems, Services, and Applications (TSSA), pp. 227–231, 30–31 October 2012
186. Veltri, L., Morabito, G., Salsano, S., Blefari-Melazzi, N., Detti, A.: Supporting information-centric functionality in software defined networks. In: 2012 I.E. International Conference on Communications (ICC), pp. 6645–6650, 10–15 June 2012
187. Bennesby, R., Fonseca, P., Mota, E., Passito, A: An inter-AS routing component for software-defined networks. In: 2012 I.E. Network Operations and Management Symposium (NOMS), pp. 138–145, 16–20 April 2012
188. Sonkoly, B., Gulyas, A., Nemeth, F., Czentye, J., Kurucz, K., Novak, B., Vaszkun, G.: On QoS support to Ofelia and OpenFlow. In: 2012 European Workshop on Software Defined Networking (EWSDN), pp. 109–113, 25–26 October 2012

189. Egilmez, H.E., Dane, S.T., Bagci, K.T., Tekalp, A.M.: OpenQoS: an OpenFlow controller design for multimedia delivery with end-to-end Quality of Service over Software-Defined Networks. In: 2012 Asia-Pacific Signal & Information Processing Association Annual Summit and Conference (APSIPA ASC), pp. 1–8, 3–6 December 2012
190. Kind, M., Westphal, F., Gladisch, A., Topp, S.: SplitArchitecture: applying the Software Defined Networking concept to carrier networks. In: 2012 World Telecommunications Congress (WTC), pp. 1–6, 5–6 March 2012
191. Ruth, P., Mandal, A., Xin, Y., Baldine, I., Heerman, C., Chase, J.: Dynamic network provisioning for data intensive applications in the cloud. In: 2012 I.E. 8th International Conference on E-Science (e-Science), pp. 1–2, 8–12 October 2012
192. Fonseca, P., Bennesby, R., Mota, E., Passito, A.: A replication component for resilient OpenFlow-based networking. In: 2012 I.E. Network Operations and Management Symposium (NOMS), pp. 933–939, 16–20 April 2012
193. Paul, S., Jain, R.: OpenADN: mobile apps on global clouds using OpenFlow and Software Defined Networking. In: 2012 I.E. Globecom Workshops (GC Wkshps), pp. 719–723, 3–7 December 2012
194. de Oliveira Silva, F., Goncalves, M.A., de Souza Pereira, J.H., Pasquini, R., Rosa, P.F., Kofuji, S.T.: On the analysis of multicast traffic over the Entity Title Architecture. In: 2012 18th IEEE International Conference on Networks (ICON), , pp. 30–35, 12–14 December 2012
195. Devlic, A., John, W., Skoldstrom, P.: A use-case based analysis of network management functions in the ONF SDN model. In: 2012 European Workshop on Software Defined Networking (EWSDN), pp. 85–90, 25–26 October 2012
196. Kawai, E.: Can SDN Help HPC?. In: 2012 IEEE/IPSJ 12th International Symposium on Applications and the Internet (SAINT), pp. 210–210, 16–20 July 2012
197. Monga, I., Pouyoul, E., Guok, C.: Software-Defined Networking for big-data science—architectural models from campus to the WAN. In: 2012 SC Companion High Performance Computing, Networking, Storage and Analysis (SCC), pp. 1629–1635, 10–16 November 2012
198. Das, S., Parulkar, G., McKeown, N.: Why OpenFlow/SDN can succeed where GMPLS failed. In: 2012 38th European Conference and Exhibition on Optical Communications (ECOC), pp. 1–3, 16–20 September 2012
199. Matias, J., Tornero, B., Mendiola, A., Jacob, E., Toledo, N.: Implementing layer 2 network virtualization using OpenFlow: challenges and solutions. In: 2012 European Workshop on Software Defined Networking (EWSDN), pp. 30–35, 25–26 October 2012
200. Naudts, B., Kind, M., Westphal, F., Verbrugge, S., Colle, D., Pickavet, M.: Techno-economic analysis of Software Defined Networking as architecture for the virtualization of a mobile network. In: 2012 European Workshop on Software Defined Networking (EWSDN), pp. 67–72, 25–26 October 2012
201. Kempf, J., Johansson, B., Pettersson, S., Luning, H., Nilsson, T.: Moving the mobile Evolved Packet Core to the cloud. In: 2012 I.E. 8th International Conference on Wireless and Mobile Computing, Networking and Communications (WiMob), pp. 784–791, 8–10 October 2012
202. Jeong, K., Kim, J., Kim, Y.-T.: QoS-aware network operating system for software defined networking with generalized OpenFlows. In: 2012 I.E. Network Operations and Management Symposium (NOMS), pp. 1167–1174, 16–20 April 2012
203. Bifulco, R., Canonico, R., Brunner, M., Hasselmeyer, P., Mir, F.: A practical experience in designing an OpenFlow controller. In: 2012 European Workshop on Software Defined Networking (EWSDN), pp. 61–66, 25–26 October 2012
204. Kang, M., Park, J., Choi, J.-Y., Nam, K.-H., Shin, M.-K.: Process algebraic specification of Software Defined Networks. In: 2012 Fourth International Conference on Computational Intelligence, Communication Systems and Networks (CICSyN), pp. 359–363, 24–26 July 2012
205. Lin, H., Sun, L., Fan, Y., Guo, S.: Apply embedded openflow MPLS technology on wireless Openflow—OpenRoads. In: 2012 2nd International Conference on Consumer Electronics, Communications and Networks (CECNet), pp. 916–919, 21–23 April 2012

206. Boughzala, B., Ben Ali, R., Lemay, M., Lemieux, Y., Cherkaoui, O.: OpenFlow supporting inter-domain virtual machine migration. In: 2011 Eighth International Conference on Wireless and Optical Communications Networks (WOCN), pp. 1–7, 24–26 May 2011
207. Scott-Hayward, S., O'Callaghan, G., Sezer, S.: SDN security: a survey in 2013 I.E. SDN for Future Networks and Services (SDN4FNS).
208. Jarraya, Y., Madi, T., Debbabi, M.: A survey and a layered taxonomy of Software-Defined Networking. IEEE Commun. Surv. Tut. **16**(1) (2014)
209. Mendonca, M., Nunes, B.A.A., Nguyen, X.-N., Obraczka, K., Turletti, T.: A survey of Software-Defined Networking: past, present, and future of programmable networks. hal-00825087, version 2 (2013)
210. Kreutz, D., Ramos, F.M.V., Verissimo, P., Rothenberg, C.E., Azodolmolky, S., Uhlig, S.: Software-Defined Networking: a comprehensive survey. Proc. IEEE **104**, 14–76 (2014)
211. Hu, F., Hao, Q., Bao, K.: A survey on Software-Defined Network (SDN) and OpenFlow: from concept to implementation. IEEE Commun. Surv. Tut. (2014)
212. Casado, M., Garfinkel, T., Akella, A., Freedman, M.J., Boneh, D., McKeown, N., Shenker, S.: SANE: a protection architecture for enterprise networks. In: Proceedings of the 15th Conference on USENIX Security Symposium, vol. 15, ser. USENIX-SS'06, Berkeley, CA, USA (2006)
213. Feamster, N., Balakrishnan, H., Rexford, J., Shaikh, A., van der Merwe, J.: The case for separating routing from routers. In: Proceedings of the ACM SIGCOMM Workshop on Future Directions in Network Architecture, ser. FDNA'04. ACM, New York, NY, USA, pp. 5–12 (2004)
214. Casado, M., Freedman, M.J., Pettit, J., Luo, J., McKeown, N., Shenker, S.: Ethane: taking control of the enterprise. In: SIGCOMM, pp. 1–12 (2007)
215. McKeown, N., Anderson, T., Balakrishnan, H., Parulkar, G.M., Peterson, L.L., Rexford, J., Shenker, S., Turner, J.S.: OpenFlow: enabling innovation in campus networks. Comput. Commun. Rev. **38**(2), 69–74 (2008)
216. Gude, N., Koponen, T., Pettit, J., Pfaff, B., Casado, M., McKeown, N.: Scott Shenker: NOX: towards an operating system for networks. Comput. Commun. Rev. **38**(3), 105–110 (2008)
217. Benson, T., Akella, A., Maltz, D.A.: Network traffic characteristics of data centers in the wild. In: Internet Measurement Conference, pp. 267–280 (2010)
218. Mysore, R.N., Pamboris, A., Farrington, N., Huang, N., Miri, P., Radhakrishnan, S., Subramanya, V., Vahdat, A.: PortLand: a scalable fault-tolerant layer 2 data center network fabric. In: SIGCOMM, pp. 39–50 (2009)
219. Heller, B., Seetharaman, S., Mahadevan, P., Yiakoumis, Y., Sharma, P., Banerjee, S., McKeown, N.: ElasticTree: saving energy in data center networks. In: NSDI, pp. 249–264 (2010)
220. Koponen, T., Casado, M., Gude, N., Stribling, J., Poutievski, L., Zhu, M., Ramanathan, R., Iwata, Y., Inoue, H., Hama, T., Shenker, S.: Onix: a distributed control platform for large-scale production networks. In: OSDI, pp. 351–364 (2010)
221. Dobrescu, M., Egi, N., Argyraki, K.J., Chun, B.-G., Fall, K.R., Iannaccone, G., Knies, A., Manesh, M., Ratnasamy, S.: RouteBricks: exploiting parallelism to scale software routers. In: SOSP, pp. 15–28 (2009)
222. Han, S., Jang, K., Park, K., Moon, S.B.: PacketShader: a GPU-accelerated software router. In: SIGCOMM, pp. 195–206 (2010)
223. Curtis, A.R., Mogul, J.C., Tourrilhes, J., Yalagandula, P., Sharma, P., Banerjee, S.: DevoFlow: scaling flow management for high-performance networks. In: SIGCOMM, pp. 254–265 (2011)
224. Farrington, N., Porter, G., Radhakrishnan, S., Bazzaz, H.H., Subramanya, V., Fainman, Y., Papen, G., Vahdat, A.: Helios: a hybrid electrical/optical switch architecture for modular data centers. In: SIGCOMM, pp. 339–350 (2010)
225. Lantz, B., Heller, B., McKeown, N.: A network in a laptop: rapid prototyping for software-defined networks. In: HotNets, p. 19 (2010)

226. Sherwood, R., Gibb, G., Yap, K.-K., Appenzeller, G., Casado, M., McKeown, N., Parulkar, G.: FlowVisor: a network virtualization. Layer. Tech. Rep. OPENFLOW-TR-2009-01, OpenFlow Consortium, October 2009
227. Guo, C., Lu, G., Wang, H.J., Yang, S., Kong, C., Sun, P., Wu, W., Zhang, Y.: SecondNet: a data center network virtualization architecture with bandwidth guarantees. In: CoNEXT, p. 15 (2010)
228. Sherwood, R., Gibb, G., Yap, K.-K., Appenzeller, G., Casado, M., McKeown, N., Parulkar, G.M.: Can the production network be the Testbed? In: OSDI, pp. 365–378 (2010)
229. Ganjali, Y., Tootoonchian, A.: HyperFlow: a distributed control plane for OpenFlow. In: INM/WREN (2008)
230. Foster, N., Harrison, R., Freedman, M.J., Monsanto, C., Rexford, J., Story, A., Walker, D.: Frenetic: a network programming language. In: ICFP, pp. 279–291 (2011)
231. Reitblatt, M., Foster, N., Rexford, J., Schlesinger, C., Walker, D.: Abstractions for network update. In: SIGCOMM, pp. 323–334 (2012)
232. Kazemian, P., Varghese, G., McKeown, N.: Header space analysis: static checking for networks. In: NSDI, pp. 113–126 (2012)
233. Sherwood, R., Chan, M., Adam Covington, G., Gibb, G., Flajslik, M., Handigol, N., Huang, T.-Y., Kazemian, P., Kobayashi, M., Naous, J., Seetharaman, S., Underhill, D., Yabe, T., Yap, K.-K., Yiakoumis, Y., Zeng, H., Appenzeller, G., Johari, R., McKeown, N., Parulkar, G. M.: Carving research slices out of your production networks with OpenFlow. Comput. Commun. Rev. **40**(1), 129–130 (2010)
234. Khurshid, A., Zou, X., Zhou, W., Caesar, M., Brighten Godfrey, P.: VeriFlow: verifying network-wide invariants in real time. In: NSDI, pp. 15–27 (2013)
235. Song, H.: Protocol-oblivious forwarding: unleash the power of SDN through a future-proof forwarding plane. In: Proceedings of the Second ACM SIGCOMM Workshop on Hot Topics in Software Defined Networking (HotSDN '13). ACM, New York, NY, USA, pp. 127–132. doi:10.1145/2491185.2491190 (2013)
236. Cabral, C.M.S., Rothenberg, C.E., Magalhes, M.F.: Reproducing real NDN experiments using mini-CCNx. In: Proceedings of the 3rd ACM SIGCOMM Workshop on Information-Centric Networking (ICN '13). ACM, New York, NY, USA, pp. 45–46. doi:10.1145/2491224.2491242 (2013)
237. Voellmy, A., Kim, H., Feamster, N.: Procera: a language for high-level reactive network control. In: Proceedings of the First Workshop on Hot Topics in Software Defined Networks (HotSDN '12). ACM, New York, NY, USA, pp. 43–48. doi:10.1145/2342441.2342451 (2012)
238. Qazi, Z.A., Tu, C.-C., Chiang, L., Miao, R., Sekar, V., Yu, M.: SIMPLE-fying middlebox policy enforcement using SDN. SIGCOMM Comput. Commun. Rev. **43**(4), 27–38 (2013). doi:10.1145/2534169.2486022
239. Fayazbakhsh, S.K., Sekar, V., Yu, M., Mogul, J.C.: FlowTags: enforcing network-wide policies in the presence of dynamic middlebox actions. In: Proceedings of the Second ACM SIGCOMM Workshop on Hot Topics in Software Defined Networking (HotSDN '13). ACM, New York, NY, USA, pp. 19–24. doi:10.1145/2491185.2491203 (2013)
240. Monaco, M., Michel, O., Keller, E.: Applying operating system principles to SDN controller design. In: Proceedings of the Twelfth ACM Workshop on Hot Topics in Networks (HotNets-XII). ACM, New York, NY, USA, Article 2, 7 pages. doi:10.1145/2535771.2535789 (2013)
241. McGeer, R.: A correct, zero-overhead protocol for network updates. In: Proceedings of the Second ACM SIGCOMM Workshop on Hot Topics in Software Defined Networking (HotSDN '13). ACM, New York, NY, USA, pp. 161–162. doi:10.1145/2491185.2491217 (2013)
242. Gupta, M., Sommers, J., Barford, P.: Fast, accurate simulation for SDN prototyping. In: Proceedings of the Second ACM SIGCOMM Workshop on Hot Topics in Software Defined Networking (HotSDN '13). ACM, New York, NY, USA, pp. 31–36. doi:10.1145/2491185.2491202 (2013)

243. Kuzniar, M., Peresini, P., Canini, M., Venzano, D., Kostic, D.: A SOFT way for openflow switch interoperability testing. In: Proceedings of the 8th International Conference on Emerging Networking Experiments and Technologies (CoNEXT '12). ACM, New York, NY, USA, pp. 265–276. doi:10.1145/2413176.2413207 (2012)
244. Vishnoi, A., Poddar, R., Mann, V., Bhattacharya, S., Effective switch memory management in OpenFlow networks. In: Proceedings of the 8th ACM International Conference on Distributed Event-Based Systems (DEBS '14). ACM, New York, NY, USA, pp. 177–188. doi:10.1145/2611286.2611301 (2014)
245. Haw, R., Hong, C.S., Lee, S.: An efficient content delivery framework for SDN based LTE network. In: Proceedings of the 8th International Conference on Ubiquitous Information Management and Communication (ICUIMC '14). ACM, New York, NY, USA, Article 71, 6 pages. doi:10.1145/2557977.2558087 (2014)
246. Voellmy, A., Wang, J., Yang, Y.R., Ford, B., Hudak, P.: Maple: simplifying SDN programming using algorithmic policies. SIGCOMM Comput. Commun. Rev. **43**(4), 87–98 (2013). doi:10.1145/2534169.2486030
247. Choi, T., Kang, S., Yoon, S., Yang, S., Song, S., Park, H.: SuVMF: software-defined unified virtual monitoring function for SDN-based large-scale networks. In: Proceedings of the Ninth International Conference on Future Internet Technologies (CFI '14). ACM, New York, NY, USA, Article 4, 6 pages. doi:10.1145/2619287.2619299 (2014)
248. Nelson, T., Ferguson, A.D., Scheer, M.J.G., Krishnamurthi, S.: Tierless programming and reasoning for software-defined networks. In: Proceedings of the 11th USENIX Conference on Networked Systems Design and Implementation (NSDI'14). USENIX Association, Berkeley, CA, USA, pp. 519-531 (2014)
249. Erickson, D.: The beacon openflow controller. In: Proceedings of the Second ACM SIGCOMM Workshop on Hot Topics in Software Defined Networking (HotSDN '13). ACM, New York, NY, USA, pp. 13–18. doi:10.1145/2491185.2491189 (2013)
250. Bozakov, Z., Papadimitriou, P.: AutoSlice: automated and scalable slicing for software-defined networks. In: Proceedings of the 2012 ACM Conference on CoNEXT Student Workshop (CoNEXT Student '12). ACM, New York, NY, USA, pp. 3–4. doi:10.1145/2413247.2413251 (2012)
251. Reitblatt, M., Canini, M., Guha, A., Foster, N.: FatTire: declarative fault tolerance for software-defined networks. In: Proceedings of the Second ACM SIGCOMM Workshop on Hot Topics in Software Defined Networking</em > (HotSDN '13). ACM, New York, NY, USA, pp. 109–114. doi:10.1145/2491185.2491187 (2013)
252. Porras, P., Shin, S., Yegneswaran, V., Fong, M., Tyson, M., Gu, G.: A security enforcement kernel for OpenFlow networks. In: Proceedings of the First Workshop on Hot Topics in Software Defined Networks (HotSDN '12). ACM, New York, NY, USA, pp. 121–126. doi:10.1145/2342441.2342466 (2012)
253. Monsanto, C., Foster, N., Harrison, R., Walker, D.: A compiler and run-time system for network programming languages. In: Proceedings of the 39th Annual ACM SIGPLAN-SIGACT Symposium on Principles of Programming Languages (POPL '12). ACM, New York, NY, USA, pp. 217–230. doi:10.1145/2103656.2103685 (2012)
254. Georgopoulos, P., Elkhatib, Y., Broadbent, M., Mu, M., Race, N.: Towards network-wide QoE fairness using openflow-assisted adaptive video streaming. In: Proceedings of the 2013 ACM SIGCOMM Workshop on Future Human-Centric Multimedia Networking (FhMN '13). ACM, New York, NY, USA, pp. 15–20. doi:10.1145/2491172.2491181 (2013)
255. Handigol, N., Heller, B., Jeyakumar, V., Mazires, D., McKeown, N.: Where is the debugger for my software-defined network?. In: Proceedings of the First Workshop on Hot Topics in Software Defined Networks (HotSDN '12). ACM, New York, NY, USA, pp. 55–60. doi:10.1145/2342441.2342453 (2012)
256. Vanbever, L., Reich, J., Benson, T., Foster, N., Rexford, J.: HotSwap: correct and efficient controller upgrades for software-defined networks. In: Proceedings of the Second ACM SIGCOMM Workshop on Hot Topics in Software Defined Networking (HotSDN '13). ACM, New York, NY, USA, pp. 133–138. doi:10.1145/2491185.2491194 (2013)

257. Anderson, C.J., Foster, N., Guha, A., Jeannin, J.-B., Kozen, D., Schlesinger, C., Walker, D.: NetKAT: semantic foundations for networks. In: Proceedings of the 41st ACM SIGPLAN-SIGACT Symposium on Principles of Programming Languages (POPL '14). ACM, New York, NY, USA, pp. 113–126. doi:10.1145/2535838.2535862 (2014)
258. Qazi, Z.A., Lee, J., Jin, T., Bellala, G., Arndt, M., Noubir, G.: Application-awareness in SDN. SIGCOMM Comput. Commun. Rev. **43**(4), 487–488 (2013). doi:10.1145/2534169.2491700
259. Khan, K.R., Ahmed, Z., Ahmed, S., Syed, A., Khayam, S.A.: Rapid and scalable isp service delivery through a programmable middlebox. SIGCOMM Comput. Commun. Rev. **44**(3), 31–37 (2014). doi:10.1145/2656877.2656882
260. Jafarian, J.H., Al-Shaer, E., Duan, Q.: Openflow random host mutation: transparent moving target defense using software defined networking. In: Proceedings of the First Workshop on Hot Topics in Software Defined Networks (HotSDN '12). ACM, New York, NY, USA, pp. 127–132. doi:10.1145/2342441.2342467 (2012)
261. Dixit, A., Hao, F., Mukherjee, S., Lakshman, T.V., Kompella, R.: Towards an elastic distributed SDN controller. SIGCOMM Comput. Commun. Rev. **43**(4), 7–12 (2013). doi:10.1145/2534169.2491193
262. Yu, Z., Li, M., Liu, Y., Li, X. GatorCloud: a fine-grained and dynamic resourcesharing architecture for multiple cloud services. In: Proceedings of the 2014 ACM International Workshop on Software-Defined Ecosystems (BigSystem '14). ACM, New York, NY, USA, pp. 13–20. doi:10.1145/2609441.2609640 (2014)
263. Sun, W., Ricci, R.: Fast and flexible: parallel packet processing with GPUs and click. In: Proceedings of the Ninth ACM/IEEE Symposium on Architectures for Networking and Communications Systems (ANCS '13). IEEE Press, Piscataway, NJ, USA, pp. 25–36 (2013)
264. Vestin, J., Dely, P., Kassler, A., Bayer, N., Einsiedler, H., Peylo, C.: CloudMAC: towards software defined WLANs. SIGMOBILE Mob. Comput. Commun. Rev. **16**(4), 42–45 (2013). doi:10.1145/2436196.2436217
265. Hong, C.-Y., Kandula, S., Mahajan, R., Zhang, M., Gill, V., Nanduri, M., Wattenhofer, R.: Achieving high utilization with software-driven WAN. SIGCOMM Comput. Commun. Rev. **43**(4), 15–26 (2013). doi:10.1145/2534169.2486012
266. Nelson, T., Guha, A., Dougherty, D.J., Fisler, K., Krishnamurthi, S.: A balance of power: expressive, analyzable controller programming. In: Proceedings of the Second ACM SIGCOMM workshop on Hot Topics in Software Defined Networking (HotSDN '13). ACM, New York, NY, USA, pp. 79–84. doi:10.1145/2491185.2491201 (2013)
267. Ghobadi, M., Yeganeh, S.H., Ganjali, Y.: Rethinking end-to-end congestion control in software-defined networks. In: Proceedings of the 11th ACM Workshop on Hot Topics in Networks (HotNets-XI). ACM, New York, NY, USA, pp. 61–66. doi:10.1145/2390231.2390242 (2012)
268. Sezer, S., Scott-Hayward, S., Chouhan, P.K., Fraser, B., Lake, D., Finnegan, J., Viljoen, N., Miller, M., Rao, N.: Are we ready for SDN? Implementation challenges for software-defined networks. IEEE Commun. Mag. **51**(7), 36–43 (2013). doi:10.1109/MCOM.2013.6553676
269. Hakiria, A., Gokhale, A., Berthoua, P., Schmidt, D.C., Thierrya. G.: Software-defined Networking: challenges and research opportunities for FutureInternet. Preprint submitted to Elsevier 30 July 2014
270. Yeganeh, S.H., Tootoonchian, A., Ganjali, Y.: On scalability of software-defined networking. IEEE Commun. Mag. **51**(2), 136–141 (2013). doi:10.1109/MCOM.2013.6461198
271. Tootoonchian, A. et al.: On controller performance in Software-Defined Networks. In: Proc. USENIX Hot-ICE'12, p. 10 (2012)
272. Yu, M. et al.: Scalable flow-based networking with DIFANE. In: Proc. ACM SIGCOMM 2010 Conf., pp. 351–362 (2010)
273. Banafa, A.: Software-Defined Networking (SDN): an opportunity ? a distinguished tenured faculty in Heald College, 30 March 2014
274. Yegulalp, S.: five sdn benefits enterprises should consider, 12, July 2013. http://www.networkcomputing.com/networking/five-sdn-benefits-enterprises-should-consider/a/d-id/1234292? Accessed 12 September 2014

275. Kepes, B., SDN meets the real-world: implementation benefits and challenges, Gigaom Research, 29 May 2014. http://www.nuagenetworks.net/wp-content/uploads/2014/06/Gigaom-Research-SDN-Meets-the-Real-World-Final.pdf. Accessed 12 September 2014
276. Li, L.E., Mao, Z.M., Rexford, J.: Toward Software-Defined Cellularnetworks. In: Proceedings of IEEE EWSDN (2012)
277. Jin, X., Li, L.E., Vanbever, L., Rexford, J.: SoftCell: Scalable and FlexibleCellular Core Network Architecture. In: Proceedings of ACM CoNEXT2013 (2013)
278. Feamster, N., Rexford, J., Zegura, E.: The road to SDN: an intellectual history of programmable networks. SIGCOMM Comput. Commun. Rev. **44**(2), 87–98 (2014)
279. Levin, D., Wundsam, A., Heller, B., Handigol, N., Feldmann, A.: Logically centralized?: state distribution trade-offs in software defined networks. In: Proceedings of the First Workshop on Hot Topics in Software Defined Networks (HotSDN '12). ACM, New York, NY, USA (2012)
280. Kreutz, D., Ramos, F.M.V., Verissimo, P.: Towards secure and dependable software-defined networks. In: Proceedings of the Second ACM SIGCOMM Workshop on Hot Topics in Software Defined Networking (HotSDN '13). ACM, New York, NY, USA (2013)
281. Jammala, M., Singha, T., Shamia, A., Asal, R., Li, Y.: Software-Defined Networking: state of the art and research challenges. Submitted for review and possible publication in Elsevier's Journal of Computer Networks (2014)
282. Kim, Y.-J., He, K., Thottan, M., Deshpande, J.G.: Self-configurable and scalable utility communications enabled by software-defined networks. In: Proceedings of the 5th International Conference On Future Energy Systems (e-Energy '14). ACM, New York, NY, USA (2014)
283. Voellmy, A., Wang, J.: Scalable software defined network controllers. SIGCOMM Comput. Commun. Rev. **42**(4), 289–290 (2012)
284. IBM, Software-Defined Networking: a new paradigm for virtual, dynamic, flexible networking, October 2012
285. Sun, L., Suzuki, K., Yasunobu, C., Hatano, Y., Shimonishi, H.: A network management solution based on OpenFlow towards new challenges of multitenant data centers. In: Proceedings, 2012 Ninth Asia Pacific Symposium on Information and Telecommunication Technologies (APSITT), pp. 1–6, 5–9 November 2012
286. Pries, R., Jarschel, M., Goll, S.: On the usability of OpenFlow in data center environments. In: Proceedings, 2012 I.E. International Conference on Communications (ICC), pp. 5533–5537, 10–15 June 2012
287. HP: Realizing the power of SDN with HP virtual application networks. http://h17007.www1.hp.com/docs/interopny/4AA4-3871ENW.pdf (2012)
288. Brocade communications systems: network transformation with software-defined networking and Ethernet fabrics, CA, USA (2012)
289. Nunes, B.A.A., Mendonca, M., Nguyen, X.-N., Obraczka, K., Turletti, T.: A survey of Software-Defined Networking: past, present, and future of programmable networks. IEEE Commun. Surv. Tut. (2014)
290. Scott, C., Wundsam, A., Raghavan, B., Panda, A., Or, A., Lai, J., Huang, E., Liu, Z., El-Hassany, A., Whitlock, S., Acharya, H.B., Zarifis, K., Shenker, S.: Troubleshooting blackbox SDN control software with minimal causal sequences. In: Proceedings of the 2014 ACM Conference on SIGCOMM (SIGCOMM '14). ACM, New York, NY, USA (2014)
291. Kotronis, V., Dimitropoulos, X., Ager, B.: Outsourcing the routing control logic: better internet routing based on SDN principles. In: Proceedings of the 11th ACM Workshop on Hot Topics in Networks (HotNets-XI). ACM, New York, NY, USA (2012)
292. Heller, B., Scott, C., McKeown, N., Shenker, S., Wundsam, A., Zeng, H., Whitlock, S., Jeyakumar, V., Handigol, N., McCauley, J., Zarifis, K., Kazemian, P.: Leveraging SDN layering to systematically troubleshoot networks. In: Proceedings of the Second ACM SIGCOMM Workshop on Hot Topics in Software Defined Networking (HotSDN '13). ACM, New York, NY, USA (2013)

293. Agarwal, K., Rozner, E., Dixon, C., Carter, J. SDN traceroute: tracing SDN forwarding without changing network behavior. In: Proceedings of the Third Workshop on Hot Topics in Software Defined Networking (HotSDN '14). ACM, New York, NY, USA (2014)
294. Jin, D., Nicol, D.M.: Parallel simulation of software defined networks. In: Proceedings of the 2013 ACM SIGSIM Conference on Principles of Advanced Discrete Simulation (SIGSIM-PADS '13). ACM, New York, NY, USA (2013)
295. Farhadi, A., Du, P., Nakao, A.: User-defined actions for SDN. In: Proceedings of the Ninth International Conference on Future Internet Technologies (CFI '14). ACM, New York, NY (2014)
296. Na, T., Kim, J.W.: Inter-connection automation for OF@TEIN multi-point international OpenFlow islands. In: Proceedings of the Ninth International Conference on Future Internet Technologies (CFI '14). ACM, New York, NY, USA (2014)
297. Sivaraman, V., Moors, T., Gharakheili, H.H., Ong, D., Matthews, J., Russell, C. Virtualizing the access network via open APIs. In: Proceedings of the Ninth ACM Conference on Emerging Networking Experiments and Technologies (CoNEXT '13). ACM, New York, NY, USA (2013)
298. Ashton, M. et al.: Ten Things to Look for in an SDN Controller, Technical Report (2013)
299. Yazıcı, V., Sunay, O., Ercan, A.O.: Controlling a Software-Defined Network via distributed controllers. NEM Summit, Istanbul, Turkey. http://faculty.ozyegin.edu.tr/aliercan/files/2012/10/YaziciNEM12.pdf, October (2012)
300. Macapuna, C.A.B., Rothenberg, C.E., Magalhaes, M.F.: In-PacketBloom filter-based datacenter networking with distributed OpenFlow controllers. In: IEEE 2010GLOBECOM Workshops, pp. 584–588, 6–10 December 2010
301. Marsan, C.D.: IAB Panel Debates Management Benefits, SecurityChallenges of Software-Defined Networking. IETF Journal, October 2012
302. Cai, Z., Cox, A.L., Ng, T.S.E.: Maestro: a system for scalable OpenFlow control. Rice University Technical Report TR10-08, December 2010
303. Cai, Z., Cox, A.L., Ng, T.S.E., Maestro: balancing fairness, latency, and throughput in the OpenFlow control plane. Rice University Technical Report TR11-07, December 2011
304. Heller, B., Sherwood, R., McKeown, N.: The controller placement problem. In: First workshop on hot topics in Software-Defined Networks, pp. 7–12 (2012)
305. Hu, Y.-N., Wang,W.-D., et al.: On the placement of controllers in Software-Defined Networks. , October 2012
306. Alsmadi, I.: The integration of access control levels based on SDN. Int. J. High Perform. Comput. Netw. **9**(4), 281–290 (2016). doi:10.1504/IJHPCN.2016.077820
307. Aleroud, A., Alsmadi, I.: Identifying DoS attacks on Software Defined Networks: a relation context approach. In: NOMS (2016)
308. Alsmadi, I., Munakami, M., Xu, D.: Model-based testing of SDN firewalls: a case study. In: Proceedings of the Second International Conference on Trustworthy Systems and Their Applications (TSA'15), Taiwan, July 2015
309. Alsmadi, I., Xu, D.: Security of Software Defined Networks: a survey. Comput. Secur. **53**, 79–108 (2015)
310. Baker, C., Anjum, A., Hill, R., Bessis, N., Kiani, S.L.: Improving cloud datacenter scalability, agility and performance using OpenFlow. In: Proceedings, 2012 Fourth International Conference on Intelligent Networking and Collaborative Systems (INCoS), pp. 20–27, 19–21 September 2012

Index

I.M. Alsmadi et al. (eds.), *Information Fusion for Cyber-Security Analytics*, Studies in Computational Intelligence 691, DOI 10.1007/978-3-319-44257-0

Z

9783319442563